T0075316

The Gastro-Archeologist

Jeremy Woodward

The Gastro-Archeologist

Revealing the Mysteries of the Intestine and its Diseases

Jeremy Woodward
Addenbrooke's Hospital
Cambridge, UK

Illustrations by
Charlie E. Manning
Oakington, UK

ISBN 978-3-030-62623-5 ISBN 978-3-030-62621-1 (eBook)
https://doi.org/10.1007/978-3-030-62621-1

This Springer imprint is published by the registered company Springer Nature Switzerland AG.
The registered company address is: Gewerbestrasse 11, 6330 Cham, Switzerland

Preface

I have worked as a gastroenterologist—a doctor specialising in guts—over the last 25 years. During this time, I have witnessed dramatic advances in our understanding of diseases and our ability to treat them. At the same time, I have seen how there is a growing desire amongst non-specialists to understand how our bodies work. Thanks to the Internet and the 'popular science' press, not only have terms such as 'DNA' and 'protein' become familiar to those without a scientific background, but also the general concepts behind them. Over these years, the nature of my job has also changed. A doctor is appropriately no longer an authoritarian figure, expecting the patient to dutifully take their medicine with little in the way of explanation, but is now primarily an educator.

At the time of writing, the only treatment available for coeliac disease is the lifelong and absolute avoidance of eating gluten—there is no other 'medicine' for it. In order for patients to accept such a change in their lifestyle, my prescription therefore has to be the imparting of sufficient knowledge to understand not only *how* but *why* to follow such a diet. For 2 hours every month over the last 8 years, a dietitian and I have done exactly that in an informal group setting with patients newly diagnosed with coeliac disease. The benefits of this approach have been frankly extraordinary (and have saved substantial healthcare resources) simply by empowering patients to look after themselves. The overview of coeliac disease that I provide in these sessions is necessarily very superficial. However, I often get asked interesting and challenging questions that lead me to realise that quite a few people wish to understand the condition in more depth.

Working as I do in Cambridge I actually have quite a few molecular biologists as patients, but how does one begin to describe to those people without any prior knowledge of biology the complexities of intestinal immunology? It is a subject that popular science authors have assiduously avoided and is notoriously challenging to teach even to medical students. However, a few hilarious sessions at local Coeliac UK patient groups and one occasion with catering managers showed me that it is not altogether impossible—albeit with the use of oven gloves, scrunched up paper balls attached to a length of string and audience members participating in role play! This is where this book started—but as with all such projects it took on a life of its own and I found myself describing the history of the gut and the immune system in order to explain coeliac disease before embracing other conditions such as food allergies and Crohn's disease as well.

This book is not intended as a textbook, as I have avoided any jargon unless absolutely necessary. Neither is it really 'popular science'—it covers far too much ground for that. I hope that it will be accessible in most part to people with relatively little biological knowledge, but will equally entertain those who also have some or even a lot of science behind them as it may present well-known concepts in an unusual or interesting way. To the experts reading this book, I apologise in advance for the degree to which I have simplified complex processes. However, the view is often clearer when free of intricate details. Just as a long-distance path can be completed very enjoyably in stages and each part may differ in its appeal, so one should not feel bad about not completing the journey in this book to its conclusion. The history of the gut and the workings of the immune system make for perfectly interesting stories in their own right, but I have kept them together in order to provide the complete story (so far) behind the diseases. Many unknowns remain for us to uncover in these conditions as I point out along the way but I have made every effort to make sure that the stories I relate here are as accurate and up to date as possible.

Cambridge, UK Jeremy Woodward

Acknowledgements

There are so many people who have contributed to the telling of this story (often without knowing that they have) that I simply cannot acknowledge them all here, but my gratitude to them is none the less for that. I list those without whom I would probably never have started, or finished.

Jane, my extraordinary, amazing wife, has supported me, fed me and generally looked after me whilst I was helplessly 'in the zone' ignoring everything and everyone around me for hours on end whilst staring at this laptop screen. I cannot begin to express my gratitude for this indulgence. Or indeed for the life that we share with our five fabulous children, Josh, Toby, James, Jack and Tom.

My parents, the kindest and most generous folk that I know, undoubtedly shaped this book through me. My father, John, instilled in me my captivation and enthusiasm for life in all its forms and my mother, Anne's revulsion of snakes, moths and butterflies, might have had something to do with my fascination for those lifeforms in particular! My first publication at the age of 14 (in 'Camping and Caravanning World'—much to the hilarity of my academic colleagues) was written with my father on the reptiles of the UK, with photographs taken by him. Memories of developing and printing black and white photographs in the attic of our home in Sidcup remain as a magical experience.

Bob Allan was my gastroenterologist mentor in Birmingham, and I owe him a huge debt of gratitude for passing on his love of the subject, as well as to John Owen, Eric Jenkinson and Graham Anderson in the thymic biology laboratory, for opening up the amazing vistas of immunology research.

My long-suffering patients and my equally forgiving students are of course my teachers and I continuously learn from them. Extra special thanks to those of them who have sense-checked and proofread for me. Equally my colleagues who have acted as sounding boards for my ideas, and others who unwittingly started my writing in ways that that they will never know. Stephen Moss, Dunecan Massey and Andrew Butler in particular helped me to shape my ideas through stimulating discussion. Matt Mason, University of Cambridge Physiologist, rigorously corrected my spelling, grammar and schoolboy howlers for which I am forever indebted to him (despite the need for substantial rewriting!).

All at Springer Verlag have been enormously patient with me and I am very grateful to them for taking on this project—with special thanks to Phillipp Berg, Ulrike Daechert, Tanja Weyandt, Parthiban Kannan and Birke Dalia for getting it into production.

Introduction

The primal importance of our guts has long been reflected in our beliefs, our language and our culture. From antiquity, gastrointestinal organs have been considered to originate emotions and temperament. Today, we talk of 'gut feelings' and use the word 'visceral' to mean something derived from deep feelings rather than rational thought.[1] The words for 'bowels' and 'mercy' even share the same Semitic word root for 'something deep within'[2] signifying the origin of the sense of compassion (and incidentally leading to some rather humorous interpretations of Bible passages). Similarly, the use of 'guts' to represent courage dates back to at least the Middle Ages and persists in most modern cultures—in Finland the word 'Sisu' (which quite literally translates as 'guts') has become a popular word to define a stoic tendency in the national characteristic.[3] The symbolic disembowelling of criminals guilty of treason in

---

[1] The Norwegian psychologist Gerda Boyesen (1922–2005) even developed a branch of alternative medicine—'Biodynamic psychotherapy'—on her ideas that linked the emotions to the gastrointestinal tract. She would use the sounds made by the intestines to interpret moods and emotions and called the process '*psychoperistalsis*'.

[2] Confusion apparently arose during translation of the Septuagint Bible into Greek with usage of splanchnois (**σπλαγχνοις**) for words originating from the Semitic root resh/chet/mim (**רחם**) meaning 'from deep within' and used for both compassion (emanating from a deep feeling) and bowels in Hebrew. This is the possible explanation for verses such as Genesis 43:30 'And Joseph made haste for his bowels did yearn upon his brother' and Song of Solomon 5:4 'my beloved put in his hand by the hole in the door and my bowels were moved for him'.

[3] Interestingly, the Finnish word Sisu also derives from a word meaning 'inside' or 'interior'—Sisus. Whilst a part of the culture for hundreds of years, its first appearance in the English language probably dates back to this excerpt from *Time* magazine in January 1940: *'The Finns have something they call Sisu. It is a compound of bravado and bravery, of ferocity and tenacity, of the ability to keep fighting after most people would have quit, and to fight with the will to win. The Finns translate Sisu as "the Finnish spirit" but it is a much more* ***gutful*** *word than that.'*

medieval England presumably shares its origin with this intrinsic notion of the viscera housing the strength of spirit.[4]

And yet, there are few sights more shocking to us than that of the human body, stripped of its cosmetically enhanced and muscularly defined outer layers, with its inner workings fully on display. The guts (Fig. 1), filling the abdominal cavity in a squirming mass of wormlike tubing, bring about the most intense feelings of revulsion, not just in the squeamish or faint-hearted.

Why intestines should engender this reaction—more than for instance the heart or the lungs—may be explicable on many levels. It is not due to unfamiliarity—at first inspection our intestines resemble those of many other mammals on display at the butcher's shop or roadkill neglected beside our highways. The intestines often appear to have a mind of their own—writhing and wriggling in coordinated peristalsis completely independently of conscious control. Indeed, we are so rarely aware of the scale of the activity inside our abdominal cavity that this surprising revelation may endow the intestines with alien qualities that makes it feel that they are not actually even a part of us.

Ultimately however, I suspect that the distaste with which we relate to our own viscera has more to do with how we see ourselves in the natural world, set apart from other forms of life by our superior intellectual development. Our bodies—and our guts in particular—serve as a great leveller to diminish our conceit. Remove all the external superficial trappings of human beings and we are seen to be much the same as other animals. Even apparently simple creatures contain a gut which is often very similar in appearance to ours. The reaction that we have to seeing how we work on the inside is perhaps therefore akin to the shock experienced in futuristic fantasies when the skin of the major character peels back to expose electronic circuitry and reveals them as a mere android machine.[5]

Underlying this realisation is the tacit understanding that we all must have—that just as the robot's wiring is essential to its functioning—so the intestine is the basic necessity of life. We are fundamentally beholden to our aesthetically distasteful innards. All that we are has come through the wall of

[4] The revolting torture of 'hanging, drawing and quartering' involved hanging the victim for several minutes until near the point of death and then cutting open the abdomen and eviscerating them whilst still alive prior to decapitation and cutting the body into four parts. The practice was a statutory penalty for treason in England from 1351 and only removed from the statute books by the Act of Forfeiture in 1870.

[5] I entirely accept that I might be overstating the case here and that the reason that we are revolted by our guts is that they are messy and smelly and if you can see them then it probably means that the person they belong to has died unpleasantly!

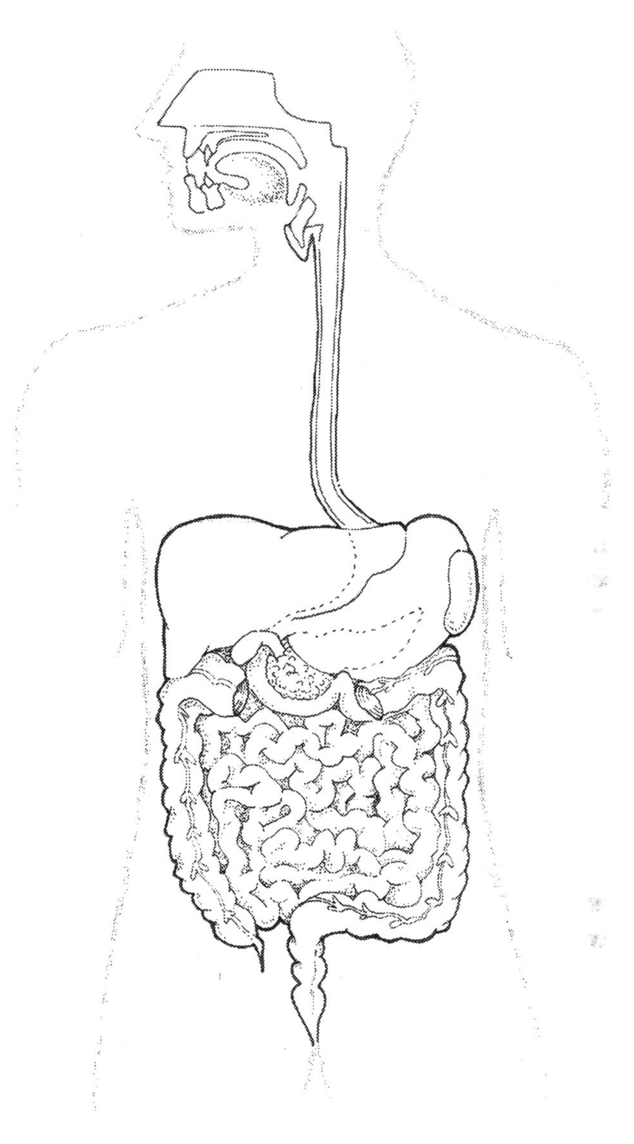

Fig. 1 The human gastrointestinal tract—from mouth to anus

the gut. The molecules that make us, the energy that drives us, even the fluid that hydrates us—all have been ingested, digested to basic constituents in the gut lumen, selectively absorbed through the lining of the intestine, repackaged and exported to the rest of the body.[6]

[6] Except some of the oxygen molecules which of course come through the lungs.

I number myself amongst the fortunate few privileged enough to spend their days working intimately with guts and the human beings that live around them. As a gastroenterologist, a physician specialising in internal medicine, I study the ways our intestines work, how they malfunction and how we interact with them—and them with us. In so far as we feed our guts and empty them on a regular basis (at least in part as a response to instructions from them, whether we are aware of it or not), we arguably interact more with our gastrointestinal tract than any other organ system. Over the last 25 years, my ongoing relationship with the gastrointestinal tract continues to awe and astonish me every day, on every level. I have come to appreciate the extraordinary beauty of the gut—I have to admit for instance, to lingering during endoscopic procedures that examine the small intestinal lining to admire the finger-like 'villi' wafting in the current like a bed of sea anemones in a rock pool. I still feel genuine excitement whilst watching 'peristalsis'—the periodic contractions of the tube that propel its contents along it—whether recorded on an oesophageal pressure tracing or observed directly in a newly transplanted intestine in the operating theatre.

I have come to learn that our guts do indeed sometimes behave as if they have a mind of their own—we now call it the 'gut-brain'—and harbour so many other microorganisms within themselves that they are truly in large part alien to us. However, it is the amazing complexity and our nascent understanding (that still frankly amounts to little more than staggering ignorance) of the workings of the gut that fascinates me the most.

Sometimes it is simply a matter of seeing things differently. Early in my career I stumbled upon a metaphorical viewpoint that allowed an unfolding of the landscape below me as if standing on a rocky outcrop or having climbed above the treeline on a mountainside. I cannot claim to be the first or the only person to have enjoyed this view—it has always been there and like so many ideas it is obvious to the point of trivial once pointed out. However, the landscape that it opens up is always changing as, despite remaining mostly in darkness, discoveries begin to illuminate more and more of the vista. This is my invitation to you to join with me on the rocky outcrop to share the splendid view it affords (Fig. 1).

This book is split into three parts of which the first two describe the route to our viewpoint. In Part 1, our path follows a journey from the most basic concepts of life that prophesy the evolution of guts. We will then follow the way in which guts have dictated the paths that life has taken. In Part 2, we retrace our steps to see how guts have hosted the development of 'immunity'—not merely a defence system but the rules dictating interactions within organisms and with others. Having arrived with an understanding of how the gut has in large part written the story *of life itself, we will be able to see how the story has been laid down in layers over time. Just as an archaeologist would make little sense of a jumble of artefacts muddled together from different ages and learns much more by painstakingly recording the separate strata in which they are found, so it is with the gut. Seen as a whole, the functions of our intestines are bewilderingly complex. But visualise the individual layers over time and place them in an evolutionary context and suddenly everything begins to make more sense. This is my metaphorical viewpoint—I call it '**Gastro-Archeology**'.*

Diverticula

As with all journeys we will find along our way points of potential distraction. These will be labelled in a box as 'diverticulum'—a word which derives from the Latin word meaning a 'bypath' (from *devertere*, to turn aside). Diverticula (*sing.* diverticulum) are a common finding in the gastrointestinal tract and are simple outpouchings on the side of the tube. Whilst they can occur anywhere from the gullet to the rectum, they are commonly found in the human colon where their numbers increase over time. The terminology often causes confusion—'Diverticulosis' is the condition of having (colonic) diverticula (as do most people in the UK over the age of about 40 years), but 'diverticulitis' is the presence of infection or inflammation in a diverticulum. I tend to illustrate this by considering the human appendix. We are all born with this particular diverticulum, but relatively few people develop an infection within it and experience 'appendicitis'.

Whilst substantial distractions will be located in the diverticula, it is not unusual to find other points of interest on any journey and additional facts and historical comments will be found as footnotes.

Contents

About the Author

Jeremy Woodward is a consultant gastroenterologist working in Addenbrooke's Hospital in Cambridge. He specialises in diseases of the small intestine and clinical nutrition, which extends from tube-feeding techniques to intestinal transplantation. He lectures on gastroenterology and nutrition in the Medical School in Cambridge University where he is also a communications skills facilitator, and he teaches gastrointestinal physiology to undergraduates in Christ's College (where his own career started). An enthusiastic wildlife observer and photographer, he particularly enjoys studying moths in his back garden and hunting reptiles when on holiday in warmer climes.

Illustrator

Charlie E. Manning comes from a long line of creative folk. Educated in Saffron Walden, Essex, and in Cambridge, she has worked within the art industry in roles such as college technician, art and antiquities custodian and art materials advisor and retail for over 20 years. She chose to refresh some of her skills and studied illustration and visual communications with the Open College of the Arts. Whilst pursuing various creative avenues, she enjoys trying to incorporate recycling wherever possible, being a history nerd, folklore enthusiast, appreciator of traditional crafts and an avid plant collector. Working predominantly in pen and ink, gouache, coloured pencil, mixed media collage and Lino printing, she takes inspiration from the natural world and can be found recharging her batteries outdoors, preferably in the dappled shade of woodland amongst the flora and fauna, taking lots of photographs and possibly hugging a tree. Currently, she is dwelling near the Fenlands of East Anglia. She is a devoted auntie to her nieces and nephews and works part-time in floristry.

Part I

Selection of the Fattest

The transition of the 'Enterocene' (Gut Age) to the 'Anthropocene' (or Brain Age) (Reprinted with permission from CartoonStock: www.cartoonstock.com)

> '*...one can perhaps even view an animal as nothing more than a group of cells clustered around a gastro-intestinal tract, differentiated for and dedicated to the task of keeping that gut full.*'
>
> Wayne Becker[1]

---

[1] Wayne Becker, Professor of Botany at University of Wisconsin. This quote comes from an unpublished set of course notes taken by a student that I found on the internet. The cartoon is by Patrick Hardin from Michigan. For our purposes the human's thought bubble should probably read 'I know I shouldn't, and I really don't need it, but that cream puff is just *too* tempting'. Or be replaced by Homer Simpson holding a doughnut...

Introduction

One of the greatest threats to human longevity is an evolutionary omission. Until now life has never existed in an environment where food was so plentiful that it was present in excess. As a result, we have not really required mechanisms to shed weight but only to conserve or gain. However, thanks to our developed brains, we can now manipulate our environment to be able to provide sufficient to feed us all. Dire warnings of global catastrophe will probably have to await the effects of climate change rather than population growth. Sadly, we have singularly failed to develop the societal mechanisms that allow us to share and to prevent the disintegration of kinship into conflict based on need or greed. We therefore live in a World where poverty and plenty co-exist—and those who have most suffer from the effects of excess.[2]

With the advent of human society, life on earth has now entered a period where our intelligence has allowed us to adapt our environment rather than to it. This 'Brain Age' has now even been credited with its own geological epoch—the 'Anthropocene'.[3] However, up until this point, it is the relative lack of food throughout time that has driven competition. The perpetual need to feed the gut, and the animal through it, has perhaps been the most potent selection process on the planet. Whether the 'Anthropocene' started around 12,000 years ago with the spread of settlements and farming or more recently in the nuclear age is debated. For our current purposes it is immaterial as our journey through the 'gut age' spans from the invention of guts around 600 million years ago up until now.

However, we will need to set off on our path considerably earlier than this. Life itself is born out of a set of temporary solutions to irresolvable conflicts and long before the first gut came into being, its future existence was prophesied by the necessary compromises. I promise that we will not linger over-long in the inhospitable environment at the dawn of life but we do have a lot of ground to cover—so we had better get going!

[2]The prevalence of obesity in the UK increased from 15% in 1993 to 26% in 2014. Even higher rates of obesity were reported in other developed countries such as the USA (35%) and New Zealand. Obesity accounts for approximately 44% of all cases of diabetes, and 23% of ischaemic heart disease. It is now (at the time of writing) the fifth largest cause of death worldwide, and one in five deaths in North America can be related to obesity. Tragically, it is not just the wealthy and the greedy corners of society that suffer from the complications of excess—cheap highly calorific but poorly nutritious foods are often consumed in poor societies leading to obesity co-existing alongside under-nutrition.

[3]The term Anthropocene is attributed to Paul J Crutzen. He is an atmospheric chemist who shared the Nobel prize for chemistry in 1995 for his work, particularly relating to the formation and depletion of ozone.

1

The Invention of Eating

Summary In which we see the cell membrane as the defining boundary of life but also as a confining barrier that must be overcome to allow substances in and out of the cell. This is the 'containment paradox'. Single-celled animals ultimately developed the ability to bring components from outside the membrane into their substance through mechanisms such as 'endocytosis' and 'phagocytosis'. This was the 'invention' of eating and significantly enhanced the organism's capacity to assimilate nutrients. The ability to ingest particles had enormous significance for the evolution of life. Whole bacteria that came to live as 'endosymbionts' within the cell took on the energy conversion processes from the surrounding surface membrane (where this previously occurred) and enabled it to take on additional roles. Importantly, phagocytosis led to the ingestion of whole living organisms. This was to alter the relationship between lifeforms with consequences that would lead ultimately to the requirement of both a gut and an immune system.

J. Woodward, *The Gastro-Archeologist*, https://doi.org/10.1007/978-3-030-62621-1_1

The sketch is inspired by a painting by Caspar David Friedrich 1774–1840 which is entitled 'The Wanderer (or 'hiker') in the Mists'. It was painted in 1818 and now hangs in the Kunsthalle in Hamburg. It has always enchanted me since I first saw it used as the cover illustration for the Penguin Classics edition of Friedrich Nietzsche's Ecce Homo and I have to admit I found it far more inspirational than the text. I love the portrayal of an emerging vista from a new viewpoint

The Containment Paradox

The first stage of our journey will take us from the beginnings of life to a momentous event in its history—we could call it the' invention' of eating. Although this will take over 2 billion years, we should not hurry to get started as we need first to establish one or two of the basic principles of life that ultimately made eating necessary.

All life exists inside a bubble. Its membrane is a highly organised two-dimensional fluid enclosing a three-dimensional space called the cell, the smallest independent unit of life.[1] This liquid skin just 4 millionths of a millimetre across,[2] is the sole interface between the vagaries of the outside World and the relatively constant and chemically different internal milieu required by life. Its essential structure is so basic that if you break it up into its constituent molecules it will spontaneously re-assemble itself, simply because oil and water do not mix.

Fat molecules—also known as 'lipids'—are chains of Carbon and Hydrogen atoms joined together. They repel water—hence when washing up after a roast dinner we see round globules of fat floating in the water in the sink. However, if we add a detergent such as washing up liquid, the globules vanish leaving a white emulsion of the fat in the water. The way in which detergents do this is that their molecules each have one end that dissolves in fat and the other in water. They are thence able to bridge—and mix—the two otherwise incompatible fluids. Likewise, the molecules that make up the cell membrane have a fat chain on one end and a water-soluble molecule (usually phosphate) at the other end—and are therefore called 'phospholipids'. When mixed with water, such chemicals spontaneously align themselves to have their lipid portions in proximity to each other and as far away from the water as possible. These molecules might for instance orientate themselves in a ball with the lipid ends all pointing inwards and the water-soluble ends pointing outwards.

Alternatively, if they enclosed a bubble of water they would be pointing the other way, but this would leave the fat soluble ends in contact with outside water and this would be unstable. However, if on top of this skin around the bubble there was *another* layer of phospholipids pointing the other way, then they have created a molecular sandwich—a lipid 'bilayer' with water on both sides separated by a fatty barrier.[3] Thus, we define the boundary of life—the double layered cell membrane (Fig. 1.1).

[1] Cell (from Latin—cella—meaning 'small room'). The concept of the cell literally becomes the living room of life. Different types of cell are identified by the suffix '-cyte' from the Greek word 'Kytos' meaning 'container'.

[2] The thickness of the membrane can be estimated on the basis of an experiment performed by the American diplomat and polymath Benjamin Franklin on a pond in Clapham Common in the early 1770s! He poured a teaspoon of oil onto the pond and estimated that it covered about half an acre when it spread out—from this you can estimate the thickness of a single layer of molecules. The experiment was not designed for this purpose though but to see if there was any truth in the efficacy of pouring oil on water to calm choppy seas.

[3] The bubbles that are made by blowing through a film of washing up liquid between our fingers or in children's bubble toys are also a lipid bilayer, but the exact opposite of the cell membrane. Here the double layer is made up of the fat-loving ends pointing outwards from the membrane with a thin layer of water sandwiched in between the water-loving ends in the middle.

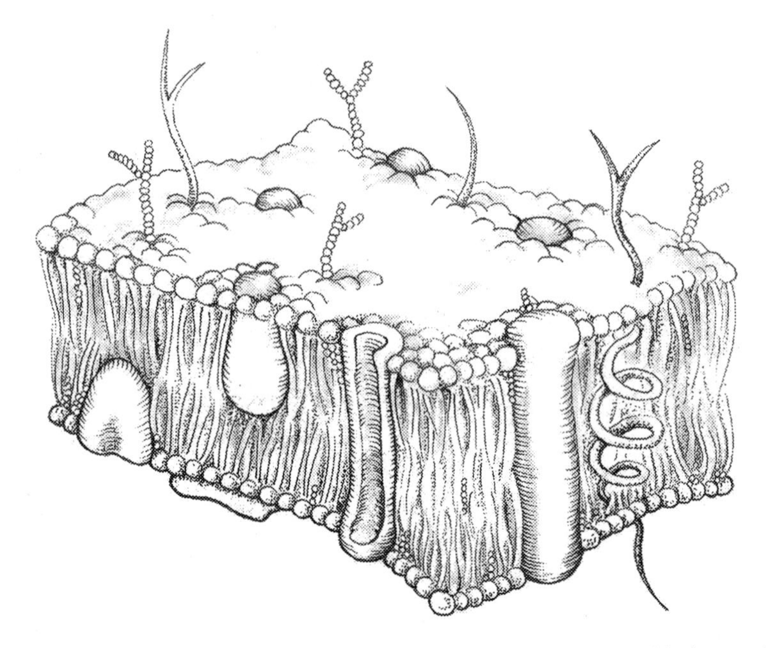

Fig. 1.1 The cell membrane—the boundary that defines life. A double layer of lipids pointing inwards with water soluble ends on each 'out'side, with proteins embedded or traversing it to provide passages and means of interaction with the outside world

The importance of the cell membrane cannot be overstated. Imagine it as the geographical borders of a country. It can control the flow in and out of the region it encloses. The environment on either side of the boundary can be very different, akin perhaps to crossing the border between two countries as different as North and South Korea. Trade can take place across the border and commodities can be exchanged by bartering. It can communicate by sending and receiving messages far outside its boundaries. The boundary permits the area inside to differ from that on the other side. The borders have created an 'entity' that can have its own individual 'id-entity'.

One can therefore understand the importance of the membrane in confining and defining the living space of the cell. It creates an enclosed environment which can be maintained stable regardless of the changes occurring outside. Toxins can be excluded or nutrients selectively accumulated across the boundary. Optimum conditions can be set within the cell for biochemical functions, such as the acidity and salinity. Enclosure by the membrane increases the local concentration of chemicals to encourage them to interact rather than simply diffusing away into the surrounding environment and this speeds up the reactions of life.

Containing life within the confines of the membrane and the creation of an 'entity' has further implications in the context of that essential pre-requisite of life—the means of reproducing itself. Self-replicating molecules exist within all living organisms as the basis of the 'genetic code' in the form of nucleic acids—DNA and RNA. The information encrypted in the four-letter alphabet of these vast molecules can be translated (by special cellular machinery) into the long chains of amino acids which form proteins. Proteins are the nano-machines that form the engines of life. They can be made of chains as short as 20 amino acids joined together (the hormone, insulin has just 51) to many thousands such as the appropriately named 'Titin' which is made of nearly 30,000.[4] Shorter chains of amino acids are called 'peptides' rather than 'proteins'. Unlike simple peptides, proteins are defined by their ability to form three dimensional structures which gives them their functional capabilities. For instance, they can catalyse chemical reactions (such proteins are called 'enzymes') by attracting molecules together into a cleft or 'active site' where they can react; they can act as messengers between cells (such as insulin) or they can build up structures (such as the proteins in hair, or titin, which effectively acts like a spring in muscles).

Faults, or mutations, that occur in the genetic code or its translation into proteins might be catastrophic, but occasionally result in an improved version of the protein. This is a fundamental component of genetic adaptation and is critical to allow organisms to respond to change. Regardless of the outcome the membrane serves initially to limit the damage—or the benefit—of any such mutation to the single unit in which it occurred. The change relates only to the single cell and its progeny. If a harmful mutation leads to the death of a single celled organism and its progeny, it will not affect others of its kind. Whilst it could be argued (as Richard Dawkins does in 'The Selfish Gene'[5]) that natural selection can operate at the level of the replicating gene itself, it is enclosure by the cell membrane that turns the whole cell into the 'unit of evolution'.[6]

Evolution has tinkered with the cell membrane and found uses for it that extend beyond its boundary role. It has become a vibrant trading place and a

[4] All proteins can be described by a given name, or by individually naming the amino acids that make them up—given that there are over 30,000 in titin this would therefore make an extremely long name indeed (over 189,000 letters) and some have suggested that this is the longest word in the English language, or indeed any. I think it is stretching it a little though to use a chemical formula as a word.

[5] At its simplest, life could be seen as the emergence of self-replicating molecules. Such a concept was promulgated by Richard Dawkins in 'The Selfish Gene' in 1976 in which he envisaged the genetic material as the being of life on which evolutionary selection worked, and the organism around it as the mere container or vehicle.

[6] A broader perception is that of David Hull who added the necessity for the 'interactor' to the 'replicator': 'A process is a selection process because of the interplay between replication and interaction. The structure of replicators is differentially perpetuated because of the relative success of the interactors of which the replicators are part. In order to perform the functions they do, both replicators and interactors must be discreet individuals which come into existence and cease to exist'.

chemical crossroad of reactions. A vast array of proteins embed themselves within the membrane and stretch out into the external medium or span across it. They function as passive channels or active pumps to alter the concentration of chemicals such as salts on either side, or to receive messages from outside or signal to other cells. The membrane acts as a two-dimensional fluid platform in which these proteins float such that they can aggregate together and work collaboratively to assemble larger structures, just like machines in a production line comprising different parts. The immense utility of lipid bilayers as both boundaries and scaffolds for useful proteins has led to the lipid bilayer configuration fulfilling a wide number of functions within the cell itself. It has been used to create sub-compartments within the cell, each with their own contained chemical environment. In some highly specialised cells, the boundary membrane only comprises around 2% of the total lipid bilayers of the cell.

There is a downside to all of this of course. By concentrating and controlling the matter of life within the boundary of a cell membrane there still remains the issue of how the cell is able to harness the nutrients from the environment that it requires in order to power it. Somehow it needs to get the raw ingredients into the cell in a form that it can use. It requires an ability to maintain strict border controls on immigration but allow trade for the essential commodities. This compromise between containment and consumption is the first great primal conflict of life that predicts the future development of the gut and will become one underlying theme of our story as it plays out over the history of time. I call this the 'paradox' of 'containment'.

With this in mind, let us commence our journey!

Diverticulum #1.1: The energy of life

Just being alive requires huge amounts of energy. Given our current sedentary lifestyles less than a quarter of our daily energy consumption is used in physical activity but even in those who exercise regularly it is still under half. The rest is needed just to maintain our bodies—not simply in breathing or the heart pumping blood, but in the internal workings of every cell.

Plants harness the energy from sunlight to make complex 'organic' molecules (the 'chemicals of life') largely from carbon dioxide and water which results in the generation of oxygen as a by-product. In beautiful natural symmetry, animals effectively reverse this process, using chemical reactions between oxygen and complex organic molecules (from food) to generate carbon dioxide and water. However, all living organisms ultimately use the resulting energy in the same way—by first converting it into electricity and then storing it in a chemical form, just like charging a battery.

The electrical generator in cells is found on the inside of specialised lipid bilayer membranes in compartments within the cell called 'Mitochondria'. The miniscule current it produces is quite literally the 'spark of life'—just as one can imagine passing between the outstretched fingers of God and Adam in Michelangelo's depiction on the ceiling of the Sistine chapel.[7] Or perhaps (more crudely) as the lightning bolt harnessed in a B-movie depiction of Dr Frankenstein's laboratory.[8]

The phenomenon that we recognise as 'electricity' is simply a flow of negatively charged particles called electrons. We are most familiar with them passing along a copper wire surrounded by insulating material such as PVC. Instead of a wire, life forms generate the current by passing an electron along a chain of proteins embedded on one side of a cell membrane—a bit like a rugby ball being passed down a line of players out of the scrum. Instead of copper, the conductor is made up of clusters of iron and sulphur within proteins and the insulator is the membrane itself.

The 'battery' to store the energy is a chemical phosphate bond in a molecule called ATP (adenosine triphosphate). Most energy-requiring chemical processes in the cell are powered by the effects of breaking the chemical phosphate bond—it has been called the 'energy currency' of life. We all apparently create and consume our own body weight in ATP every day! The battery is 'charged' (ATP formed) by pumping hydrogen ions (carrying a single positive charge equal and opposite to that of the electron) across the membrane using the electricity of the electron transport chain. This builds up a gradient of hydrogen ions across the insulating membrane—and the energy can then be harnessed by allowing these ions to move back in through specialised channels.[9] These channels contain a protein shaped like a turbine that even rotates as a result of the flow of hydrogen ions and creates three ATP phosphate bonds on each turn. In fact, it bears remarkable similarity to the three-cylinder radial engine used in Louis Bleriot's aircraft for the first flight across the English Channel in 1909![10] The entire mechanism is akin to that of hydroelectric power with water backed up behind a dam flowing down through channels to drive a turbine. The only difference is that life forms store the energy in the battery pack of ATP—engineers are looking at similar ways of storing generated electricity!

---

[7] Michelangelo painted the ceiling of the Sistine chapel between 1508 and 1512. It depicts the Creation as told in the book of Genesis and the Creation of Adam is the centrepiece—famously as God and Adam nearly touch fingers in the giving of life. The image of God is surrounded by a red mantle—whilst many have pointed out the anatomical similarity of the shape of this ensemble to the human brain it also closely resembles the form of the uterus after giving birth. Which of these was intended is not known—is it possible that Michelangelo's genius was to combine both forms?

[8] 'Frankenstein; or the Modern Prometheus' was written by the young Mary Shelley who published it in 1818 at the age of 20. She referred to Victor Frankenstein's monster in different tellings of the tale as 'Adam' and the monster refers to himself in talking to Frankenstein as 'The Adam of your labours'. Interestingly, the actual animation of the monster is not described in her book, although she hints strongly that it was connected to Victor Frankenstein's interest in electricity.

[9] The coupling of the proton gradient to the synthesis of ATP is known as the 'Chemiosmotic hypothesis' and was first espoused by Peter D Mitchell in 1961—he was awarded the Nobel Prize in Chemistry for his work in 1978.

[10] In 1908 the Daily Mail newspaper offered £500 for anyone achieving a powered flight across the English Channel—Le Matin paper in Paris said that it could never be won. The prize was doubled to £1000 in 1909 when on 25th July, Louis Bleriot claimed it by flying his mark XI monoplane from near

The Journey Begins

Determining how long ago the first cells came into being on our planet presents a significant challenge. The history of life is written in the fossil record of the rocks of our planet, which helpfully assists in dating the era of death of the preserved form of the organism. The earliest fossils that are indubitably of bacterial forms are found in rocks that date back a 'mere' 3 billion years. However, specialised techniques are required to identify microscopic fossils and below a certain size it can be difficult to discern the structures of life forms from the crystals of the rock itself. Identification of organic molecules associated with life—such as the phospholipids of the cell membrane—may just date the availability of the chemical ingredients rather than their association into living entities.[11] Strong candidates for the earliest living things are unusual structures called 'Stromatolites' which are found in old rocks that have been made by the progressive layering of sediment. They resemble features made by modern day bacteria living in shallow waters. Such deposits found in Australia have been dated to around 3.5 billion years before the present and are thought to represent the oldest evidence of lifeforms similar to those in existence today. Nevertheless, there are hints of possibly more ancient life forms in even older rocks—up to 4.1 billion years ago. To put this into context, this is about as far back in time as any rocks can be reliably dated. The young planet, less than a billion years old—was up to this point inhospitably hot and the surface was constantly renewing itself through volcanic activity and being pulverised by meteorites. If life could form in such an environment so soon after the formation of the planet itself then perhaps it was not so difficult to create after all.

Calais to Dover in a time of 36 min. He crash-landed as he had not previously reconnoitred a landing site on the English side. Perhaps he never expected to make it across.... The radial engine, designed by Alessandro Anzani, became the prototype for all aircraft engines until long after the Second World War.

The rotating three step mechanism of the ATP synthase was first described by Paul Boyer at UCLA and confirmed by John Walker from Cambridge following elucidation of the structure of the molecule—they shared the 1997 Nobel prize in Chemistry for their achievement.

[11] These difficulties are faced by exo-biologists looking for traces of life on nearby planets and moons of the solar system, but compounded by the necessity of transporting a miniature laboratory across space, and constrained by knowledge of the molecules associated with life on earth. An example of such potential confusion is the presence—or not—of 'nanobacteria'. These are tiny structures which, by their complexity and ability to perpetuate themselves, have been considered by some to represent small organisms, about 100 times smaller than most bacteria. Nanobacteria have been implicated as the cause of certain diseases in humans—including calcifying conditions such as atherosclerosis and kidney stones. They have also been found on material emanating from other worlds (the 'Alan Hills 84001' meteorite from Mars that was found in Antarctica) and considered as possible evidence for the existence of extra-terrestrial life. Whilst the debate continues, it is most likely that these structures on Earth are simply mineral-protein complexes that are capable of self-propagation whilst the appearances on the Mars rock are now thought to be artefacts from the preparation for electron microscopy.

Fig. 1.2 The 'Lost city' under the sea: were structures such as these the original cradle of life?

Our journey begins at a remote point on our planet, some 2000 ft below the surface of the middle of the Atlantic Ocean. Here lies what has been dubbed the 'Lost City'—an eerie field of white chimney-like structures growing through the murky depths on the ocean floor (Fig. 1.2). Its discovery in December 2000 has radically changed our views on the potential origins of life. Unlike nearby volcanically generated 'black smokers' formed by the deposition of sediments from superheated water at up to 400 °C, the Lost City is produced at much lower temperatures (50–90 °C) by alkaline hydrothermal vents. Seawater percolating deep into the earth's upper mantle reacts with the particular type of rock found there (called 'Olivine') to generate heat and rise back up to the sea floor as an alkaline solution rich in minerals and hydrogen gas. (According to the NASA probe Cassini, a similar process appears to be happening currently in the oceans of Enceladus, one of the moons of Saturn).

On contacting the cold seawater, minerals such as magnesium carbonate settle out from the plume and form thin semi-porous inorganic membranes. In the tranquil stillness of the ocean depths these fragile structures can reach enormous sizes—up to 60 m tall in places. Such membranes would have been highly enriched in iron and sulphur in the oceans of the early planet—a combination that may well have been capable of generating electrical gradients through tiny pores in the structures.

The energy from such gradients generated inorganically could have powered the chemical reactions to create the organic molecules of life. In other words, unlike Frankenstein's monster, the electrical spark was not the finishing touch but the *start* of the creation of life. If life did originate at such sites as a result of inorganically generated electrical currents, then it is perhaps no coincidence that all of the proteins that make up the electron transfer chain (see Diverticulum #1.1) in all life forms use inorganic catalytic centres—made up of exactly the same iron and sulphur.

In order to leave the comfortable ancestral home of the alkaline vent, it is postulated that life-forms developed the necessary lipid bilayer membranes within the pores or pockets of the chimneys and the necessary protein pumps to create energy-generating gradients across them. Therefore 'life' would already have been quite advanced before it left home, and it is likely that if we had been able to sample the residents of the vents 3.5 billion years ago, we would have found a venerable distant relative still living there going by the name of 'LUCA', whom I shall now introduce to you.

The Tree of Life

When tracing our family origins, most of us are able to go back three generations with ease and if we are lucky to have access to the records we can then follow certain lines back a few more. Evolutionary biologists have taken this to the ultimate length in postulating that all current life can be traced back to single predecessors, or groups of genetically similar individuals. In other words, that there is a tree of life[12] that can be traced back to a single 'last common ancestor' at the base of the trunk.

[12] Charles Darwin put this much more eloquently—"As buds give rise by growth to fresh buds, and these, if vigorous, branch out and overtop on all sides many a feebler branch, so by generation I believe it has been with the great Tree of Life, which fills with its dead and broken branches the crust of the earth, and

LUCA is the 'Last Common Universal Ancestor' for all currently existing life forms. Since the first complete bacterial DNA was decoded in 1995, the full genomes of over 100,000 different species have been published. It is possible to attempt to recreate the genetic make-up of LUCA by determining which genes have been preserved in 'primitive' life forms that are still in existence. Around 350 such different genes may be so old that they have been present since the time of this ancient ancestor. Significantly, these genes all encode proteins that would be required for life to exist at the alkaline hydrothermal vents—being able to utilise carbon dioxide and hydrogen, and appearing to associate with iron and sulphur.

Prior to 1977, all life forms were considered to fall into one of only two major categories depending on the complexity of the structures within the cell—either bacteria ('prokaryotes') or 'eukaryotes' which included all more advanced life forms including multicellular organisms, such as plants and animals. The two cell types are easily distinguishable by the presence or absence of internal membrane components—particularly a cell 'nucleus'. Prokaryotes have simple, often circular loops of DNA floating freely within the cell, whereas eukaryotes have their genetic material contained within a membrane-enclosed structure called the nucleus. They also contain a whole host of other such membrane-bound structures called 'organelles' by early microscopists.[13] These include the 'mitochondria' which generate the energy currency molecule, ATP within the cell (from the electron transport chain proteins embedded within its inner membrane) and 'chloroplasts' in plant cells where a similar process occurs, powered by light. All prokaryotic organisms comprise a single cell, whereas eukaryotes can be unicellular such as amoebae, or constructed of many cells (multicellular) such as ourselves—each of us is made up of about 30 trillion!

---

covers the surface with its ever-branching and beautiful ramifications." Darwin, *The Origin of Species* (1872), 104f. Darwin was a student of theology at Christ's College in Cambridge between 1828 and 1831 (having failed to show any interest or aptitude in medical studies at Edinburgh) and embarked on the legendary voyage of the Beagle shortly after leaving Cambridge. His student rooms have been restored and are well worth a visit, as is the statue of him sitting on a bench made by Anthony Smith in gardens in the college recreated with species identified from his voyage.

[13] Robert Brown described the first eukaryotic organelle—the nucleus—in 1833. Like Darwin, he also dropped out of medical studies in Edinburgh, as he was more interested in botany. He also described 'Brownian motion'—the random motion of particles suspended in fluid due to collision with fast moving molecules. Although mitochondria were first spotted inside the cell in the 1850s, their definitive description fell to the German pathologist, Richard Altmann in 1890. He called them 'bioblasts' and thought that they were 'elementary organisms' living independently within the cell. He would have been gratified to know that this is concurrent with present thinking, even though his theory was initially ridiculed (as all the best are!). Mitochondria are now thought to have evolved from formerly free-living bacteria. The term 'mitochondrion' was coined in 1898 by Benda, from the Greek for 'thread' and 'granule' after the appearance of the mitochondria in the tail of the sperm.

However, we now know that the sapling of life had not two, but three main branches. A group of bacteria previously thought to be extremely primitive were found to have significant enough differences to classify them separately as an equal prokaryotic domain called the 'Archaea'. This branch of life is named after the Greek word meaning 'beginning' or 'origin' from which words such as 'archeology' are derived. Indeed, the similarities between the archaeans and eukaryotes appear greater than those between the archaeans and the bacteria.

Some investigators consider that the persistence of primitive traits within eukaryotes may lead us to discover that it is our domain that bears the most resemblance to LUCA.[14] The apparent simplicity of the other two branches may represent evolutionary fine tuning by cutting back redundant mechanisms—perhaps through separate common ancestors living in high temperature environments where RNA turns out to be unstable. Nevertheless, we have already travelled a great distance from our starting point—the first sign of 'eukaryotic' life appears only about 2 billion years ago. It has actually taken us longer to get there from our first stop at the hydrothermal vents than it took for LUCA to appear after the birth of the planet.

The Nourishment of Life

Cells require nutrients to provide energy and the necessary ingredients for the machinery of life. The 'Containment Paradox' is that the cell closes off the outside World but still has to find ways of getting what it needs and also disposing of waste products across the barrier of the membrane.

The membrane is not a hard wall but a fluid interface and some substances can pass across it quite easily. These include important gases such as hydrogen, carbon dioxide and oxygen. Being made of lipid, chemicals that dissolve in fat can also cross the membrane with relative ease. However, water-soluble chemicals and large molecules such as proteins are blocked. The cell has got around this problem by developing specialised proteins that are embedded within the membrane. Some of these proteins span the membrane and are capable of selectively pumping particular chemicals (including charged particles called

[14] Carl Woese (1928–2012), along with George Fox at the University of Illinois pioneered the use of 16s ribosomal RNA to understand the relationship of different bacterial and archaeal species over time. 16s ribosomal RNA is a structural nucleotide that makes up part of the 'ribosome' a subcellular structure responsible for the translation of the nucleic acid genetic code into proteins—effectively the 'protein factory' within cells. It mutates very slowly due to its critically important role and therefore changes in the 16s rRNA can be used to define different species. In common with many other original thinkers in science who challenge the established paradigm, Woese was considered to be something of a crank—but later recognised as a 'scarred revolutionary' as his theory became accepted by the end of the 1980s.

ions) in or out. Other 'transporters' can simply stick to larger organic molecules such as glucose or amino acids and ferry them across both layers of the membrane into the cell. Some proteins simply make 'pores' or holes in the membrane to selectively channel small components across the membrane. However, this is a risky business as such pores need to be carefully controlled—through molecular size or electrical charge—to prevent equilibration of the internal compartment of the cell with the outside world.

Whilst such mechanisms suffice for simple bacteria and archaeans that are capable of feasting on chemicals abundant in their immediate environment, more advanced cells have greater requirements and need to become specialist and bulk importers. In order to understand how they do this we have to think a bit like rocket scientists!

Imagine for a moment the International Space Station. To keep the astronauts alive inside it requires an absolute exclusion of its internal environment from the vacuum of space. A single hole in the outer fabric and all the air could be lost. Similarly, whilst tiny pores can selectively control the movement of some substances across the membrane, a large enough hole leads to dilution with water and leakage of the living cell contents and extinguishes life. Getting things in and out of the space station without creating a hole is a challenge similar to the cellular containment paradox. As we all know however, astronauts are able to enter and exit the spacecraft using an 'air lock'. This is a small enclosed area which can be opened on either side—but not both at the same time! Rotating doors of hotel lobbies work similarly.

The rocket scientist is sadly limited by the necessary rigidity of the materials required to build the spaceship. Cells on the other hand enjoy the benefits of a fluid surrounding layer—which can not only patch up any spontaneous holes that form in it, but also allow it to form sophisticated 'air locks' to bring substances in and out. It can do this in a number of ways. The first is called 'endocytosis' which literally means 'bringing into the cell'. Those old enough to remember them should at this stage visualise a 'lava lamp' from the 1970s.

Endocytosis works by a dimple or pit forming on the cell surface, gradually enlarging into a cavity and then forming a bubble that breaks off internally from the cell membrane. The fluid membrane has simply closed behind it and never breaks the seal. However, unlike the air lock in a space station, the cell cannot simply afford to open the bubble containing a bit of the outside world and allow its contents inside. This is because (with the exception of science fiction films) space is just empty and does not contain harmful substances that could run amok inside the spaceship. Unfortunately for cells this is not the case and they must keep the outside world separate from their internal workings. So, whilst the cell has cleverly managed to 'ingest' some of its

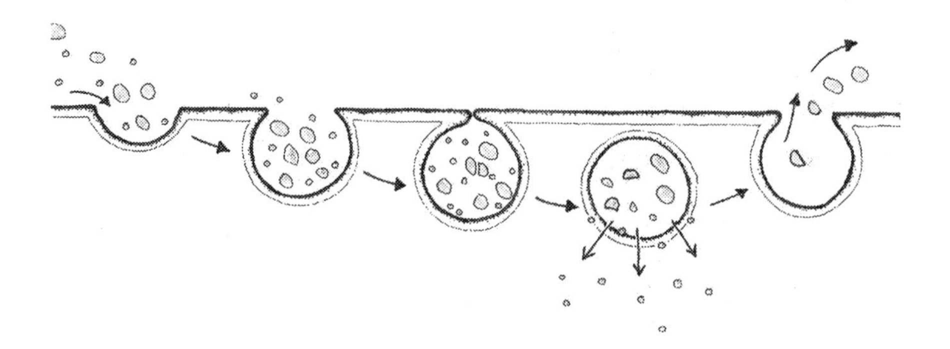

Fig. 1.3 Solving the containment paradox through 'endocytosis'. The membrane buds internally and encloses some of the outside fluid within it to bring it into the cell as a 'vesicle' or bubble. Desired nutrients can be selectively transported across the membrane of the vesicle into the substance of the cell, and the unwanted debris expelled by the reverse process of 'exocytosis'

surrounding fluid, it is not really any further forward as it still has to get what it wants out of the bubble and into the substance of the cell (Fig. 1.3).

We will park this conundrum for a moment and look briefly at how cells manage to export stuff to the outside instead. As we have seen, the cell has developed uses for the lipid bilayer structure beyond its boundary role. Next to the nucleus is a structure that looks like a stack of pancakes layered on top of each other, named after the nineteenth century Italian biologist Camillo Golgi[15] who first described it. This is effectively the sorting and distribution centre for the cell and sends out newly synthesised proteins within membrane-contained bubbles (or 'vesicles') to various destinations within or outside the cell. These vesicles can simply reverse the endocytosis process and merge with the cell membrane to discharge their contents (predictably called '*exo*cytosis'). Additionally, the lipid bilayer of the vesicle has inserted itself into the cell membrane and therefore any proteins that were formed within it or sticking out of it are now part of the outer cell membrane. This is how the various pumps, channels and signalling proteins come to reside there.

[15] Camillo Golgi (1843–1926) was an Italian pathologist who studied the nervous system, having invented a stain that demonstrated nerve fibres and for which he received the Nobel prize for his work in 1905. The stain also demonstrated the intracellular apparatus that bears his name and is his most lasting legacy.

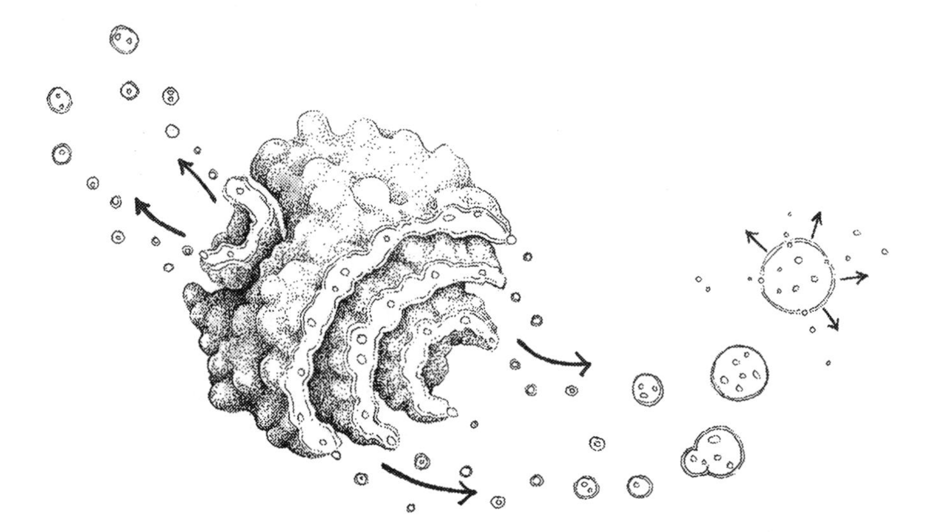

Fig. 1.4 The Golgi Body—the cell's sorting and distribution centre. Vesicles are formed with proteins embedded into the membrane or packaged inside. Upon fusion with the cell membrane the packaged proteins are released into the surroundings by 'exocytosis'. By the same process, the proteins embedded in the vesicle membrane become part of the cell's boundary membrane where they may act as transporters or channels to allow specific molecules into or out of the cell, or interact with other cells. Special vesicles called 'lysosomes' are filled with digestive enzymes that can then merge with the vesicles formed by endocytosis to digest the contents without damaging the cell substance, as they remain compartmentalised

Some of the vesicles created inside the cell are not destined for the outer cell membrane. These are called 'lysosomes' (from the Latin, *lysis* = to loosen and *soma* = body). They may contain chemicals such as digestive enzymes that would be unwise to let loose within the cell itself and much safer to keep enclosed in a bubble—for obvious reasons. The cell is now in an ideal situation to harness the contents of the bubbles created by endocytosis. When lysosomes collide with these endocytic bubbles within the cell, they fuse together and mix their internal contents and their fluid membranes. This allows the cell to effectively 'inject' not only chemicals into the bubble to digest and liberate its contents, but also special pumps and transporters into its membrane to selectively pump out the desired nutrients into the cell. The remainder and the unwanted bits of the outside world then go back to the cell membrane and discharge by 'exocytosis' (Fig. 1.4).

So far, so good. However, nature loves to tinker and we can perhaps now envisage further ways in which this process of getting things in and out of the cell can be significantly enhanced. Consider the case of a particular molecule that is really valuable to a cell (see our Diverticulum #1.2 about iron below).

Rather than just randomly enclosing a bubble of the outside fluid into the cell in the hope that it will contain some of the prized nutrient, specific receptor proteins can be placed into the membrane that can stick and hold onto it. Even better, the cell can extend its range by freeing up the receptor molecules from the surface to send them out into the fluid medium as 'remote probes' to capture the nutrient. The cell then creates new 'docking' molecules on its surface that recognise the free receptors laden with their cargo in order to bring them into the cell by endocytosis.

I have unashamedly but very significantly simplified the process of endocytosis in my description, which requires a multitude of different steps and proteins to effect. However, despite its complexity, there are enormous benefits to the cell in doing so. The quantity of trade with the outside world is dramatically increased, and the cell can 'cherry pick' what it wants by the use of specific surface 'docking' receptors. Previously, the processing or digestion of foods to liberate their nutrients in a usable form could only take place *outside* the cell, where the results would be equally available to neighbouring competitors. Now the cell, by bringing them *inside* and processing them internally can keep all of the fruits of its labours to itself. This process is so important that in active cells endocytosis leads to 50% of the surface cell membrane being recycled every hour. It also comes as no great surprise, given the complexity of the process, that it took around a billion years of evolution to crack the problem of the 'containment paradox' by ingestion.

Diverticulum #1.2: The Great Iron Rush

There is a huge demand from all living organisms for iron in view of its central importance in providing the energy to power the cell. Unfortunately iron in the environment is usually present as an insoluble form—'ferric' (Fe^{3+}) rather than a soluble 'ferrous' (Fe^{2+}) ion. Acidification of the environment to convert ferric iron to its ferrous form is one strategy employed by organisms to take up iron—notably plants that live in calcareous soils. Bacteria bring the ferric iron into the cell before converting it to ferrous iron. They do so by the secretion of chemicals (called 'siderophores') into the environment in order to capture the iron, and then bring it back into the cell by docking with cell surface receptors specific to the siderophore. Bacteria compete by producing siderophores with greater and greater affinity for iron—effectively inventing molecules that can snatch it from each other. This original arms race has gone to incredible lengths with such extraordinary affinities for iron that some chemicals can even grab it from the air! Unfortunately, once outside the cell membrane, the iron-bound siderophore is available to any organism and some bacteria cheat by simply producing a surface receptor for the siderophores produced by other organisms in order to steal the iron! Iron is of such importance to bacteria that one of the first defences of animals to bacterial infection is to try to remove all the available iron from the environment. A molecule secreted in human saliva, breast milk and tears called lactoferrin binds iron to prevent bacteria from having it. However, rather than producing chemicals with greater affinity for iron to snatch it back from the bacteria, some animals simply produce proteins that stick to the bacterial siderophore to prevent the bacteria from having it. Such molecules include 'siderocalin' in humans, and this is also one of the many functions of the commonest protein in the blood—albumin. One bacterial siderophore—called 'desferrioxamine'—is used as a drug in human medicine to reduce toxic levels of iron in the body. Animals also respond to an active infection by locking iron away into stores in the tissues rather than allowing it to continue circulating in the blood where it may be accessible for bacteria. When infection or inflammation are prolonged this also deprives the host organism of the iron it needs to make blood cells and can result in an 'anaemia of chronic disease'.

The Invention of Eating

We have looked at 'ingestion' by forming *inward* invaginations of the surface membrane to suck substances inside the cell by endocytosis. There is of course another way to internalise a portion of the outside world—by sending protrusions *outwards* instead. Being freely able to change its shape, an animal cell can mould itself around a droplet of liquid or a particle and engulf it by then merging the membranes together on the other side. Just as with endocytosis, this process results in the inclusion of a bubble containing part of the outside world inside the cell. However, it allows a much larger area to be enclosed and importantly it also allows particulates to be engulfed, disassembled and stripped of their assets. This process is known as 'phagocytosis'[16]—derived from the Greek word to eat—or more colourfully—to devour. Just as with endocytosis, particles may be identified by specific chemicals on their surfaces that dock with receptor proteins on the surface of the cell, allowing cells to select their food. In other words, they are effectively 'tasting' it (Fig. 1.5).

The advent of phagocytosis was clearly a momentous turning point in the pathway of life. It is an extraordinarily complex mechanism that requires substantial intracellular machinery and a considerable amount of energy to power it. However, the potential gains and consequences are dramatic. Being able to

[16]The originator of the term 'phagocytosis' was the Ukrainian biologist Elie Metchnikoff who noticed accumulations of cells around thorns that he pinned into starfish larvae in 1882. He postulated that these cells were attempting to engulf and destroy the foreign bodies. For this work he was awarded the Nobel prize in 1908 and is known as 'The father of natural immunity'. He also thought that aging could be prevented by drinking sour milk every day and attributed the longevity of Bulgarian peasants to this. This theory has not been substantiated to date! However, the sour milk theory may just be one of the first steps towards our current enthusiasm for 'probiotic' remedies.

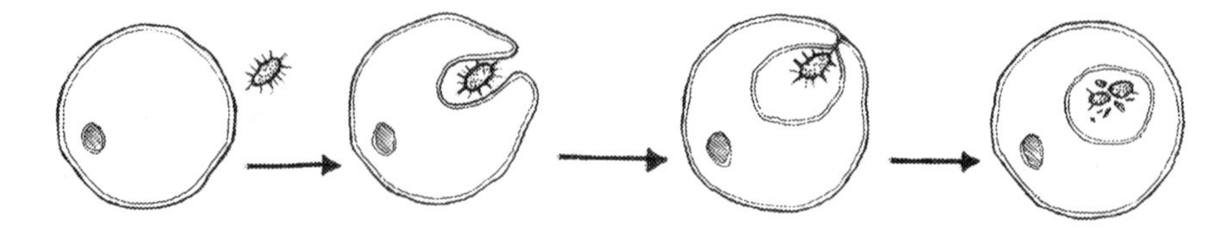

Fig. 1.5 Phagocytosis—the invention of 'eating'. The ability to ingest particles requires significant complexity of mechanisms and structures within the cell and took hundreds of millions of years to evolve, but was a turning point in the evolution of animal life

consume a large particle and digest it internally provided significantly greater sources of nutrients—and being locked away inside the cell, prevented competitors from accessing it. It also fundamentally altered the relationships between organisms in other ways.

Bacteria had probably already learnt the ability to compete aggressively with each other. They might for instance secrete digestive chemicals to 'dissolve' a neighbouring cell. Unfortunately, this allows other nearby cells to benefit from the proceeds, unlike phagocytosis which internalises the food. Evidence for direct bacterial invasion into other bacteria in order to 'digest' them from the inside is remarkably scant—only one example has been reliably demonstrated. However, one species gets as far as burrowing between an outer and inner membrane in other bacteria where it kills the cell.[17] With the invention of 'eating', such 'cheating' could become much more widespread and possible on a much larger scale. Organisms could now save energy on complex organic synthetic processes by allowing others to do so...and then devour them and steal their work.

Phagocytosis was to irrevocably alter the predator-prey interaction as a result of live prey becoming consumed within the cell.[18] Once ingested, the lifeform would be subject to the forces of natural selection and evolve escape mechanisms. Survival within the cell, having left the phagocytic vacuole could damage or destroy the cell and result in a form of 'disease'. This became so successful that many bacteria currently recognised as causing infections even encourage the cell to eat them as a way of gaining entry.

[17] The delightfully named 'Vampirococcus' adheres to other bacteria then digests them from outside. Bdellovibrio burrows through the outer surface membrane of bacteria that are double membraned and secretes digestive chemicals from its protected niche to destroy the cell. Only Daptobacter, described in 1986 has been reliably shown to invade other bacteria.

[18] Whilst we have words for most forms of eating, based on the suffixes '–vory' (such as carnivore, omnivore etc) or '–phagy' being the respective Latin or Greek roots, interestingly there is no word that expresses eating something alive. Perhaps 'Vitavory' would be the appropriate expression.

The Commune of Life

Let us take a closer look at some of the bean-like structures within the eukaryotic cell that define its nature—the mitochondria and chloroplasts. They are known as the 'powerhouses' of the cell,[19] where the spark of life is converted into chemical energy. The electron transport chain of proteins (see Diverticulum #1.1) is found on the inner membrane of the mitochondrion (a mitochondrion, unlike its host, is surrounded by two membranes). The 'chloroplasts' in plant cells contain the green chlorophyll pigment that is central to the conversion of light to chemical energy through the process of photosynthesis. Extraordinarily, these organelles contain their own self-contained genetic code, kept separate from that of the cell. However, the DNA differs slightly from that of the nuclear DNA of the eukaryotic cell by being shaped into a simple circular structure rather than tightly packed with proteins into the structures known as 'chromosomes'. The DNA code of the mitochondrion also differs from nuclear DNA in its use of a different initiation code for translation into proteins and slight variations in the meaning of the code.[20] Mitochondria and chloroplasts are capable of dividing independently within the eukaryotic cell. All in all, they strongly resemble bacteria—in the case of chloroplasts an ancient and simple group of photosynthetic bacteria called the '*cyanobacteria*'. Mitochondria almost certainly arose from a different precursor, an oxygen metabolising proteo-bacterium possibly related to modern bacterial species called '*Rickettsia*'.[21] In common with some forms of bacteria they also have two cell membranes surrounding them with a small space between.

[19] Philip Siekevitz (1918–2009) first called the mitochondrion the 'powerhouse of the cell' in 1957.

[20] The 'words' of the DNA code are each made up of three letters. Each of the three letter combinations codes for a different amino acid—these are then added together in a chain to make a protein. The sequence of letters that codes for that polypeptide is called a 'gene'. There are differences in the code between different types of organisms. Eukaryote codes have a set starting point and therefore always see the same three letter words and the sentence always translates into the same sequence of amino acids to make up the protein. Bacteria also always start the protein with a variant of the amino acid methionine called N-formyl methionine. Mitochondria also initiate proteins with this molecule, but eukaryotes and archaeans do not. This supports current evolutionary theories of the relationship of bacteria with mitochondria, and of eukaryotes with archaeans.

[21] Rickettsia is a genus of bacteria that can only live within other cells. That this genus includes *R. prowazeckii* which is the causative organism of louse-borne epidemic Typhus fever is perhaps pertinent to our story. Both of its discoverers—HT Rickett and SJM Prowazek—succumbed to the disease which is undoubtedly one of the greatest epidemic scourges of mankind. It was responsible for about 3 million deaths in Eastern Europe after the First World War but has been responsible for changing the course of history on many occasions—for instance it led to the surrender of Prague to the French in 1741 when the Austrian army was decimated by the infection. The Nobel prize in 1928 went to Charles Nicolle for the discovery of its mode of transmission.

Diverticulum #1.3: Maternal inheritance

The offspring of sexual reproduction inherit half of their genes from each parent. However, the gametes (or 'seeds') are of unequal sizes and in the case of mammals, the mitochondria in the sperm are packed into the tail which is shed on fertilising the egg (ovum). Therefore none of the mitochondria are inherited from the male, but all come from the mother. As the mitochondria have their own DNA, this means that some genes are only transmitted down the maternal line. Mutations in the mitochondrial DNA can cause devastating disease given their importance in cellular energy processes and include conditions such as 'mitochondrial myopathy' leading to muscle weakness, or 'Leber's Hereditary optic neuropathy' causing blindness. These conditions are transmitted only from mother to offspring. As mitochondrial DNA mutates only slowly, analysis and comparison of the DNA from different individuals and species allows a matrilineal tree of descent to be generated. In this way, the last common female ancestor—called 'Mitochondrial Eve'—for all living human beings has been estimated to have lived only around 160 thousand years ago.

The inescapable conclusion, now firmly established beyond hypothesis, is that of 'endosymbiosis'—literally 'inside collaboration'.[22] At some stage in the genesis of eukaryotes, a bacterium has been ingested and survived within the cell as a partner or 'symbiont'. There is considerable debate still raging about whether or not this took place by phagocytosis. Some have suggested that the outer membrane of the mitochondrion derives from the phagocytic vacuole, however its chemical composition makes it more likely to have arisen from ingestion of a bacterium that already had two membranes. Very few examples of bacteria that can ingest by phagocytosis have ever been described—this is for the most part a hallmark feature of eukaryotes. However, there are many examples of bacteria that have others residing within them (although we don't know how they got there). It also appears that the energy required for phagocytosis and the degree of intracellular organisation necessary may indeed have only become available thanks to the presence of mitochondria that were already present in the cell. Alternative theories suggest that eukaryotes evolved from archaeans that merged with or somehow ingested a proteobacterium

[22] The first person to describe the theory of 'endosymbiosis' was the Russian botanist Constantin Mereschkowsky (1855–1921) who postulated this mechanism for the origin of chloroplasts in plants. Lynn Margulis (1938–2011) reinvigorated the endosymbiosis theory and added the possibility of endosymbiotic origin of mitochondria. Her seminal paper on the subject was published in 1967 (under the name Lynn Sagan, as she was at the time married to the astronomer, Carl Sagan) after being turned down by 15 journals. Perhaps as a result of this experience she became drawn to other ideas and theories rejected by mainstream thinking such as the 'Gaia hypothesis' that have attracted less support over the following years. She was refreshingly outspoken and unafraid to be proven wrong—but then it took many years for the scientific community to accept the evidence for endosymbiosis.

prior to the evolution of phagocytosis, but the precise mechanism is as yet unclear.

However, 'secondary' endosymbiosis also appears to have occurred between eukaryotes as a result of phagocytosis. Several species of freshwater algae known as 'cryptomonads' contain chloroplasts with four rather than two membranes. Analysis of the DNA in these organisms suggests that a type of red algae (a eukaryote rather than a bacterium) has been completely engulfed—the four membranes therefore comprise (from inside out)—those of the initial bacterium, then the outside membrane of the algal cell itself that gave it a home and finally that produced by the process of ingestion of the alga by the precursor of the cryptomonad. As if further evidence were needed, remains of a nucleus like structure is found between membranes number 2 and 3—that of the original red algal cell!

An interesting example of endosymbiosis comes from recent discoveries of organisms that still practice it. *Hatena arenicola* is a single celled eukaryote discovered on a beach west of Osaka in Japan. Its name is justly derived from the Japanese for 'enigmatic'. This organism actively feeds on green algae until it comes across a particular species called '*Nephroselmis*' that it can ingest and live with as an endosymbiont inside itself. It then stops eating and lives by photosynthesis provided by its algal lodger. Only then is it capable of division, but of the two resulting cells only one can inherit the passenger and continue photosynthesis—the other has to find and ingest its own symbiont.

Some forms of marine plankton retain chloroplasts for photosynthesis but resort to eating bacteria by phagocytosis at times of nutrient shortages just like insectivorous plants such as *Dionaea*, (the well-known Venus Fly Trap) that lives in nutrient-poor soils and enhances its diet by eating.

One of the consequences of endosymbiosis is that genetic material from the incorporated organism is retained intact within the cell. Over time, genes have been transferred from the passenger DNA into the nucleus leaving only a reduced genome encoding a small number of proteins (just 37 in human mitochondria). This allows the eukaryote to control the replication of its mitochondrial lodgers but more importantly reduces the possibility of the accumulation of deleterious mutations in the mitochondrial genome as a result of many rounds of asexual reproduction (see 'Muller's ratchet' in Chap. 2).

It is also likely that eukaryotes have at times been able to incorporate or 'steal' genes from bacteria that were ingested as food. This form of information theft is known as 'horizontal gene transmission' as opposed to 'vertical' transmission (with reference to the tree of life) where genes are just passed from parent to offspring. This is a direct demonstration of how the invention of eating might have actually led to an acceleration in the rate of evolution.

Finally, the presence of mitochondria within the cell allowed eukaryotes to use the cell membrane for a variety of different purposes. In many forms of bacteria, the inner cell membrane is where the all-important electron transport chain machinery resides (as indeed in mitochondria). No longer using the surface membrane for essential energy conversion therefore opens up the possibility of large-scale endocytosis and phagocytosis without compromising the cell's power supply.

The Benefits of Being Big

Organisms tend to evolve over time to become larger. This is known as 'Cope's Rule'[23] and has been observed across a wide range of different animals in comparison to known ancestors from the fossil record. The reasons for this are not clear and one can equally consider many selective advantages in remaining small. However, it was only with the arrival of eukaryotic single-celled creatures that size really first began to matter.[24] It is entirely possible that it was simply the invention of eating itself that led to the explosive escalation in the scale of life. Whilst impressive feats of phagocytic gorging have been observed by amoebae there has to be a size beyond which one cell cannot engulf another. The predator-prey relationship therefore favours larger size even at the single-celled level. There are of course other benefits in being big—it may improve the ability to move faster or over greater distances, and there may be metabolic advantages such as increased storage capacity to pack reserves of excess food for when times are hard. These presumably usually outweigh the benefits of being small (such as hiding and escaping predation).

Thus, the cells of life forms are defined, yet confined by their boundary membrane. This is the 'containment paradox'. Whilst maintaining a relatively

[23] Edward Drinker Cope (1840–1897) was a critic of Darwinian natural selection and a proponent of Lamarckism—namely the inheritance of acquired characteristics. A prolific author (he published over 1400 papers over 40 years) and theorist on paleontology, his professional life was characterised by his rivalry with Charles Othniel Marsh and on his deathbed he left his skull to medical science in a challenge to his great rival to compare the size of their brains!

[24] To a certain extent this is mirrored in the bacterial world. The small size of Bdellovibrio is necessary in order to gain entry to the space between the two bacterial membranes. However, there are plentiful examples of bacterial sticking together in filaments after replication to increase their size and introducing predatory bacteria into culture systems appears to select for these larger forms—even in the absence of phagocytosis.

constant internal environment separated from the outside, cells need to be able to import and export the chemicals needed to support life. A variety of different selective transmembrane pores, channels and transporters evolved in order to do so. However, the 'invention of eating' by phagocytosis permitted the ingestion and subsequent internal digestion of particles, substantially increasing the ability to bring nutrients into the cell. The evolution of phagocytosis required considerable re-organisation of the structure of the cell and investment of energy that would in all probability have been provided by mitochondria. Given that mitochondria are believed to be derived from endosymbiotic bacteria that came to reside within the cell, it is something of a 'chicken or egg' problem to work out how they got there if not by phagocytosis! Some simpler form of eating that permitted whole organisms to be ingested, perhaps by unequal cell fusion would need to be invoked.

Nevertheless, the invention of eating would appear to have changed the landscape of life. It permitted a greater supply of nutrients to power the cell and freed up the surface membrane for uses other than energy conversion. It locked food sources inside the cell where they could be digested internally and not lost to competitors. It also altered the fragile relationships between organisms as whole, living cells could now be ingested. This significant step up in the predator-prey interaction was to have significant consequences as we shall see throughout this story.

2

The Gut Revolution

Summary In which we see how animals initially found safety in number and greater collective size through aggregation as colonies. The primordial cell blue print for multicellular animals was probably the 'choanoflagellate' due to its polarisation with a flagellum and feeding apparatus at one end, allowing it to replicate side by side yet retain its nutritive abilities. Early multicellular organisms may have digested food outside their body in a primitive 'exo-gut'. However, true multicellularity (rather than colonial collections of individual cells) requires the ability to keep and regulate a separate internal environment much as we have seen with single cells. Palisades of lining cells called 'epithelia' came to perform a similar containment function in multicellular animals as membranes do for single cells. Such compartmentalisation requires the existence of a specialised gut for digestion and nutrient assimilation. The ability of the gut cells to provide nutrients to the rest of the organism constitutes the 'gut revolution' which permitted the diversification of specialised cell types that no longer had to feed themselves. The evolution of an efficient 'sealed' gut cavity may have triggered rapid increases in animal diversity such as the 'Cambrian' explosion from which originated the majority of animal forms that we currently recognise.

[1] Ernst Haeckel (1834–1919), arguably the greatest German natural philosopher of the nineteenth century and a contemporary of Charles Darwin. His great interest in embryology and the observation that animal embryos are superficially similar during stages of their embryonic development led to his best-known legacy—the theory that 'Ontogeny recapitulates Phylogeny'. This theory—first promulgated by Etienne Serres in the 1820s—suggested that during their embryonic development ('ontogeny') animals pass through modified adult stages of other primitive life forms ('phylogeny'). Haeckel developed and popularised this theory and extrapolated widely from it—including his political beliefs that human society could be understood in basic biological principles. This led to his formation of a political move-

J. Woodward, *The Gastro-Archeologist*, https://doi.org/10.1007/978-3-030-62621-1_2

'No way here'

'The gut is the oldest of all the organs…the result of the first division of labour among the homogeneous cells of the earliest multicellular animal body was the formation of an alimentary cavity'.

Ernst Haeckel[1]

ment—the 'Monist league' and it is postulated that the Nazi ideology took some of its ideas from Haeckel's work. Part of Haeckel's output was discredited in 1868 when it was clear that the illustration of three embryos from different animals used to stress their similarity were in fact from the same woodcut. Contrary to popular belief, he was not brought to task and dismissed from the University of Jena even though it has entered science folk-lore as the first instance of scientific fraud. The 'Recapitulation' theory in the form espoused by Haeckel is unsupportable for numerous reasons, however it is clear that animal embryos do pass through similar stages in their ontogeny that frequently parallel phylogeny and can provide some useful insights into developmental biology.

Size Matters

Having started our journey with the probable origins of life at the alkaline hydrothermal vents as long ago as 4 billion years we have seen how it took half of the time that life has existed on this planet just to overcome the containment paradox, with the invention of eating around 2 billion years ago. We now have to travel yet a further billion years until we encounter the first 'multicellular' eukaryotic organisms (in other words, made up of more than one cell) at around 1 billion years ago. In this part of our journey we meet the first such multicellular organisms and the earliest structures that could be considered to be 'guts'...

Size really matters in nature. We have already seen how the invention of eating meant that there was an advantage to being larger. However, this comes with a price. Fundamentally, as animals get larger the ratio of their surface area to volume reduces. For a simple sphere, the surface area increases proportionally to the width of the sphere multiplied by itself, but the volume increases by this much multiplied again by the width. This rule was described by Galileo as the 'square-cube law'[2] and dictates many key biological observations.

For instance, it has been invoked to explain phenomena such as the rapid heart rate and huge appetites of small mammals, the large size of elephant ears, the increase in bulk of Swedish elk in more northerly latitudes and the enormous proportions attained by some herbivorous dinosaurs in the Cretaceous period.[3] For our purposes however, the relevance is that simple

[2] 'The Discourses and Mathematical Discussions Relating to Two New Sciences' was Galileo's final work, completed and published in 1638. His previous work promoting the Copernican theory of the Earth and other planets orbiting the sun was considered heretical by the Catholic Church and banned from publication along with all subsequent works by him—appearing on the 'Index of Forbidden Works' until 1835. Fortunately 'Two New Sciences' was published in Holland by the House of Elzevir, seemingly safe from the Roman Inquisition. In it, he describes his theories by discourse between three individuals thought to represent his own viewpoints at different stages in his life—Simplicio, Sagredo and Salviati. He describes the problems associated with simply scaling up structures by pointing out that a horse will break bones if dropped from a small height compared to a grasshopper that will survive falling from a tower.

[3] Bergmann's Rule (from Carl Bergmann, 1814–1865) is that the colder the climate, the larger the animal. Originally used to compare related animal species, it generally holds true for individuals within a species such as the elk. The reason is that the surface area to volume ratio reduces with increasing size and therefore the *proportionate* heat loss is less for larger animals. The huge bulk attained by some Jurassic period dinosaurs such as *Apatosarus* may have led to 'thermal inertia' whereby the relatively low heat loss as a result of the small surface area in comparison to volume would have led to a relatively constant temperature without the animal taking any active measures to control it. Small warm-blooded animals have high metabolic rates for their body size partly as a result of needing to generate extra heat to balance the relatively increased heat loss—which translates into faster heart rates. There is some question as to how this translates into life expectancy as larger animals also live longer. Astonishingly, the total number of heartbeats in a lifetime—about 1.1 billion—is the same for a 2 g shrew that lives 14 months as for a

organisms derive their nutrients across their surface *area*—but the amount that they require depends on the *volume* of metabolically active tissue. As organisms evolve to be larger, there simply comes a point beyond which they can grow no further, because just like Napoleon's army marching on Moscow, it outgrows its logistics supply (as he is quoted as saying—'an army marches on its stomach'[4]). This imposes strict limits on both the size of a cell and of multicellular organisms that are comprised simply of many cells of the same type. Of course, nature always finds ways of breaking the 'laws' that we make for it. Some consider an unfertilised bird's egg to be just a single cell (think of the size of an Ostrich egg!) but this is somewhat stretching the definition as the yolk in avian eggs is largely outside the cell itself. However, there are vast unicellular organisms inhabiting the deep ocean floors known as Xenophyophores. These amoeba-like creatures live at depths of up to 6 miles below the ocean surface. They can reach sizes of up to 25 cm across (about 25,000 times larger than an average mammalian cell), but remain bound by a single plasma membrane and are therefore single celled. Relatively little is known about these benthic monsters as they are exceptionally fragile and difficult to bring to the surface to study. However, they buck the size limitations in various ways. Firstly, they have many nuclei (this is called a 'syncytium' or 'giant cell'). Each nucleus is able to respond individually—switching on the genes for particular proteins necessary in its immediate vicinity—so that the whole cell is capable of reacting differently in separate parts. In some ways this is a little like a single country but with different regional assemblies (or as some Europeans might visualise the European Union!).

Secondly, Xenophyophores develop an outer casing of concreted sediment and waste products that it shapes into long fragile tubes up to 25 mm long but only 0.5–1 mm across. These tubes contain thin extensions of the cell and serve to maximise the surface area of the membrane in order to permit adequate nutrient absorption for the volume attained. Such a strategy is only possible in the uniquely still environment inhabited by these organisms and it is difficult to envisage such an opportunity elsewhere on the planet. However this is by no means the size record for a single cell—at one stage during its life cycle, the slime mold, *Physarum polycephalum* produces an enormous,

100,000 kg blue whale which lives for 80 years, and the amount of oxygen consumed by each gram of tissue before it dies is about the same.

[4] There is no evidence that Napoleon Bonaparte ever said this. However, it has also been attributed to Frederick the Great as well as the Greek philosopher and physician, Aelius Galen (129–200 AD) whose theories were to dominate western medicine for the following 1300 years.

superficially spreading, mass of slime up to 2 m across. It is still a single cell—again a syncytium with lots of separate nuclei—and takes nutrients across its cell membrane by phagocytosis. In this instance it manages to evade the square-cube law by spreading itself very thinly in order to maintain the necessary high surface area to volume ratio.

Safety in Numbers

With the rare exceptions of 'syncytia' that only achieve greatness by extreme morphological adaption, the maximum size of individual cells is highly constrained. The largest diameter attained by a prokaryote—the bacterium *Thiomargarita namibiensis*—is about the same as the largest mononuclear eukaryote, an amoeba, at around 0.76 mm. Individual spherical cells could realistically grow no larger as a result of the square-cube law. However, if the progeny of cell division were to aggregate together then they could no longer be consumed by other single cells because joined together, they would be too large. Such a strategy could perhaps be likened to that of the Duke of Wellington's redcoat soldiers forming defensive squares at the Battle of Waterloo in 1815.[5]

In this way, the earliest multicellular animals most likely formed colonies as aggregates of identical cells for mutual protection but still spent large parts of their life cycles individually as single cells. Despite being a major step, such a change in the pathway of life was not a crossroads but a slip road that was capable of re-joining the main carriageway, as some organisms subsequently lost this ability and reverted to being single-celled. Simple colonial multicellularity where the cells remain identical and can live together or independently has probably evolved as many as 46 times in the history of life—and is exhibited by various different types of organism including fungi, plants and

[5] It is typical that the British identify Wellington with this military strategy—however, the 'hollow square' had long been a successful tactic used by infantry when faced with cavalry charges and was first recorded in use by the Romans at the battle of Carrhae in 53 BC (where they were crushingly defeated!). It particularly came into its own with muzzle-loading firearms that took time to reload, as each side could be made up of 3–4 lines of soldiers that could each fire in turn. Ideally infantry would wait until approaching cavalry were around 20 m away where the bodies of downed cavalrymen and horses would build up a barrier against further charges. Using feints or dummy attacks the cavalry would attempt to get the infantry to discharge their weapons early which would then leave them effectively defenceless. Cannon fire was too inaccurate to decimate square formations unless brought perilously close or the squares were sufficiently large. The strategy continued to be used in battles until the invention of repeating weapons made any type of line formation vulnerable.

animals despite diverging from each other whilst still at the single-celled stage of their evolution.

One colonial organism that has helped in the study of the development of multicellularity is called *Salpingoeca rosetta* which comprises around 50 cells arranged in a rosette (Fig. 2.1). The cells that make up *Salpingoeca* belong to a group known as 'choano-flagellates' (from *Khoane* meaning 'funnel' in Greek), due to a specialised collar that assists them in trapping bacteria for phagocytosis. This collar surrounds a flagellum—the whip-like structure that can be used for propulsion or to generate a current in the water—at one end of the cell. Crucially, when the cells of *Salpingoeca* divide, they stay connected together side by side, rather than forming colonies through the aggregation of genetically disparate individuals. This ability of cells to divide but stay connected together afterwards would later become of fundamental importance to animal development in the formation of embryos. It appears that the many surface molecules required to link the cells together are present in related species that spend all of their lives alone as single cells. It is therefore likely that these molecules served some purpose in sensing their surrounding prior to their role in adhering cells to each other.

Notably, although gaining benefit from being clustered together, colonies of choanoflagellates such as *Salpingoeca* are still made up of individuals that can survive independently—and importantly the food each cell takes in is solely for its own benefit. These colonies are not really teams of interdependent cells but individuals coming together for mutual benefit when it suits them. As we all know though, teams work best when individuals develop their own specialisations to become expert in different areas and should be so much more than just the sum of the individual parts!

The Same....But Different

Unlike choanoflagellate colonies, multicellular organisms that we are familiar with comprise many different types of specialised cell rather than a multitude of identical ones. We humans are collections of over 250 different cell types—for instance cells that make bone, constitute blood, skin, muscle or nerve. Yet each cell of the body is basically genetically identical[6] as it carries the same DNA as all of the others, and we know that it is the genetic code that tells

[6] With the exception of the cells involved in sexual reproduction of course—the ova and sperm—or else we would all be identical clones!

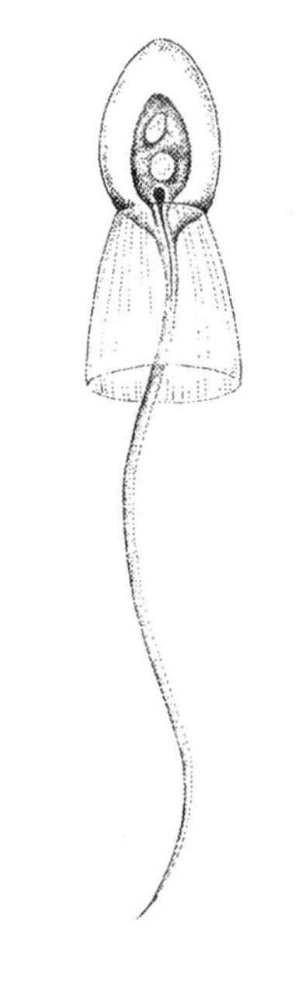

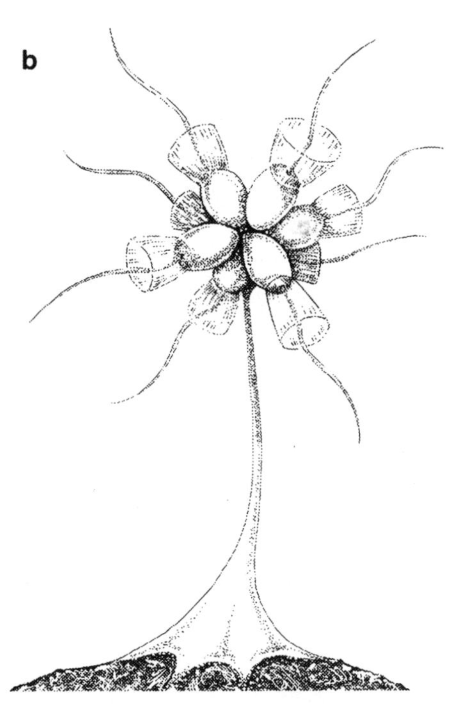

Fig. 2.1 (**a, b**) The primal choanoflagellate (above), so named for its collar (choano) and whip-like flagellum that act in concert to draw particulates to the end of the cell where they can be trapped and ingested. The colonial choanoflagellate organism *Salpingoeca* (below) made of an aggregate of single cells, but exhibiting a degree of cell specialisation by formation of a stalk. Choanoflagellates are thought to represent the precursor cells of later multicellular animals as they have polarised their feeding mechanism to one end, thus allowing them to line up side by side without compromising their ability to ingest nutrients, unlike amoebae that require their whole surface area

each cell to become what it is and to behave in the way it does. How then can the same instruction booklet tell one cell (for instance) to make insulin in the pancreas and another to sense light in the eye?

The answer to this conundrum is that cellular DNA carries a large encyclopaedia of genetic instructions which are not all used within every cell. Which genes are 'expressed'—translated into proteins—is dependent on a wide range of different regulatory factors and allows cells to respond to varied stimuli by the production of different proteins. Techniques of altering the genes expressed by the cell are described as 'epigenetic' (from 'above' or 'over' genetic) mechanisms. It turns out that the amount of DNA that actually encodes proteins varies tremendously between different species—we humans have around 19,000–20,000 coding genes but this makes up only 1.5% of our total DNA. The other 98.5% appears to have no coding function. On the other hand, *Utricularia gibba* (a rather interesting carnivorous bladderwort plant that captures and digests prey such as water fleas in its underwater bladder traps) has almost exactly the opposite ratios with 97% of its DNA encoding proteins. The spare DNA was once considered to be just 'junk' DNA but is now known to contain sequences that control or regulate the expression of the genes into proteins.

Every part of the DNA that actually encodes a protein will have certain recognisable 'non-coding' sequences neighbouring the gene. Specific protein molecules may identify and attach to these segments of DNA and thence promote or repress the reading of the gene to make the protein. Such mechanisms can regulate genes on a minute-to-minute basis and allow the cell to respond rapidly to environmental changes. Clearly cells need longer-term methods if the effects are to the last for the life of the cell and their progeny and one way in which this can be done is by chemically altering specific letters in the code.

The process of 'methylation' (the simple addition of a single carbon atom-containing molecule) acts to 'silence' the gene and stop it from being translated into a protein. Gene methylation is easily reversible but can be made permanent within the cell and even be passed on to daughter cells after division. A more sophisticated method of controlling gene expression occurs solely in eukaryotes which package their DNA by winding it around proteins called histones. When tightly wound, the DNA cannot be 'read' but if the particular histone protein is chemically modified with an attachment (this time usually with a two-carbon atom group, called 'acetylation'), then it no longer interacts so closely with the DNA, allowing it to unwind and permit gene expression to occur in that segment.

Hence, by the different *expression* of genes, cells with the same genetic blueprint can become completely different—a process given the name of differen*tiation.* These changes may be made effectively irreversible and passed onto the cellular offspring of the differentiated cell. In a sense therefore, although multicellular organisms comprise many different types of cell, they can be grouped into three broad categories—the **germ cells** that carry the genetic information between generations (which usually have most of the epigenetic 'slate wiped clean' on fertilisation in order to allow all genes equal rights of expression); the '**somatic' (or 'body') cells** that are the specialised end-result of differentiation; and the **stem cells** that retain the ability to become different types of somatic cell. The latter may be limited to particular lineages only—for instance the stem cells of the mammalian intestine can only change into a limited number of different cell types within the gut lining. Loss of gene regulation is one of the underlying causes of cancers—the uncontrolled replication of individual cells.

We now come to a rather exciting juncture in our journey as we have everything in place for the arrival of the first guts. But—if you can bear the suspense—we will first pay a visit to a rather hapless little creature in a cul-de-sac[7] of evolution.

The Exo-gut: 'No Way Here'

I recall a classic scramble route on Honister Crags in the English Lake District that I climbed when I was (much) younger. The route was complex and I kept the guidebook in my back pocket so that I could consult it regularly. At one stage it directed me along a narrow ledge to identify a rock that protruded 'like a cannon'. Edging gingerly on tip toe and fingertip around the sheer cliff face, I found the rock and fumbled precariously for the guide with one hand whilst clinging on desperately above the dizzying drop with the other. What I read next has stayed with me ever since—'There is no way here……now retrace your steps'. Taking my life once again in my hands I inched back to safety and made sure I had read the rest of the route before continuing—and I now read exam papers, cookery recipes as well as scrambling guides to the end before embarking on such ventures. However, 'no way

[7] 'Cul-de-sac' is of course a French expression that we have adopted in English. A 'Cul' in French is actually any blind-ending organ and was used in medieval times for the vagina (leading to some interesting street names in the Parisian red light district) but latterly for the bottom (which is of course not blind-ending!). It has recently taken on a meaning in French slang as a word for 'pornography'. Cul-de-sac therefore translates as 'bum end of the bag'!

here' reminds me of *Trichoplax*, the Betamax video equivalent of animal evolution.

Trichoplax adhaerens is a 'one of a kind'—all animals are classed in related groups or 'phyla' (sing. 'phylum') of which there are in total just 34 (or 35—experts cannot quite decide on this!). *Trichoplax* was the only member of the phylum Placozoa and Placozoa *was* the only phylum with only one member, until 2017 since when two other species have been identified. All biologists have their favourite animal, and in the true British fashion of favouring the 'also-ran', this is mine. It is a tiny creature made up of only a few thousand cells and it measures 1–2 mm across. It looks and behaves like a large version of an amoeba that has forgotten that it is now multicellular. It was first found in an aquarium in the late nineteenth century[8] but was subsequently identified in warm waters where it lives on the shore line. It can perform the neat trick of passing through solid structures like a ghost as long as they are porous—if pushed through a sieve and divided into single cells they will all come together again, and if the cells from two different *Trichoplax* are combined then the resulting animals will also remain of mixed origin. It is not however a simple colonial animal, being comprised of five different types of differentiated cell that make up two layers separated by one huge jelly like middle. It has no sensory nerves or muscle cells and moves simply by the vibration of tiny hairs on its surface. It has no front or rear and is not bilaterally symmetrical, but it does have well-defined upper and lower surfaces.

Trichoplax has developed not one but two unique ways of feeding—one by each layer. In its upper surface, it can simply exude mucus slime to trap small creatures and bacteria which can then be drawn back in between the cells and phagocytosed into them. However, it is the lower layer that has evolved a means of digesting larger creatures. Once the Trichoplax has passed over its prey, the bottom surface invaginates into a cavity—again much as if it was behaving like a single cell ingesting by endocytosis. Specialised gland cells then secrete digestive chemicals into the cavity outside the animal to digest the food that can then be absorbed through the surface lining cells. It has effectively created an outside or 'exo'-gut around its prey to digest it outside the body (Fig. 2.2). Whilst *Trichoplax* demonstrates many features that seem to predict the future evolution of animals, it

[8] *Trichoplax* was discovered in an aquarium in Graz, Austria in 1883 by the German marine biologist Franz Schutze who was also one of the first to describe Xenophoryphores. *Hoilungia* was described as a separate species in 2017 on the basis of genetic differences from Trichoplax, and *Polyplacatoma* was discovered free-living in the Mediterranean Sea in 2019.

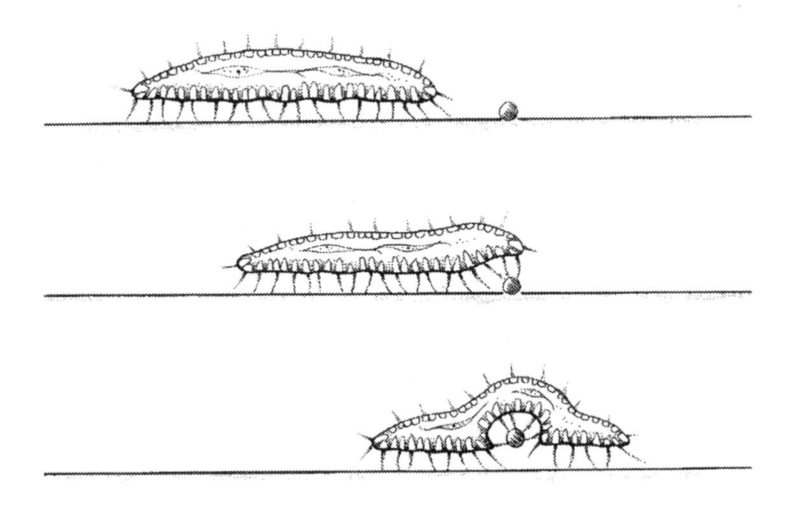

Fig. 2.2 Trichoplax adhaerens feeding. The animal passes over particulates which it senses and surrounds, then secretes enzymes to digest it so that the nutrients can be absorbed. This is the 'exo' or external gut, an inefficient digestive mechanism as it risks losing nutrients and digestive chemicals into its surroundings

appears to have occupied an isolated backwater in the pathway of life. Perhaps this is further proof for Galileo that you cannot simply scale up an amoeba, even if it is made of separate cells!

However, there are some fundamental features of *Trichoplax* that bear closer inspection. Firstly, the containment paradox is at play again with the exo-*gut* which is bound by the same limitations of efficiency as experienced by single cells that secrete enzymes into the medium to digest outside the cell membrane. The nutrients and the enzymes can simply diffuse away into the medium. Instead, the future for multicellular organisms would be exo-*digestion* but within a cavity entirely contained within their own body plan (the gut) in order to maximise efficient capture of nutrients and recycling of enzymes. This is the multicellular equivalent of endocytosis. Secondly, *Trichoplax* has two layers of cells. This allows one layer to develop nutritive functions and the other to interact with the outside environment—normally by shutting it out in forming a skin. However, this is a major step—it requires some cells to give up feeding altogether whilst relying on others to nourish them—and *Trichoplax* never quite achieves this as both layers appear to feed independently. To do so requires a revolution akin to the societal changes that have allowed me to sit here and write this whilst fellow humans produce the food that sustains me.

First Guts

Having travelled over 3 billion years from the origin of life we now find ourselves at a point between 800 and 600 million years ago when the first truly multicellular organisms arose. Their first convincing fossil traces date from about 600 million years ago, as the earth was thawing from a global ice age. This is in some ways the 'dark ages' of life as many of the key developments happened in quick succession and left little trace—and as a result there is quite heated controversy within the 'evo-devo' (*evo*lutionary *dev*elopmental biology) science community about the sequence of events. What is generally accepted is that the first truly multicellular animals resembled modern day sponges and arose from choanoflagellate-like cells. The major reason for this is that uniquely amongst single-celled organisms they had developed a specialised feeding mechanism at one end (the eponymous collar surrounding the flagellum) that would not be compromised if they were joined together side by side, whereas others would lose crucial surface area by doing so.

Sponges[9] have a basic body plan of a cavity lined with cells that look very much like choanoflagellates and are called 'collar cells' because they have a similar arrangement of a central propeller-like 'flagellum' surrounded by tiny projections forming a 'collar' that traps bacteria and small particulates.[10] The outside lining of the sponge is made of specialised cells called 'pinacocytes' (from the Greek 'pina' = to drink and 'cyte' = cell) that are flattened to form a skin, and are also capable of adhering to a surface. The pinacocytes also secrete a jelly-like substance between the two layers.

The sponge gains the benefit of larger size as the flagella of the collar cells all beat together to generate a current in the water that carries food particles through tiny pores in the outside layer and into the body cavity. This is clearly much more efficient than being dependent on random collision or actively seeking out tiny prey by swimming. However, in trying to achieve a larger size, animals are at a distinct disadvantage compared to plant cells.

[9] Henry James Clark (1826–1873), an American natural scientist who started his career in botany but later developed an interest in zoology was reportedly the first to notice the similarity between the collar cells of sponges and choanoflagellates although some claim that the French biologist, Dujardin made the same association about 25 years earlier in 1841. Clark however considered the sponges to be colonial choanoflagellates like *Salpingoeca*, and this was later disproven by Ernst Haeckel's studies.

[10] Sponges comprise the phylum 'Porifera' (derived from Greek = 'carrying pores' from their sieve-like structure). It is the oldest subdivision of multicellular animals with fossil traces dating back to the Ediacaran era around 600 million years ago. Currently over 8000 different existing species of sponge have been described. A subtype called 'Demosponges' form the soft variety that has been used by humans for cleaning and padding purposes and have been substantially over-harvested, leading to the use of artificial substitutes. Recent uses include reusable feminine sanitary products. Humans are not the only species to use sponges as a tool—it has also been noted as a recent development in a group of bottlenose dolphins studied in Australia which use hard sponges to disturb sediment and dislodge fishes.

The reasons for this is that having overcome the containment paradox through phagocytosis, animals are now entirely dependent on maintaining a membrane that is sufficiently fluid to facilitate it - that lacks a rigid structure. Plants on the other hand, by using sunlight to power the binding of simple chemicals together into more complex ones, have been able to surround their individual cells with strong, thick walls that can support enormous structures such as trees, on land, unsupported by water, over 100 m tall![11] It was the jelly like interior, sandwiched between the two body layers that allowed the first aquatic multicellular animals the structural stability in order to grow. Later cell specialisation would generate the hard structures made of silica or calcium carbonate that act as skeletal scaffolds—that when laid down and compressed over time became rocks such as flint or chalk (Fig. 2.3).

So—we now have our first primitive gut, or 'proto-gut' of sorts. It is a simple body cavity that concentrates and selects appropriately sized food particles within it, lined by specialised cells that ingest them. The first gut however probably still only nourishes itself—it has not yet learnt how to share with the rest of the organism.

Diverticulum #2.1: Our Selfish Guts

A key step in the evolution of animals came when cells that absorb nutrients in the gut lining learnt how to share them with the rest of the organism. Presumably this could only come about when the efficiency of digestion allowed for excess nutrients to be released from the food. These nutrients need to be provided in the correct forms and quantities for the non-gut cells to thrive.

Perhaps not surprisingly, the cells of our guts still take up nutrients directly from the food passing through them rather than just from the blood which supplies all the tissues of the body. Starvation or bypassing the gut by feeding directly into the blood stream ('intravenous nutrition') results in damage to the intestinal lining that can affect its function. Interestingly the favourite fuel of the cells lining the small intestine is an amino acid called glutamine. It may be that in this way enterocytes do not use up all the glucose which is a favourite energy source of many other cells. Other reasons include the extra nitrogen provided in glutamine for rapidly dividing cells. When we try to restore a deficiency of glutamine in the body by giving it as a food supplement, we see very little of it being shared with the rest of the body as the gut cells take it all up!

Similarly, the cells lining the large bowel (colon) derive a large part of their nutrition from small fatty acid molecules that are produced by the bacteria that live there. When the bowel is disconnected, we see a 'diversion colitis' that can cause bleeding and diarrhoea, due to the lack of nutrients available for the lining cells.

[11] The tallest tree in the world was discovered in 2006 and has been called 'Hyperion'. A specimen of *Sequoi sempervirens*, the giant Californian redwood, it has been measured at 115 m high with a circumference of over 30 m and a weight of 1 million kg. Its exact location is kept secret.

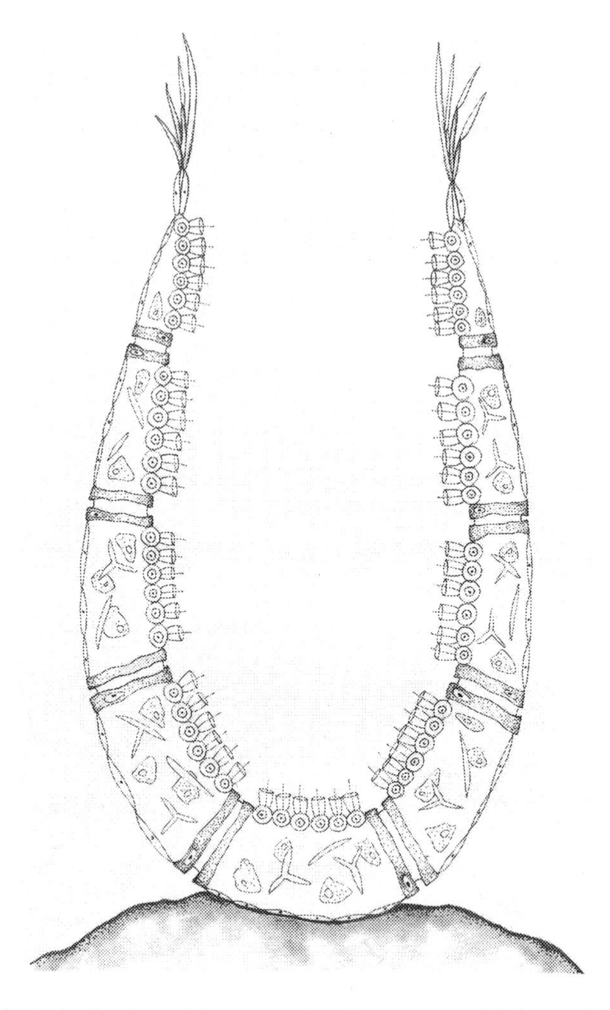

Fig. 2.3 The basic body plan of the sponge—a 'proto gut'. This is a cavity acting like a sieve with fluid and particulates drawn in through pores by the co-ordinated beating of flagellae. Note the similarity of the gut lining cells to the choanoflagellates

Sex and the Gut

Organisms made up of different cell types face the problem that only some will have a nutritive function (such as the collar cells in sponges) whilst others have given up this role in order to perform other functions. The different parts of the animal therefore need to work collaboratively and in particular the cells that get the food need to pass it around! Indeed, it is likely that the earliest animal cells did not show any such significant degree of altruism but that different cells simply developed their own particular ways of feeding. For instance, the surface cells of sponges are still capable of phagocytosis and ingest the larger particles that get stuck in the small pores rather than filtering into the gut cavity. Furthermore, the middle jelly layer provides a rich environment for bacteria to grow in—many of these develop a 'symbiotic' relationship with the host sponge and generate nutrients that can be used by both cell layers—or they can be eaten when excess to requirements! It is estimated that up to 80% of the energy requirements of some sponges are provided by symbiotic bacteria which can make up as much as one third of the mass of the sponge. Within this rich middle layer, cells called 'amoebocytes' are found which resemble amoebae and exhibit phagocytosis.

A crucial step then in developing multicellular animals from colonial species is the degree of collaboration between different cell types that allows some to lose their ability to feed altogether and to rely entirely upon a specialised gut for nourishment. This only comes about when the 'unit of evolution'—on which natural selection acts—is not the single cell, but the whole organism made up of mutually interdependent cells. And this only happens when the animal can reproduce itself as an animal—true to form including all of its different constituent cell types in the correct relative positions, numbers and relationships to each other. Whilst basic organisms such as sponges can simply spawn by budding or breaking off fragments that grow separately, complex creatures need the body parts to develop in sequence. This requires the ability to grow as an embryo from an egg with all the cell fates and the body plan pre-programmed into the DNA code.

Diverticulum #2.2: The need for Sex

Sexual reproduction probably evolved around 1 billion years ago in eukaryotic cells at a time prior to the first multicellular organisms. Most—but not all—animals have two sets of genetic information, one from each parent. Thus humans have 46 chromosomes (bundles of DNA) of which there are 22 pairs and separate X and Y sex-determining chromosomes. During cell reproduction these can swap (crossover) genes creating a mixture of both parents' information on each of the sister chromosomes.

The benefits of sexual reproduction stem from the ability to disseminate advantageous genetic combinations which can arise as two individuals' genomes come together, while helping to prevent deleterious mutations from accumulating in successive generations (deleterious mutations are assumed to be more likely than beneficial ones!). It is therefore relevant that species that have the ability to reproduce either way prefer asexual reproduction when there is little environmental change and sexual reproduction when there is a need to adapt quickly. It has been suggested that the advent of infection has led to a 'mutation race' between parasites and the host, with the prize going to whichever can adapt most rapidly to the latest challenge provided by the other. Hence those species most prone to parasitic infection will likely gain most from sexual reproduction. This is the so-called 'Red Queen Hypothesis' named after the character in Lewis Carroll's *Alice in Wonderland* who needs to run faster and faster just to stay in the same place.

Having two copies of all the genetic instructions in cells is useful for repair of any damage to DNA as it provides a reference template and it may well have been this function itself that actually led initially to the development of sexual reproduction. A second copy also allows nature to tinker with the duplicate to see if it can come up with something better whilst not doing away with the original—we will see this again repeatedly throughout our journey.

Asexual reproduction on the other hand can generate large numbers of offspring quickly. However, with no way of correcting or selecting out the deleterious mutations that inevitably result, these build up over time to the significant detriment of the population—this is called 'Muller's Ratchet'.[12]

We can make one last simple observation from our brief foray into the fascinating world of sponges. Their spermatozoa are formed from collar cells which retain the propeller-like flagellum to allow them to swim to another sponge in order to fertilise an ovum (which is derived from an amoebocyte). We should never judge solely by appearances, but it would seem that the form—and swimming behaviour—of animal spermatozoa has barely changed throughout all 600 million years of evolution thereafter (and that includes us). We might also note in passing that the ancient gut lining once played a role in sexual reproduction by itself producing the male spermatozoa.

Containment Revisited

We have seen how in order to 'define' life it required containment within the cell membrane to provide a separate and different internal environment. With the advent of multicellular organisms with surface layers and an 'inside' middle layer we are now in a similar situation. The jelly in the sponge (and creatures such as the jellyfish) is composed of molecules that hold water. The

[12] Herman Joseph Muller (1890–1967) was an American geneticist whose work on genetic mutations caused by radiation led to him being awarded the Nobel Prize in Physiology or Medicine in 1946. He was somewhat peripatetic in life—his left wing political tendencies led to his involvement with a reactionary publication called 'The Spark' at a politically sensitive time in 1930s USA which led to him working initially in Berlin and then following the rise of Nazism to the USSR. His work fell into disrepute with Stalin and he returned to the USA in 1940 via Edinburgh and Madrid.

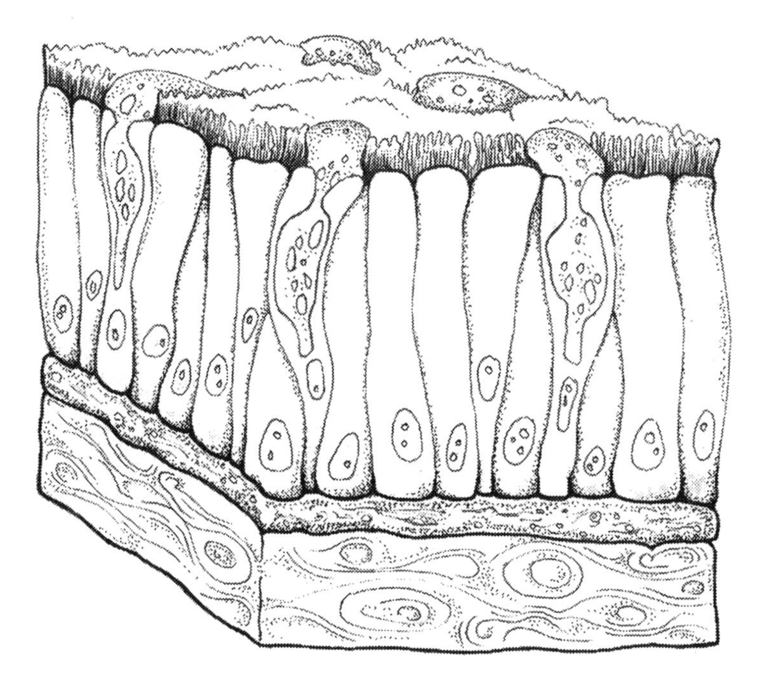

Fig. 2.4 A simple epithelium. A single layer of cells sitting on a basement membrane. The cells making up this epithelium are columnar in shape, as found in the gut lining. The epithelium provides 'containment' to the multicellular organism in the same way as the membrane does for the single cell

turgor in the jelly and therefore the structural support that it can provide depends on its state of hydration. Whilst one might think that this would be constant in an aquatic environment, this may well not be the case as it can depend on the relative amount of salt present in the gel and in the surrounding water—the salinity (which can vary). Water will move from areas of low salt concentration to areas of high salt concentration—which is why pouring table salt onto freshly spilt red wine is effective at drawing it out of a carpet. There are therefore benefits for organisms in regulating the composition of their middle layers, just as for the cell in maintaining its internal environment. Too much sodium and water and it swells up, too little and it desiccates and shrivels. This once again requires containment, and whereas the membrane is the solution for the cell, the answer for the multicellular organism goes by the name of 'epithelium'[13] (Fig. 2.4).

[13] The word 'epithelium' is a complete misnomer as is the case with many biological terms—it means 'overlying (epi) the nipple (=Greek, thele)', for some reason that is entirely obscure and has nothing to do with its true meaning.

An epithelium is basically a sheet of cells that has several characteristics—the cells all sit on a basement layer of connective tissue; they are all joined together to form a barrier and they are often 'polarised' such that one end of the cell behaves very differently from the other. Epithelia can be made of several layers of cells—such as our skin—or just a single layer of cells as in our intestines or the lining of our blood vessels, but can be made up of more than one type of cell. They basically 'contain' the internal milieu inside our bodies and control what passes in or out just as the cell membrane does for the individual cell.

The cell layers in *Trichoplax* are not complete epithelia as the cells do not sit on a basement membrane—and this is why the upper layer can feed in the way it does by exuding mucus between the cells and drawing it back into the organism. Whilst it has long been contentious, there is now clear evidence that sponges do have established epithelial barriers, but this was always likely given that they can exist in freshwater as well as sea. Therefore, epithelia evolved as an important feature along with the first multicellular animals.

However, the 'containment paradox' is once again at play. Now that epithelia are present, the organism has closed off the outside world and requires a way of transporting substances into itself. A variety of different routes are available—from opening up the junctions between cells in a rather non-specific way, to selective 'transcytosis' whereby the endocytic vesicles (that we first encountered in Chap. 1) travel with a cargo inside a 'bubble' from one side of the cell to be discharged on the other. It is also possible for epithelial cells to selectively place pumps or channels across the membrane in both the end of the cell facing into the gut or the other end facing the inside of the animal. This can allow for selected substances to be actively pumped across the barrier made by the cell junctions and into the interior of the animal. Probably the first and most important of these pumps among animals is one that exchanges sodium ions (Na^+) for another important ion, Potassium (or K^+: see Diverticulum #2.3). This pump is present in all cell membranes to keep the sodium content inside the cell low compared to the outside (in our case the blood—which has a threefold higher concentration). Some have even gone so far as to suggest that the maintenance of higher salinity in the fluid outside the cells of our bodies reflects the marine environment in which life evolved, even though it falls a long way short of the sodium content of modern seawater.

Now perhaps 3 billion years from its beginnings, animal life had overcome numerous obstacles. Food and feeding were the fundamental drivers behind the problems and their solutions. The barrier of containment at the cellular level had been resolved by the invention of eating, which led amongst other things to the advantage of size; the limitations of scale were then met through multicellularity; this required a body plan with specialisation of different cell types including a 'middle layer' to provide structural integrity in view of the necessary 'softness' of animal cells, and epithelia to control the internal environment. This all led to the requirement for greater feeding efficiency through the creation of a gut. Just as the agricultural revolution freed people from the land and led to the diversification of human society (artists, musicians, scientists, doctors—even evolutionary biologists!) so the 'Gut Revolution' freed cells from the need to derive their own nourishment and to specialise in different ways.

Diverticulum #2.3: Sodium/potassium exchange across cell membranes

One of the ways in which cells differ internally compared to the external environment is in the concentration of salts—sodium and potassium. This is achieved through a specialised membrane pump. Three positively charged sodium ions (Na+) are driven out of the cell in exchange for two potassium ions (K^+) that are allowed in. This requires the energy of an ATP phosphate bond to power it (see Chap. 1) hence the pump is called the 'Na^+/K^+ ATPase'. In exchanging unequal amounts of charge across the membrane it generates a tiny electrical difference between the inside and the outside of the cell which also pushes negatively charged chloride ions (Cl^-) out of the cell. In this way it reduces the amount of sodium and chloride ('salt') in the cell. A number of consequences follow. The difference in sodium concentration inside and outside the cell means that opening a channel would allow sodium to pass back quickly into the cell down its concentration gradient (helped by the electrical charge difference). This flow can be used to bring useful molecules (such as glucose) into the cell using protein transporters across the membrane that bind to both. However, fluid also travels with ions and the Na^+/K^+ ATPase is a fundamental means of controlling the internal volume of cells. The fully functional forms of this pump are first found in multicellular animals and it is most likely that it originated to control the internal environment of the animal by altering sodium concentrations across epithelia rather than the cell itself. The importance of this single molecule is such that it is estimated to consume approximately 40% of all our basic energy expenditure.

The Gut Revolution and the First Explosion of Life

Our understanding of the evolution of life over the last 600 million years or so relies on the fossil record, which is unfortunately not a complete archive. It depends significantly on the environmental conditions present at the time of an organism's demise and the presence of hard body parts. Rarely are soft parts of animals well preserved and therefore fossils are over-represented by those creatures which have developed hard shells or skeletons. Nevertheless, under the right circumstances soft creatures can be preserved as fossils, such as those found in the Ediacaran hills north of Adelaide in Australia.[14] These rare 'Ediacaran fauna' represent the first global expansion of large multicellular organisms around 575 million years ago, a time that is also known as the 'Avalon explosion'—of life.

In 1957, a 16 year old schoolboy called Roger Mason was climbing in a quarry in the Charnwood Forest near his home in Leicestershire when he came across a fossil that resembled a leaf. He took a 'rubbing' of the fossil and showed it to his father, who was acquainted with a geologist who identified it as the earliest (then) known fossil of a multicellular organism and the first evidence of such life in the Ediacaran strata. Such fossils are of such rarity that this location in England turns out to be the only place in the whole of Western Europe where they are found. The fossil has been named '*Charnia masoni*' and its discoverer went on to become a professor of geology.

Charnia is symbolic of the enigma represented by all the Ediacaran age fossils subsequently identified around the world. Initially thought to be a plant as a result of its frond-like appearance, it is believed to have existed in deep waters beyond the reach of light and therefore was more likely to be a type of animal. Exactly what kind of animal is a mystery as it bears little or no resemblance to any existing creatures. Furthermore, it does not appear to have any obvious gut aperture or feeding apparatus. Similarly, other life-forms from this era preserved as fossils exhibit bizarre shapes, quite alien to modern organisms. Theories about their nature still range widely from being types of lichen to weird animals that were composed of fluid filled sacs under pressure called 'pneu' organisms, a bit like quilted mattresses. Some have suggested that these creatures were simply nourished by symbiosis, providing a home to bacteria that could supply adequate nutrients. Some grew to enormous sizes such as *Dickinsonia* which has left fossils up to 1.4 m across. The latter organisms can be seen in some strata to have left trails behind them—possibly a sign of grazing behaviour on the algal mats that would have existed in shallow waters at the time. It has been suggested that this organism had a gut opening along its underside and therefore acted a bit like a 'hoover' and may represent a giant extinct *Trichoplax*-like organism. However, the general paucity of such 'trace fossils'—tracks or marks left in the sediment by animals moving—suggests that Ediacaran creatures were largely sedentary. The bizarre organisms of the Ediacaran era may indeed represent groups of animals that no longer exist—an entire extinct animal kingdom—which has been named the

[14] The Ediacaran hills are in the north of the Flinders range about 400 miles north of Adelaide in Australia. Their name probably derives from the aboriginal 'Yata Takarra' meaning 'stony ground'. Similar Ediacaran age rocks have been found in China (Doushanto formation), Newfoundland (Avalon peninsular), Sonora in Mexico and in Namibia as well as Charnwood forest in Leicestershire.

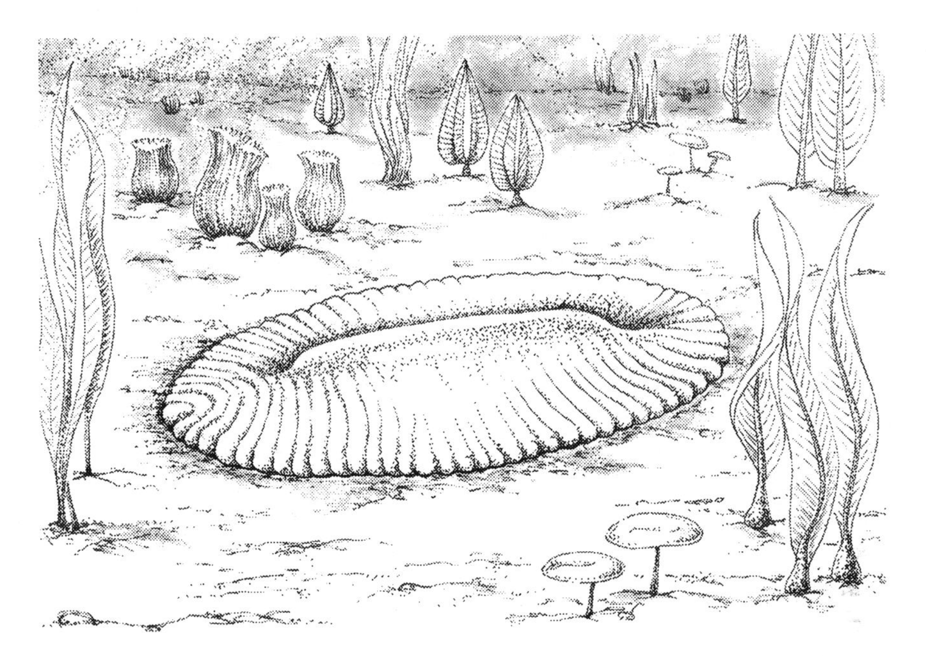

Fig. 2.5 Dickensonia in the landscape of the Ediacaran sea floor. Was this perhaps an enormous Trichoplax-like creature?

'Vendobionta' after the 'Vendian', a previous name for this geological period[15] (Fig. 2.5).

Very few 'Ediacaran' creatures are seen in the fossil record more recently than 540 million years ago. Only our old friends, the single-celled protozoans such as the choanoflagellates and multicellular sponges, are consistently present from this time onwards. Why the other types of creatures failed to prosper is a mystery and theories of the cause of their extinction abound—from climate change to the development of eyesight improving the ability of predators to hunt. However, one striking feature of the extinct types of creature from this first explosion of life is the lack of an efficient enclosed gut capable of digestion—it appears that all were nourished by symbiosis or phagocytic filter feeding similar to the sponges. Perhaps we might be so bold at this stage in our journey along the passage of time as to propose another theory for the extinction of the Ediacaran fauna—namely the powerful potential unleashed by the evolution of a true digestive system that led to overwhelming competition.

[15] The German palentologist Adolf Seilacher (1925–2014) has contributed much to the description and understanding of the Ediacaran biota and many of the suggestions in this paragraph belong to him.

The Dawn of the Gut Age

If the first 'proto-guts' of the sponges coincided with the development of multicellularity and the first global explosion of life (in the Ediacaran era), then perhaps it was the first 'true guts' that heralded the greatest expansion of life in the history of the planet. This is known as the 'Cambrian'[16] explosion, during which most currently existing types of life form came into being. Whilst it can be surprisingly difficult to prove a causative association for any observation, it follows from the logic of the 'Gut Revolution' (whereby the evolution of a digestive tract allowed multicellular diversification) that an increase in gut efficiency could result in greater specialisation by organisms—and thereby competitive advantage. Perhaps if the Ediacaran (coinciding with the sponge proto-gut) was analogous to the British Agricultural revolution, the development of true guts of the Cambrian could be considered to parallel the much more significant social advances of the Industrial Revolution.

The Cambrian explosion has caused significant difficulties for paleontologists and evolutionary biologists who struggle to explain the sudden arrival of so many diverse creatures in the fossil record. Evolution had previously been understood as a gradual slow process of adaptation and Charles Darwin himself could propose no satisfactory explanation except for the incompleteness of the fossil record.[17] Perhaps our gut-centric perspective now provides the answer!

---

[16] The 'Cambrian' period is named from 'Cambria' a post—Roman latinised form of Cymru, the Welsh name for their country, (derived from a word meaning 'fellow country folk'). It was so named by Revd Adam Sedgwick (1785–1873). Sedgwick read theology and mathematics at Trinity College Cambridge but went on to study geology and for 55 years was 'Woodwardian' chair of Geology in Cambridge (actually 'Professor of Fossils' set up in 1728 by a Dr J Woodward—no relation!). One of his students was Charles Darwin and although they remained friends, he could not bring himself to reconcile his evangelical beliefs with the theory of natural selection—writing that 'it repudiates all reasoning from final causes; and seems to shut the door on any view…of the God of Nature as manifested in his works. From first to last it is a dish of rank materialism cleverly cooked and served up'. His legacy in Cambridge is the Sedgwick Museum of earth sciences—well worth a visit!

[17] In the sixth Edition of *On the Origin of Species*, Darwin wrote: 'to the question why we do not find rich fossiliferous deposits belonging to these assumed earliest periods prior to the Cambrian system, I can give no satisfactory answer'. The rift between 'gradualists' and 'saltationists' (from Latin, *Saltare* = to leap) in evolutionary terms predated Darwin with leading biological theorists such as Etienne Geoffroy Saint-Hilaire even proposing that new species could develop instantaneously. Darwin was very definitely in the gradualist camp—'natural selection acts solely by accumulating slight successive favourable variations, it can produce no great or sudden modification; it can act only by very short steps'. His friend Thomas Huxley warned him that his work showed an 'unnecessary difficulty in adopting *natura no facit saltum* (nature does not take leaps) so unreservedly'. Debate continued into the twentieth century with evolutionary models such as 'Punctuated Equilibrium' being proposed by Eldredge and Gould in 1972 and the 'Hopeful Monster Hypothesis' by Richard Goldschmidt in 1940. There are significant differences between these approaches—the notion of punctuated equilibrium helped to explain the stasis and lack of

Returning to the sponges, we are in fact only a relatively short step away from the first true guts. We already have a gut cavity (albeit porous), specialised cells for feeding, a middle layer to provide structural integrity and rudimentary epithelia. Furthermore, sponges evolved the first muscle-like tissue (in their middle layer) in order to augment the flow of water through the gut cavity by contracting the body to act as a sort of pump. In order to co-ordinate this contractility and pumping mechanism, rudimentary nerve cells forming a 'nerve net' also formed. From here therefore, in order to create a gut, we just need to close the pores through the walls of the sponge to create a sealed cavity, find a more efficient way of pumping or bringing food into it (as continuous flagella-generated flow is no longer possible) and an ability to secrete the enzymes needed to digest food into the gut cavity. Considering the great leaps, each of a billion years or more, it has taken to arrive at this point in our journey, it is perhaps not surprising that these tweaks only took a relatively short period of time and the Cambrian Explosion followed 'only' 30 million years after the start of the Ediacaran.

In this way, it is quite likely that the first true guts looked very similar to the flask shape that we have just described in sponges with a single opening acting as both mouth and anus. This is exactly what we can see today in the 'Cnidaria', a group of animals that includes jellyfish and corals. They take their name from the Greek word 'Knide' (nettle) in view of the 'stings' they have developed. These are specialised cells called 'cnidocytes', mostly localised to tentacles around the mouth region that contain a hard barb on a spring like mechanism that discharges when triggered by physical or chemical stimuli. The barb can penetrate other organisms and inject toxins that serve to paralyse, kill and even begin to digest them internally. Jellyfish toxins can kill large prey such as fishes, and can be so potent that they are of danger to even bigger animals such as ourselves—the sting of the box jellyfish (*Chironex fleckeri*) results in 20–40 human fatalities every year in the Philippines alone.[18]

---

large outward changes in organisms well adapted to environments (although not ruling out smaller stepwise change) with rapid changes occurring at times of adaptation to changing environments. The 'Hopeful Monster' envisaged major changes occurring over a single generation. Modern genetics have considerably assisted in understanding the magnitude of evolutionary changes as proteins coded by single genes can regulate large numbers of other genes involved in shaping the organism and permit apparently large changes to occur rapidly. Needless to say, creationists have capitalised on the Cambrian explosion to support their views!

[18] The first aid management of box jellyfish stings has recently changed. Tentacles often still remain attached to the victim and the aim is to prevent discharge of the stinger cells that have not already done so (as few as 1% may have 'fired' initially). Washing in salt water and scraping off the tentacles is no longer considered the first line treatment—washing copiously in vinegar or a special solution called 'StingNoMore' ® developed by the Hawaii based researcher Angel Yanagihara is more likely to prevent stinging. Application of heat may be preferable to icepacks.

Probably the simplest Cnidarian to understand is the green *Hydra*[19] which is a flask-like structure attached to a surface at one end with its mouth/anus at the other surrounded by tentacles. When a tentacle touches its prey the stings discharge to paralyse it, and other tentacles wrap around it and drag it into the open mouth. The gut lining (called the 'gastrodermis', from *gastro* = stomach and *dermis* = skin) has developed two specialised cell types—one to secrete digestive chemicals into the gut cavity and the other for absorption. The latter has lost the large flagellum but retained the smaller projections similar to those that make up the collar of the choanoflagellate. The prey is digested and broken up—possibly also by contractions of the gut cavity as a result of some cells developing contractile fibres similar to those found in our muscle cells. Soluble nutrients are absorbed directly but remaining particles can still be phagocytosed by the gut lining cells and further digested internally. Indigestible remnants of the prey are expelled after 2 or 3 days and the feeding cycle begins again (Fig. 2.6).

Once again we should stop for a moment to reflect on a couple of aspects. Firstly, *Hydra* has a sophisticated gut that can now nourish the entire organism by passing nutrients across the gastrodermis cells to diffuse through the jelly to its outer surface cells (although it has still retained symbiotic algal partners in its jelly layer—hence the green colour!). It was once considered that the amoebocytes shuttled nutritious particles across the middle layer, but there is precious little evidence to support such a notion. Being dependent on chemical diffusion, the jelly layer in *Hydra* is necessarily thin. Jellyfish with much thicker jelly layers have developed a highly sophisticated network of finely branching channels from the gut cavity. This is called the 'gastrovascular system' (as in some ways it resembles our vascular system of blood vessels) and serves to deliver nutrition throughout the organism, whilst still relying on chemical diffusion.

Secondly and as a consequence of having a sealed gut, *Hydra* has now developed the ability to actively feed itself. This requires the co-ordination of the tentacles to sense the surroundings, move towards prey and then carry the food into the mouth. It also needs the body to generate mixing movements and expel the 'leftovers'. In order to do all this, it has developed an integrated nervous system—a basic but relatively sophisticated nerve net largely concentrated at its 'head end'. We can now see how with its efficient gut, the

[19] The *Hydra* is named after the many serpent-headed beast of Greek mythology which guarded one of the entrances to Hades and was famously slain by Hercules in his second labour. After cutting off a head it simply grew back—as can occur with *Hydra* itself—although the mythical beast would grow two heads for each one removed!

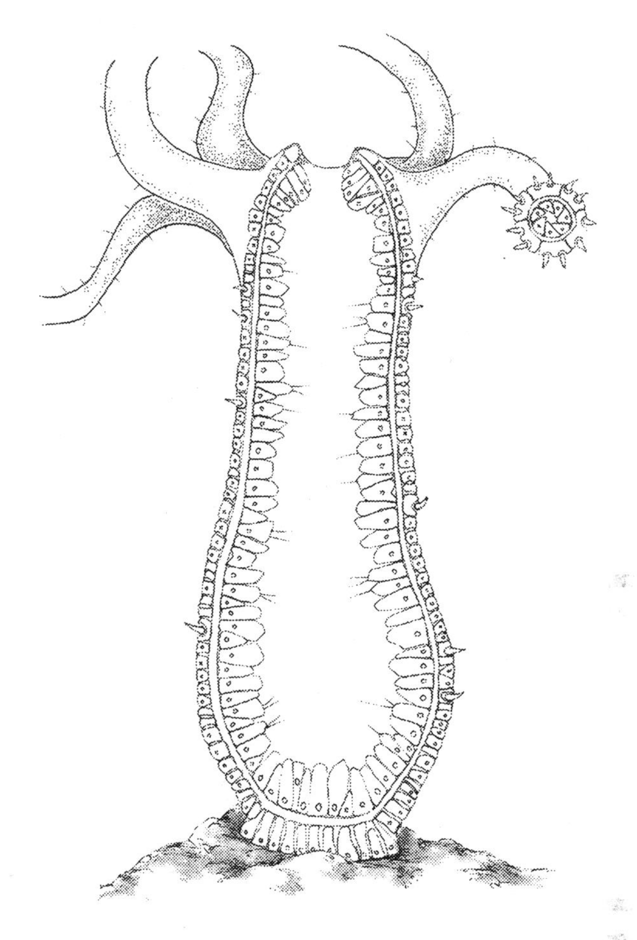

Fig. 2.6 The cnidarian body plan—a *Hydra*. The sealed body cavity acts a gut with a single opening that serves as mouth and anus

organism can invest in effective means of predation and how (and why) sensory organs, a nervous system and even appendages such as tentacles evolved. In short, it was in order to fill the gut.

Early Guts: The 'Ins and Outs'

In chemical engineering terms, the Cnidarian guts could be described as 'batch reactors'. Much like home brewing beer, all the reagents are placed into a vat and left there to react. At some stage the contents need to be emptied and the whole process started again. There are several limitations to the efficiency of this method. Firstly, every time the container is emptied all the chemicals are wasted and need to be replenished with fresh stock—for the home brewer this means the yeast, whilst for *Hydra* this includes all of the digestive enzymes poured into the gut cavity to digest the prey. Then how long should you leave the contents to stew before emptying them out and starting again? If *Hydra* expels its gut contents too soon, it might be wasting good nutrients, but there will inevitably be a diminishing return over time from hanging on to them which might disadvantage it if there are more juicy prey available. Finally, the poor *Hydra*—like a Roman who has overindulged but cannot resist the food in front of him[20]—has to vomit before he can eat again. Wouldn't it be so much better if an organism could continue to digest its food and eat more? If you are a chemical engineer, the obvious solution is the 'plug flow reactor'—also known as a 'tube reactor'. To you and me, this means that we need to invent an anus so that the gut has two ends—an in and an out. The invention of the anus was to radically improve the efficiency of digestion and alter the shape of life.

[20] The stories of Romans vomiting in order to eat more are probably overstated but this behaviour was certainly noted by Seneca and frowned upon by Cicero. It was thought to have saved Julius Caesar's life (once) when his would-be assassins waited in the latrine on the basis that he would go there to vomit but instead he purged in his room! The myth of 'vomitoria'—rooms dedicated to vomiting after meals—is completely false. The vomitoria were actually exits from large public buildings such as the Colysseum in Rome to 'disgorge' large numbers of people quickly. It is said that the 15,000 capacity of the Colysseum itself could empty through the vomitoria in under 15 min.

3

The Bowels of Existence

Summary In which we examine the first 'through guts' with the help of earthworms, and their specialisation into fore-gut, mid-gut and hind-gut regions along their length. We see how the filter feeding apparatus of early chordates became adapted to absorb oxygen from the water, and how lungs developed from the foregut. The challenges of water or salt loss required regulation through the hormone thyroxine, produced by a mucus secreting gland in the floor of the mouth that evolved into the thyroid gland. The same gland came to control the metamorphosis of larva to adult. The pancreas developed as a specialised organ for digestion and the liver for nutrient storage, both as buds from the mid-gut. We also introduce the mysterious 'typhlosole' from which the spleen may have partly evolved. Finally, we see how adaptation to life on land required developments of the mouth and feeding apparatus, and of the hind-gut to preserve water and salt.

J. Woodward, *The Gastro-Archeologist*, https://doi.org/10.1007/978-3-030-62621-1_3

Boris, the grouper

Oh wherefore, Nature, didst thou lions frame? **Bottom**

From 'A Midsummer Nights' Dream' by William Shakespeare

Making Monsters

Our journey has taken us from the dawn of life itself, through the invention of eating whilst still individual cells, to their agglomeration in communities and subsequently as multicellular organisms that defined the need for a gut. As our journey continues we see how the nature of the gut itself defines the path taken by life…

There is a saying that 'it is the exception that proves the rule'. It is in fact only half a saying. The bit that is missing is '…….in cases not excepted'.[1] In other words, if there is an exception it proves that there is actually an underlying rule to which the exception has been allowed. It is the study of exceptions, or variations, in nature that has led to many of the 'rules' of nature being discovered. One

[1] It is thought to date back to the first century BC when used in Roman Law. '*Exceptio probat regulam in casibus non exceptis*' (the exception confirms the rule in cases not excepted). This concept also underpins the Ninth Amendment of the Constitution of the United States of America—'The enumeration in the Constitution, of certain rights, shall not be construed to deny or disparage others retained by the people'. This meaning of the saying is more probable than that implied by the old English use of the word 'prove' to mean 'test'—in other words, it being the apparent exception that 'tests the rule'—as the saying occurs in many languages in some form or other.

particular such anomaly was observed by William Bateson,[2] a Fellow of St John's College in Cambridge who described in the late 1890s a rather strange bee with a leg growing out of its head in the place where an antenna should have been. He coined the term 'Homeosis' for this observation of a misplaced body part—a 'homeotic' mutation. Nearly a century later his observation became of critical importance to our understanding of the development of life forms.

Throughout the early years of the twentieth century, researchers in the USA and Russia studying the genetics of the fruit fly, *Drosophila melanogaster*, began to document several such 'homeotic' mutations that resulted in anomalous body shapes, such as the leg instead of an antenna ('antennapedia') or having four wings instead of two ('bithorax'). Breeding studies demonstrated that these major transformations were the result of mutations of single genes. Our understanding of how these genes work awaited advances in molecular genetics in the early 1980s. Researchers independently in the USA and Switzerland discovered that such homeotic genes coded for proteins that bound to DNA and acted as 'master regulators' controlling the expression of many different genes involved in embryonic development. All such genes have been found to share a short sequence of DNA called the 'homeobox' which has been highly conserved (remained the same) during evolution.

Different 'homeobox' genes control the pattern of the organism by variable expression in different parts of the developing embryo. For instance, a group of genes ('hox' genes) control the head-to-tail body pattern and are switched on in turn during development. One can perhaps visualise this best in animals made up of segments—such as the fruit fly. Each throws a complex sequence of genetic switches that control further gene expression including other homeobox genes that initiate other programmes such as 'wing' or 'leg'. In effect it is a little like pre-programmed lighting combinations in a theatre production—all the technician has to do is to press one button rather than co-ordinate all the different lights individually.

Diverticulum #3.1: Evolution and gene duplication

Whilst writing this chapter I have saved the original version (which was far too long-winded but was at least comprehensive) on my laptop in a folder and have copied it into my desktop as a working copy that I can tinker with. I know that I can always go back to the original if needs be (I have also saved a copy elsewhere in case my hard drive packs up!). So it is with evolution. Genes are frequently

[2] William Bateson (1861–1926) also first coined the term 'genetics' based on the Greek *Genno* = 'to give birth'. During his time in Cambridge he experimented with plant breeding, replicating Gregor Mendel's studies on heredity and as a result began to bridge the gap between Darwin's theory of natural selection and modern genetic-based evolutionary theory.

copied and then mutated separately. This enables the proteins that they code for to diversify and take on different functions, or just perform their original function better. If they do this really well, then the original copy may become redundant and be deleted. One way in which genes are duplicated is through sexual reproduction when the chromosomes swap genetic material. Occasionally one of the pair may fail to 'swap back' and keep hold of both copies. Sometimes the entire genome—the whole DNA code—of the organism is duplicated. This appears to have happened relatively infrequently in our history, with only two whole genome duplications occurring along the pathway from early chordates such as *Branchiostoma* (to whom I will introduce you shortly) to ourselves (we have four equivalent similar genes to every one of *Branchiostoma*). Both of these duplications in our history appear to have occurred over 500 million years ago. As a result of gene duplications and reflecting their value as master switches, we have accumulated over 200 different homeobox (pattern-controlling master switches) genes within our genome!

The enormous utility of homeobox genetics in switching and modelling pre-programmed body parts probably only came into play in the Cambrian explosion. However, the real breakthrough in homeosis was undoubtedly more subtle. Whereas Hydra has a head (mouth and tentacles) and tail (basal plate for fixing it to a surface), it has no upper or lower surface and is radially symmetrical. It really only has one axis (top to bottom) in which the position of body parts can be specified. If we add another axis (front to back) then it follows that we now also have a 'left to right' axis and we have created spatial co-ordinates for any point within the organism. This change leads to the origin of bilateral symmetry and is of such critical importance that most of the creatures that we currently identify as 'animals' from earthworms to humans are grouped together as 'bilaterians' on this basis. Now at long last, over 3 billion years into our journey and around 540 million years before the present day we have got to the point where life has the tools it needs to create a body plan—and a proper gut.

Beginning, Middle... and End

The gut fulfils in every way the above definition of what a good story requires. When I started my current job 15 years ago, I joined just two other gastroenterologists in the department and we joked about the particular areas we found most stimulating. One colleague was most interested in the 'foregut' comprising the gullet and the stomach, the other's interests lay in the 'hindgut', the large bowel or colon. For my part it was the 'midgut' that held most fascination—the small intestine and its offshoots, the liver and pancreas. It turns out that our tri-partite division of workload mimicked exactly the three part division of the gastrointestinal tract defined many millions of years previously.

Analysis of different animal classes suggests that the embryonic development of the three body layers found in vertebrates like ourselves—outside ('ecto-derm'), middle ('meso-derm') and inside ('endo-derm')—were originally controlled separately by different groups of homeobox genes along the length of the organism. These three body layers develop similarly during early embryonic development in all animals (except sponges, and perhaps Cnidaria). Just three homeobox genes are also specifically retained to programme the development of the exact same segments of the endoderm along the length of the organism that I have described above: fore (beginning), mid (middle) and hindgut (end). They have maintained their roles in all bilaterian lineages (including ourselves), with the exception that the gene initially controlling the development of the foregut development has been reassigned to the front parts of the brain.

To Understand Life, Understand the Worm[3]

Because of the proliferation of animal types and shapes that occurred during the Cambrian, we now have a choice of routes to follow at this stage in our journey. Therefore, I suggest that having arrived at this point, we take a brief stop for a picnic to look at the map. Unfortunately it is not an ideal spot as there is no Café yet at the summit of Mount Snowdon.....nor indeed a Mount Snowdon, just a shallow ocean. However, from here we can see all the paths followed by diverse forms of life spreading out towards the horizon of the present though for at least the next 100 million years they are all still underwater.

Let us start with the 'humble' worm. This is one of those words that has no exact biological equivalent—it is used to describe a long cylindrical creature without limbs. 'Worms' include animals with backbones (such as the 'slow-worm'—a lizard that has decided it is better off without legs) but most usually for invertebrates that fall into three groups—the flatworms ('platyhelminths'), the parasitic threadworms and roundworms ('nematodes' such as those that often live in our bowels and can cause itching around the bottom), and the segmented 'annelid' earthworms that we are familiar with from digging in the back garden.

[3]To paraphrase John Sulston, FRS (1942–2018). He worked with Sydney Brenner (with whom and Robert Horovitz he shared the 2002 Nobel Prize in Physiology or Medicine) on the worm *Caenorhabditis elegans* (the work he is alluding to in this quote), and he became the founding director of the Sanger Institute in Cambridge for DNA sequencing.

Worms and their guts are highly variable. The famous '*Caenorhabditis elegans*'[4] which has been used as a model organism for studying animal development is only 1 mm long and its gut is composed of only 20 cells. On the other hand, marine 'bootlace' worms (and their intestines) can grow to 55 m long.[5] Some worms have reverted to a blind ending sac-like gut similar to the cnidarians, and others such as the siboglinid 'beardworms' have done away with a gut altogether, relying entirely for their nutrition on symbiotic bacteria kept alive in a pouch. However, we will use the earthworm as our prototypical annelid worm (even though it only appears about 250 million years ago, half-way between our picnic stop and the present) to demonstrate the simple tubular gut. If you think about it, this basic body form is just a simple gut surrounded by the muscles to propel it and the nerves to co-ordinate them. In terms of 'body-plans' it is the natural successor to the Cnidarian body plan, now having a through-gut with a mouth and an anus rather than a single 'in and out' opening that serves as both. As we look around us on the ocean floor we can see plenty of evidence of worm-like forms and their burrows being present in the Cambrian period (Fig. 3.1).

Earthworms are rated as the most influential organisms on our planet of all time (apparently, we humans do not even register in the top five!)[6] and play a crucial role today in aerating soil with their burrows and digesting organic debris in the soil.[7] There may be as many as 1,750,000 earthworms per acre in fertile farming land, almost certainly weighing more than the livestock that can be supported on the surface. We might expect that their simple form contains a very ordinary gut—they are after all just a hollow tube. However, a closer look reveals some surprises.

The foregut is specialised to bring debris into the gut and comprises a muscular 'pharynx' that co-ordinates movement in the gut with the forwards

[4] *Caenorhabditis elegans* (C. elegans) is a small worm that has been extensively studied. It is effectively the first multicellular organism to be completely dissected on a cellular and molecular level, thanks to Sydney Brenner's interest dating from 1963. The development of every single cell of this organism has been mapped—the males comprise exactly 1031 cells and the more common hermaphrodite (rather than female) contains 959 cells. It was the first organism to have its complete genome sequenced in 1998, 5 years before the first human genome sequence was published in 2003. However, despite our extensive knowledge of this animal there is much still to be understood—for each of the last 5 years there have been about 1500 scientific publications a year on this species!

[5] The longest specimen of *Lineus longissimus*, the 'bootlace' worm, was washed ashore at St Andrews in Scotland after a storm in 1864.

[6] According to Christopher Lloyd in his 2009 book '*What on Earth Evolved?: 100 Species that Changed the World*'.

[7] Charles Darwin's last book was about earthworms. Darwin placed a stone in his back garden—the 'Wormstone'—at Down House in Kent and noted that it sank 2.2 mm every year due to the movement of earth by worms underneath it. He calculated from this that the worms moved a total of 18 tons of soil annually in each acre of land.

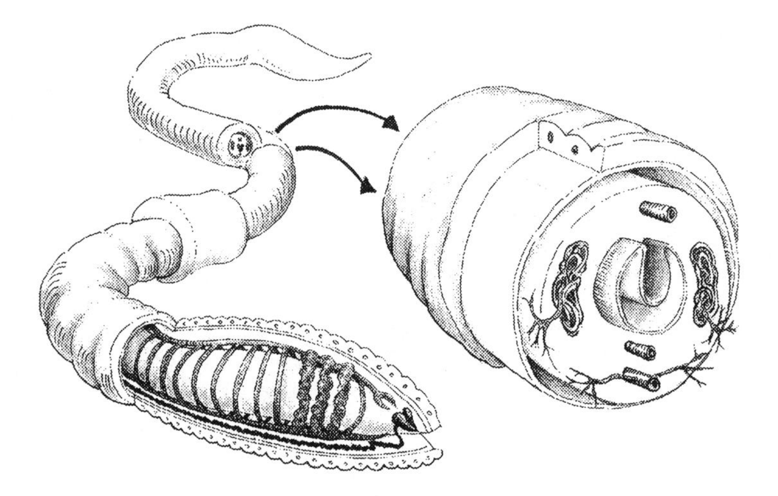

Fig. 3.1 The earthworm and its through-gut. The intestine is a straight tube that lies within the body cavity. It has its own muscle layers to aid in propulsion of its contents—an internal layer of circular muscle that surrounds it circumferentially, and an outer layer of longitudinal muscle that stretches along its length. The foregut of the worm is specialised with a muscular pharynx that sucks in debris, a storage crop and a gizzard lined with a hard surface for grinding. The rest of the gut appears to be regionally unspecialised. Note the 'typhlosole' hanging down, that is thought by some to be required for increasing the surface area for absorption. Nevertheless, the lining of the worm gut is also covered in tiny projections called 'villi' that serve the same purpose

propulsion of the worm. This is followed by an expandable chamber called the 'crop' which stores food and thus controls the rate of delivery into the intestine. The worm has no mouth or teeth to grind the food, but instead there is a 'gizzard'—an enlarged cavity beyond the crop which contains a hard surface with primitive rasping projections that perform the same function. This arrangement of the foregut is in essence similar to that of modern birds. The rest of the gut is made up of the intestine which secretes enzymes to digest the food into chemicals that it can then absorb further along. The earthworm has an enclosed circulatory blood system with vessels that surround the gut and pick up and disseminate the nutrients absorbed through it.

There are several points of interest—and even intrigue—in the worm intestine that provide useful insights into guts more like our own. Firstly, the lining cells of the gut themselves secrete the enzymes required for digestion but they do not produce the full range required to digest some of the complex organic materials found in the soil. The earthworm requires help and it employs a variety of different microbes in its gut to produce the necessary enzymes. In return it provides a comfortable environment for these bacteria, in two out-pouchings or 'diverticula' that extend on either side of the gut. A 'symbiotic' relationship with bacteria providing additional nutrients or digestive capability is a feature of all guts from the cnidarian *Hydra* through to humans—we have approximately equal numbers of human and bacterial cells in our bodies.[8]

Why keeping a colony of friendly bacteria living in the gut should be so important is unclear as animals should be perfectly capable of developing the ability to digest the same things. Indeed, some worms and insects do produce an enzyme capable of breaking down the hard cellulose structures of plant cells for which animals usually rely on their symbiotic bacterial friends. In terms of digestion it could be that the bacteria that produce the all-important enzymes mix with the food during its passage down the gut whereas the concentration of enzymes secreted by the intestinal wall will be highest around the outside of the flow of food leaving the middle part of the stream less well digested. When we look at some specialised guts we do indeed find that the part of the gut that houses the bacteria becomes a specialised chamber that has developed mechanisms for mixing them with the intestinal contents to improve digestion (for instance the ruminant stomach). Later in this book we will see why all animal multicellular life forms (without any known exception) co-exist with symbiotic bacteria.

Secondly, the intestine requires a large surface area for absorption—it does this by creating folds and also tiny little finger-like projections called 'villi'. These produce a somewhat shaggy appearance to the intestinal lining (which is exactly where the word came from as 'villus' is derived from the Latin word for 'fleece'). Villi are a feature of all intestines from earthworms to ourselves, and serve the purpose of increasing the surface area for absorption. They contain tiny thin-walled blood vessels into which the absorbed water-soluble nutrients can pass and thence be passed around the body through the circulatory system.

Thirdly, there is a large fold that hangs down in the middle of the worm intestine from about halfway along its length. Cut crossways, it looks a bit like the 'uvula', the thing that hangs down at the back of our throat, but it extends along the worm intestine and is called the 'typhlosole'. It is thought to increase the surface area for absorption, but it is unlikely to do so significantly and to be honest, nobody really knows what it is there for (and we do not have one!). We will however revisit the mysterious typhlosole at various times in our journeys to come.

[8] Popular science likes the idea that we are outnumbered in cellular terms by our bacterial companions and figures of a 10:1 ratio have been widely quoted (for instance see '*10% Human*' by Alanna Collen). Recent estimates however suggest that the ratio of human to bacterial cells in our body is closer to 1:1. No less impressive even if we are not in the minority!

Finally, there are cells around the outside of the intestine called 'chloragogen' cells that line the body cavity or 'coelom'—the space which the intestine occupies. They appear to metabolise the nutrients absorbed from the gut, converting them into forms that can be used or stored. They also accumulate and store poisons such as heavy metals absorbed from the soil to keep them from harming the worm. In many ways these functions resemble those of liver cells in mammals. It is only in the last 10 years or so that we have learnt the importance of similar 'coelomic' cells in the human that line the body cavity. They play a key role in developing both the liver and pancreas and also provide the 'pacemaker' cells that co-ordinate the nervous system of the gut and its propulsive muscular contractions.

So much then for the worm gut being just a simple tube....

Foregut First

Let us return to our Cambrian picnic stop as we now have to choose our route, surrounded as we are by a profusion of different life forms. All still exist below the surface of the water but life now no longer just inhabits the ocean floor but occupies a range of different 'ecospaces'—defined by the position in the water or on the sea floor, the mode of feeding and the degree of movement. We have already considered the fixed sponge and Cnidarian life forms with their flask-shaped guts, and the through-guts of the burrowing worms. Before us we now see a major divergence of pathways—a fork in the path. Interestingly, the lifeforms in each path are separated by the way in which they develop their guts as embryos. We can either follow one path of the 'protostomes' (molluscs, crustaceans and insects) in which the mouth is the first embryonic opening or on the other, of the 'deuterostomes' where the anus develops first.

I would suggest that we now choose the latter path as it leads us eventually to ourselves and animals like us, and therefore we will overlook the armoured trilobites shuffling like enormous woodlice across the ocean floor[9] and look up to the animals swimming in the water above us.

[9]Trilobites dominated the ocean floor from the beginning of the Cambrian for a period of around 270 million years and came to comprise over 17,000 different species. Many are similar in appearance to woodlice ('pill bugs') with a hardened protective upper surface and soft underparts that include the feeding apparatus and many pairs of legs. It is the tough upper carapace that is best preserved in the fossil record and rarely are the soft organs found in fossils. However rare finds (such as that reported in 2017 of three trilobites from Morocco) have fossilised intestines. The anatomy of the trilobite gut appeared to be of a mouth opening directly into a crop or (in some species) a cavernous stomach followed by a tapering gut. There are numerous outpouchings or diverticular paired along each side—their function was

There are great opportunities here for feeding due to the small organisms that exist by passively floating in the current and are known as 'plankton' (from the Greek word for 'wanderer'). However, in order to feast, the predator cannot simply float along with the plankton but requires the ability to swim against the current. This may require significant muscle power and for the muscles to work they need to be supported by an internal skeleton. The simplest way to do this is a stiffening rod made of cartilage running the length of the organism—called the 'notochord'.

It is useful to remember that the notochord is not-a-spine, but its precursor! Animals possessing a 'notochord' are known as 'chordates'. The majority of animals replace the notochord during development with a spine (and are hence known as 'vertebrates'). In humans the remnants of the notochord remain as the spongy centre of the discs between our vertebral bones that allow flexibility, but cause us such problems when they pop out of place and press on a nerve.

The first creatures resembling fishes possessed notochords and appeared in the fossil record about 520 million years ago (at about the same time as the trilobites) during the Cambrian Explosion. They are best identified in fossils from near Haikou in Yunan province in China, possibly the earliest being a 4 cm long creature called '*Haikouella*'. This fish-like animal had rudimentary fins and a tail, extending behind the anus. It also had tentacles around its mouth for feeding, a heart and circulatory system and a nervous system evidenced by the presence of eyes. We can glean little more detail about how it worked from the fossilised remains and instead we should perhaps look at a current surviving descendant that shows some similar features—'*Branchiostoma*', the Lancelet—to whom I promised to introduce you earlier in our journey and with whom we will spend some time (Fig. 3.2).

The old name for Branchiostoma was 'Amphioxus' which means 'pointed at both ends'. This 5–6 cm long creature resembling a small eel has used its pointed rear end to return to a sedentary (or sedimentary!) habitat by digging itself tail-first into the sea floor debris with about half of the body sticking out. Just like *Haikouella*, the lancelet has tentacles around its mouth and is a 'filter feeder', sieving the water for small organisms. How it does this is quite fascinating. The first part of the foregut—the pharynx—has developed a

either to secrete digestive enzymes or (probably) to house symbiotic bacteria. The most similar gut in currently existing creatures is found in the wonderfully named *Godzillius*. This and other 'remipedes' like it are 5 cm. long centipede-like marine creatures that use their multiple appendages to swim whilst lying on their backs. According to their fossil record, remipedes have been around for over 300 million years (hence overlapping with the trilobites), and recent genetic analysis suggests that they are most closely related to modern insects.

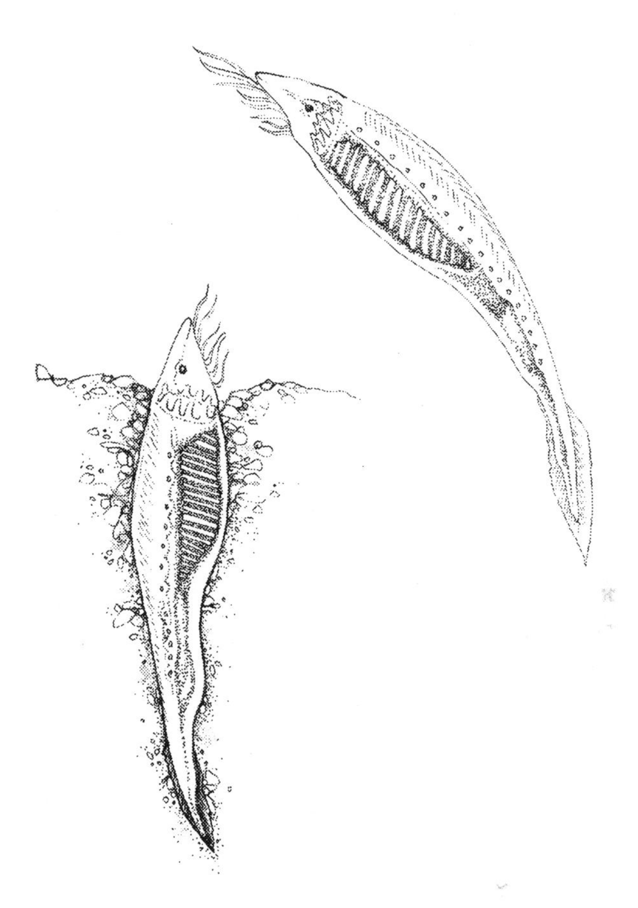

Fig. 3.2 *Branchiostoma*, the lancelet. Note the tentacles around the mouth, the circular wheel organ that draws fluid (and particles) into the foregut and the slits of the pharynx that act to filter out the particles and expel the water

uniquely specialised structure called the 'wheel organ'. It is a circular plate with a central hole and its cells are covered in little hair-like projections called 'cilia' (from the Latin for 'eyelash'). Like the flagella of the collar cells in the sponge but smaller, they similarly set up a current in the water by beating in the same direction. In the case of the wheel organ, the cilia move like a 'Mexican wave' going around a football stadium and in so doing create a whirlpool that sucks water into the gut through the central hole of the wheel organ.

However beautiful the wheel organ is to behold in action, it is the other development of the *Branchiostoma* pharynx that should interest us more—a row of oblique slits on each side which pass through the foregut wall. These pharyngeal slits act like 'baffles' in the flow to precipitate particles on the downstream surface—to visualise this one can imagine the sand building up on one side of the wooden 'groynes' or 'breakwaters' built perpendicularly from beaches into the sea to prevent coastal erosion. The pharyngeal slits capture the food particles in the slime-like mucus on their surface which are then carried along by the action of cilia wafting towards the opening of the gut. The water flowing across the slits passes out of the animal without entering the gut. Although initially devised as a means of facilitating filter feeding, the pharyngeal slits are a feature of all chordates thereafter. In many species—such as ourselves—the pharyngeal slits are only seen during embryonic stages. Their prominence in embryos contributed to the notion of nineteenth century natural philosophers that all animals 'recapitulated' the adult stages of more primitive animals during their development (see the footnote about Ernst Haeckel in Chap. 2). The foregut pharyngeal slits and the structures that they become are of such major importance in the story of life that we will stay with them for a little while.

Taking Breath

The ability of fishes to swim massively enhances their reach into the third dimension of depth instead of the two-dimensional ocean floor but requires substantial changes in form and structure. The support provided by water effectively counteracts the force of gravity, but a substantial amount of energy is required to power the fish against the current. Whilst their greater range and ability to capture food provided an ample supply of most nutrients for the first fishes, the demand for one essential nutrient now began to outstrip its supply—and that was oxygen.

Oxygen is fundamentally important in the energy-releasing reactions within the cell—it receives the electrons which have to pass along the inner mitochondrial membrane to generate the hydrogen ion gradient that powers life (as we saw in Chap. 1). Respiration can still take place without oxygen, but is much less efficient and generates only a measly 2 molecules of energy-providing ATP per glucose molecule compared to the 30 molecules of ATP generated in the presence of oxygen. Unfortunately, fresh water contains only about 4% as much life-supporting oxygen as is in the air and the diffusion rate of oxygen in water is about 10,000 times slower. Furthermore, the oxygen content of water can vary significantly depending on temperature, water quality and other chemicals present. Animals that live underwater can never take their oxygen supply for granted to the same extent as air-breathers can. They also need to find ways of extracting it from the water.

The sedentary *Branchiostoma* has a relatively low oxygen demand and simply absorbs the oxygen it needs from the water through its thin skin—amply supplied with thin-walled blood vessels across which the gas can pass. In fact this route suffices for many animals, including modern amphibians such as frogs when they are underwater. However, those animals with higher requirements as a result of greater activity required more sophisticated methods to harvest their oxygen. Predictably it was the gut—already specialised for nutrient absorption—that came up with the answer for efficient gas exchange. It did this by specialisation of the foregut. The pharyngeal slits that form the 'breakwaters' of *Branchiostoma* not only generate a large surface area, but the mechanisms derived for filter feeding pass large amounts of water over them without entering the gut. With a bit of evolutionary tweaking, these slits can be adapted to absorb oxygen. The surface cells need to lose their 'cilia' and become flattened into thin 'pavement cells'. A network of thin-walled blood vessels needs to be grown underneath them to reduce the thickness through which gases need to diffuse and the surface area needs to be substantially increased. The resulting structure—derived from the foregut—is what we recognise as 'gills'.

Breathing Underwater

In the mid nineteenth century, fossilised remains resembling teeth were found in early Cambrian rock strata dating back to about 520 million years ago. They were named 'conodonts'—meaning 'cone-shaped teeth'. Of the animals that left them there was little trace, presumably due to soft body parts decaying without leaving any impression. However, the abundance of conodonts—over a 300 million year period—meant that whichever animal they belonged to must have been prolific. The 'Conodont conundrum' was only solved by the discovery of a fossil near Edinburgh in 1982 in which for the first time the whole animal was found preserved along with conodont fossils which did indeed appear to be teeth. It was a small eel-like creature possessing a notochord and primitive gills, belonging to a group of fishes called the 'agnathans' or 'jawless' fishes. These are the only examples of vertebrates that do not have a jaw, the rest of us being classified as 'gnathostomes' (from the Greek meaning 'jaw-mouth'!). This well preserved conodont fossil tells us very little about the nature of the animal's gut (and we even have no idea whether the teeth were used for capturing prey, grinding, ripping or just filtering), but there are some existing agnathan fishes that may provide some superficial clues about the conodonts—the lampreys and hagfishes (Fig. 3.3).

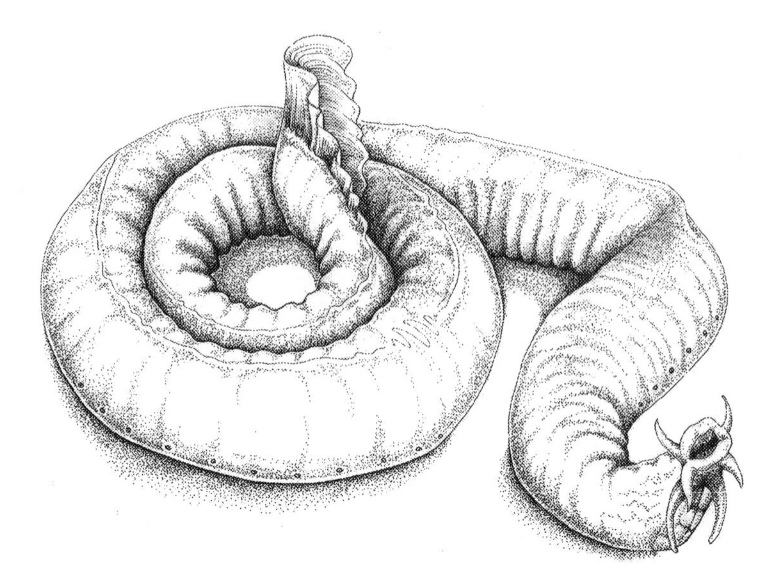

Fig. 3.3 The hagfish. A modern example of an agnathan (jawless) fish

Hagfishes are really quite bizarre creatures.[10] Eel-like and averaging around half a metre long, they feed on a variety of marine invertebrates, but also notably scavenge dead (or dying) animals whose bodies they enter and then devour from the inside. Rather than jaws they have a round sucker with two plates bearing teeth (much like the conodonts) that they can protrude and retract to rip flesh and draw it into the pharynx. The gills are contained in 5–14 (depending on the species) paired small pouches on each side of the pharynx. Rather than the feathery gills of later fishes,[11] hagfish gills comprise flat plates of cells, rather like the vanes of a car radiator. Their function appears to be something of a halfway house between the pharyngeal slits of *Branchiostoma* and the later fishes as they are still capable of absorbing soluble food nutrients as well as oxygen from the water. In this way they are also halfway between guts and lungs.

[10] Hagfishes are also notable for producing vast amounts of slime (up to 20 L!) when threatened. The slime is a defensive mechanism to prevent capture and it also clogs the gills of any predating fish leading to suffocation. Hagfish are used in cooking in Korea—not just for their meat but also for the slime which is used in the same way in recipes as egg whites are in Western cooking.

[11] The numerous feathery projections on the gills of fishes increase the surface area for absorption. For a fish the size of a sardine it can be as much as the equivalent of a side of A5 paper. However, the area of a fish gill is highly variable and depends not only on size but also on activity and oxygen availability in the environment the species inhabits. There are significant downsides to simply maximising the surface area for absorption and therefore the size of gills is strictly limited and adapted to be appropriate for the conditions.

Whilst gills are specialised for gas exchange, they are only able to do so when water is flowing across them. Cartilaginous fishes including certain species of shark have no way of keeping water moving over the gills except to keep swimming forwards—therefore they really can 'drown' in water if they are trapped in nets and stopped from moving. Later 'bony' fishes are capable of pumping water across their gills by closing the mouth and having a gill covering called the operculum that not only protects the fragile structures but also generates a second pumping chamber which allows almost continuous flow across the gills. In some ways this is similar to the 'circular breathing' that enables wind instrumentalists such as oboists to blow a continuous note through the reed whilst inhaling through the nostrils.

Diverticulum #3.2: Containment revisited (again!)

Hagfish intestines have a special trick that they share with insects. Food that is ingested is enclosed in a sort of slime cocoon that is secreted by specialised glands in the upper midgut. The cocoon is called the 'peritrophic membrane' (from 'peri' = *around* and 'trophic' = *feeding*). This enclosure serves to concentrate the digestive enzymes with the food and to contain any unwanted components of the meal. The protein matrix of the membrane has small pores through which nutrients (having been chemically broken up to be of sufficiently small size) can pass. However it excludes bacteria, parasites and larger molecules such as toxins. These are all just passed through the animal and excreted along with the membrane. Just as with other examples of overcoming the containment paradox such as endocytosis the food is 'inside' the animal and yet never enters its body tissues. The peritrophic membrane is therefore in many ways the multicellular animal equivalent of the phagosome membrane of the single cell.

Solutions to problems in nature are always compromises and the invention of fish gills for oxygen uptake is no exception. The need for a large surface area across which gases can exchange between the surrounding water and the circulating blood critically compromises the 'containment' so necessary for life. If oxygen and carbon dioxide can move freely, so then can water and salts and the circulation and internal environment of the fish will effectively equilibrate with the water. Interestingly, heat will also be lost which means that fishes can never be 'warm-blooded'. Fishes in freshwater will lose salt and absorb water and dilute their blood; fishes in saltwater will simply desiccate.

In fishes living in saltwater, salt (sodium chloride) which has been ingested with the food is specifically absorbed in the foregut, which is relatively impervious to water. This results in the fluid in the lower gut being less concentrated than seawater but equal concentration to the blood and this aids in digestion. Water and salt are then taken in by the intestine and the excess salt is excreted largely by the gills (with some help from the kidneys) using specialised molecular pumps in the membrane. Some cartilaginous fishes such as sharks have a gland within the rectum, the last part of the gut before the anus, to secrete unwanted salt.

Those fishes that live in freshwater have the opposite problem and need to rid themselves of excess water rather than salt. Their specialised kidneys excrete the equivalent of about half of their entire body content of water every hour in order to stop them swelling up like a balloon, whilst carefully retaining salts. Ah! but what then (I hear you say) about those fishes like the Atlantic salmon that spend some time in freshwater and some in saltwater? I will park that question for a short time but we will return to it a little later after we have covered a little more ground.

Clearly a major investment of energy and resources is required to overcome the breach of containment caused by leaky gills. It is perhaps the equivalent of a boat with a hole below the waterline which has to have its bilge pumps working full time (or its sailors frantically bailing out with buckets). Nevertheless, the substantial energy production supported by the uptake of oxygen clearly outweighs the enormous energy expenditure used to balance leaky containment.

Breathing Air

Around 433 million years ago (identified in the geological record as the end of the Ordovician period, immediately following the Cambrian), an unknown event resulted in rapid global cooling and a fall in water levels due to glaciation—the reverse of what is currently happening on our planet. Atmospheric oxygen levels fell from around 25% to 15% and shallow freshwater habitats dried up. The effect on marine animals was profound—this was when many species of conodonts and trilobites became extinct, along with approximately two thirds of all existing species at the time. Just as we have seen great 'explosions' of animal diversity in the Ediacaran and Cambrian periods, so there have been global 'extinctions'[12] of which this was one of the most significant.

[12] There have been numerous mass extinction events in the history of life. Five great mass extinctions are recognised, the best known being the Cretaceous-Paleogene extinction about 66 million years ago that wiped out the non-avian dinosaurs. It is widely recognised that there was a major asteroid impact with the earth at this time—evidenced by the Chicxulub crater some 110 miles wide under the Yucatan peninsula in Mexico, made by an asteroid about 9 miles across. Whether this event and its effect on the climate was the sole cause of the mass extinction is unclear. The other mass extinctions have not been linked to such cataclysmic events and may have related to climate change as a result of volcanic activity or gradual changes in atmospheric oxygen and greenhouse gases.

Some think that it was this event that led to life first emerging from the water and onto the land—effectively forced by the deteriorating environmental conditions.

The oxygen content of water depends greatly on the atmospheric oxygen concentration and the water quality. As we know from seeing a 'fish out of water', gills are not designed to work in air. However, to prove the old adage inaccurate, over 500 species of fishes *have* actually learnt how to breathe out of water. That around a third of all such air breathing species are types of catfish attests to their life in the detritus of the river bed with such poor water quality and low oxygen levels.

A variety of different techniques have been employed by fishes to breathe air, however all (with the exception of absorption through the skin) are derived as adaptations of the foregut endoderm—usually the pharynx, but occasionally even the stomach. The air-breathing organs of fishes are principally used as a supplementary means of taking up oxygen and may only be required during seasonal variations in water depth or quality. As gills have evolved to become more specialised, the number of pharyngeal slits dedicated to them has reduced over time—from the 10–14 on each side seen with the hagfish, to 5–7 in cartilaginous fishes and 3–5 in the bony fishes. The remaining pharynx behind them however adapts to produce pouches above or below the tube of the pharynx rather than to the sides (where the gills are found). It is the development of a particularly large pouch above the pharynx (towards the spine) that becomes a buoyancy chamber for bony fishes—the 'swim bladder'. This may remain attached to the pharynx by a tube that allows air to be effectively ingested into it from the water surface, or become completely separated from the gut and filled by a 'gas gland' that pumps oxygen into it. The swim bladder allows the fish to remain orientated in the water[13] and also reduces the amount of energy required to stay at a particular depth. However, swim bladders are thick-walled and necessarily impervious to gas as they are not meant for gas exchange and are therefore unlikely candidates to be the precursor of lungs. For us to see how lungs developed from the gut we will have to go to Queensland in Australia to encounter an unusual fish capable of living both on land and in the water.

[13] Swim bladder dysfunction is common in aquarium goldfish and can result in fishes sinking to the bottom of the aquarium, swimming in strange orientations or even floating upside down near the surface of the water. It is thought to be due to compression of the swim bladder by a distended intestine due to overfeeding or feeding the wrong food. Forced 3-day starvation and raising the water temperature tend to right the fish in the majority of cases. It is also the swimbladders in fish that allow them to be detected by sonar as the gas generates an acoustic echo. Sharks lack swimbladders and depend on the lift provided by their unusually-shaped heterocercal tail fins and angled heads rather than buoyancy.

Coming Ashore

'Lungfish' (*Dipnoi*) are the modern descendants of a long lineage dating back to the first bony fishes around 420 million years ago and are found in Australia, Africa and South America. A striking characteristic (apart from the fact that they have lungs and can 'breathe'!) is that their fins are attached to their skeletons by a single bone—leading to the description of them as 'lobe finned', as opposed to the majority of modern fishes that have the more familiar 'ray fins'. It was of course these bony protuberances supporting the four underside fins of the lobe-finned fishes that permitted the subsequent evolution of strong articulated limbs. They would eventually enable animals to move across the land where water no longer provided support against the force of gravity.

Diverticulum #3.3: The lungs of birds

We are in danger of being distracted by the lungs, even though we can excuse our interest as they are really part of the gut, having evolved and developed from the foregut endoderm. Avian lungs are far more advanced than ours and are incredibly efficient despite being only about half as large as of those of an equivalent sized land mammal. Our lungs are simple blind sacs that require air to be pumped in and out sequentially—much as the primitive gut of the simple Cnidarians like Hydra is with food. Birds have evolved the equivalent for gas exchange of a 'through gut'. When air is inhaled the stream is divided into left and right lungs, as in us. What happens next is more complicated. Some gas travels straight into a set of expandable posterior air sacs, found within the body cavity. They therefore contain fresh air. Other gas moves through the lung, where gas exchange occurs, and the stale gas enters some anterior air-sacs. When the bird exhales, the stale air from the anterior sacs leaves the body, but fresh air from the posterior sacs now passes through the lung tissue, and gas exchange continues. Just as with the bony fishes, we now see a pressurised system that permits constant flow over the gas exchanging tissues, and there is no mixing of fresh air with oxygen depleted air—unlike in our own lungs.

Of the existing species of lungfish, only the Australian lungfish[14] is able to breathe both through its gills underwater and also using its lungs on land—but can only survive a small number of days out of water. The other five

[14] The Queensland lungfish (*Neoceradotus forsteri*) is found in the Burnett and Mary river systems in Queensland. Fossilised remains dating back 100 million years have been found in New South Wales that are almost identical to the existing species and therefore it truly is a 'living fossil'. The Queensland lungfish can live for many years—a specimen in Chicago called 'Granddad' was first put on display in 1933 and died in February 2017! The Queensland lungfish even has a township named after it—Ceradotus—in Queensland. Lungfish are known to have the largest genome (amount of DNA) of any animals—up to 33 million base pairings.

species of lungfish have gills but they are small and unable to fully support underwater breathing—they still have to come to the surface for air. The lungs that develop in these fishes differ from swim bladders by developing on the opposite side—the under surface—of the foregut and are divided up by numerous thin walls that have a rich blood supply for oxygen exchange. Anatomically and structurally therefore these air-breathing sacs resemble the lungs of terrestrial vertebrates, but unlike them they lack a tube to connect them to the gut—the trachea—but bud directly from it.

There are distinct advantages of breathing air rather than water. Of course, the oxygen content of air is significantly greater than water and much less variable and dependent on temperature. As we have seen the gills are extremely 'leaky' and threaten containment and the maintenance of a stable internal environment. There is little such risk when breathing air. However, water is still a major concern. Surface tension tends to make water leak out of the capillaries and into the lung air spaces (thereby diminishing gas exchange by effectively increasing the barrier thickness), and evaporation of water into the gas which is breathed out can lead to significant losses in dry climates. In order to reduce such risks, the lungs secrete phospholipid molecules very similar to those of the cell membrane to act as a 'surfactant' and reduce water surface tension.

Nevertheless, there are also disadvantages to breathing air. The high oxygen content can be toxic due to the formation of 'reactive oxygen species', oxygen molecules that are capable of reacting with and damaging important organic molecules.[15]

In order to survive in an air-breathing environment, organisms therefore require an ability to quench these molecules using 'anti-oxidants' to prevent damage by oxygen. Marine organisms are able to concentrate minerals such as selenium and iodine to use as constituents of anti-oxidant enzymes. However, once on land these elements are more limited. Plants have evolved effective anti-oxidant organic molecules such as polyphenols (which include the substance that turns brown in bananas and apples when exposed to the air), and vitamins C and E. Animals either evolved their own anti-oxidants (for instance, most animals with the notable exception of guinea pigs and ourselves can make vitamin C) or gained the ability to recycle these compounds from plants.

[15] Reactive oxygen species are examples of 'free radicals', familiar to consumers of cosmetic anti-aging creams and those who take health food anti-oxidant supplements.

Lingering with Lungfish

We shall stay for a while in the agreeable climate of Northern Australia with the lungfish as, while we have been distracted by breathing and developments of the foregut, midguts have also come on really quite a long way in the preceding 100 million years.

One of the problems facing the gastrointestinal tract was the large number of different functions its cells were now being required to perform—and they were becoming more advanced and specialised all the time. This thin layer of cells lining the gut had to secrete enzymes to digest the food, to secrete mucus to lubricate it passing through and protect the gut epithelium, to absorb the nutrients, then metabolise them within the cell, detoxifying poisons and changing organic nutrients into forms that could be safely exported around the body for use in other tissues. It would be much more efficient for cells to be individually specialised for these functions rather than 'jacks of all trades'. Furthermore, not all of these tasks were required throughout the entire length of the gut. For instance there is little point in secreting digestive enzymes near the end of the gut—this is much better localised to the beginning of the intestinal tube as it allows longer for digestion to occur as the food passes along the gut. Therefore, cells began to take on separate roles of absorption and secretion. Those secreting digestive enzymes now also had no need to be directly exposed to the flow of food, and their presence in the gut tube would merely take up valuable surface area which could be better used by absorptive cells. As a result the enzyme-secreting cells became localised together in a gland—the pancreas—at the upper end of the intestine. In the lungfish, the pancreas is embedded within the stomach wall, but in later animals it becomes a separate organ, lying behind the stomach and connected to the gut by a tube through which the digestive enzymes pass. It is notable however that even humans occasionally have tiny little islands of pancreatic tissue in the stomach lining—called 'pancreatic rests'—that attest to the evolutionary origin of the pancreas.

Similarly the liver develops from a tube—the bile duct—sprouting from the midgut. The earliest structure identifiable as a liver is found as a diverticulum of the mid gut in *Branchiostoma*. Livers have so many roles that it is difficult to know their original purpose in separating from the intestine. They metabolise chemicals absorbed from the gut—detoxifying poisonous substances to excrete them in a harmless form back into the gut, and converting nutrients into other molecules required for different processes in the body. Livers assist in the absorption of fat by secreting bile into the gut, via the gallbladder in humans—bile salts have water-soluble and fat-soluble ends thereby acting as detergents. Livers also provide storage capacity for energy in the form of carbohydrates and fats. Enterocytes in the gut lining have retained many of the metabolic functions found in the liver—such as metabolising and detoxifying chemicals, including some that we use as medications. They also provide a means of excreting toxic metals which are retained within the enterocyte and shed into the lumen to pass out with the faeces. This is in fact the only route by which we can get rid of any iron that is excess to requirements and is somewhat reminiscent of the 'chloragogen' cells surrounding the earthworm gut which lock away toxic metals from the soil such as cadmium. Given that the enterocytes can do much of what the liver cells do, the driving factor for separating the liver from the intestine was most likely the need for a site of energy storage. Any such function in the intestinal lining would dramatically reduce its surface area for absorption, whereas the liver can change its shape and size without significantly compromising its function.

The need for such an energy store arose from animals developing 'eating habits'. Grazers or predators of ubiquitous but low energy foods have a constant stream of nutrients entering the body through the gut. However, predators that have to select or hunt their prey have interruptions in their energy supply—we replicate this by 'eating at mealtimes' although modern food availability has led to more of a grazing habit through constant 'snacking'. When carbohydrates are not available, the body breaks down protein at a rapid rate, and as there are no protein 'stores' in the body, muscle wasting occurs. We therefore keep a short term carbohydrate supply in the liver in the form of 'glycogen'[16] which is made up of glucose molecules joined together. These glucose molecules can be released to smooth out energy delivery between meals. Glycogen stores can last up to a day in normal situations, but athletes can exhaust their glycogen in about 100 min of exercise. This leads to the sensation of 'hitting the wall' in long distance runners and is also the reason why athletes 'carb load' prior to competitions. Only with longer periods of starvation—over 24 h—do we gradually adapt to breaking down more fat

[16] Glycogen was discovered in 1857 by Claude Bernard (1813–1878) who called it '*la matiere glycogene*'—hence its name in Engish. Bernard was undoubtedly one of the greatest physiologists of history. It was he who coined the term 'milieu interieure' to describe the consistent internal environment of cells and organisms, maintained by the process of 'homeostasis'. He also laid the foundations of medical experimentation such as controlled trials. Glycogen is the animal equivalent of starch, produced in plants as a similar store of glucose molecules connected together.

rather than predominantly carbohydrate in order to conserve protein. Hence there is a tidal 'ebb and flow' of glucose going from the gut in to the liver as a result of eating and then being released from the liver in between meals in order to provide a seamless energy supply.

The emancipation of the liver from the gut means that all the nutrients absorbed from the gut have to be transported to it for metabolism. This is accomplished by a specialised enclosed blood circuit linking the intestine to the liver such that the entire blood flow of the gut passes through the liver before finding its way back to the heart. This 'portal' system (from the Latin word 'to carry') is unusual in many regards. It has very small blood vessels at both ends—in the gut the capillaries draining the villi, and in the liver, the tiny vessels that percolate the blood through the plates of liver cells to allow maximum uptake of nutrients. As a result, the blood flow in this system is under low pressure and can be very sluggish. In some species it actually needs a bit of help—the hagfish for instance has an accessory portal heart to improve the flow.

Gut Senses

When cells with specialised functions were localised within the intestinal wall they could directly sense the presence of food or specific nutrients in the gut contents and respond accordingly. However, when the cells became separated from the gut in the liver and pancreas, some form of communication with the gut was required to know what was happening there. For instance, the liver needs to know whether to store or release glucose depending on whether a meal has just been eaten. This is accomplished by using a chemical signal—specialised cells in the intestine sense the presence of glucose and secrete a protein called 'insulin'[17] which is carried in the portal blood to the liver where it delivers its message to the cells via specific cell surface receptors. Such chemical messengers are called 'hormones'. Insulin

[17] The hormone Insulin was discovered following its purification by Frederick Banting—a Canadian surgeon—and his medical student Charles Best in 1922, working in the laboratory of John Macleod—a professor of diabetes. Banting donated half of his prize to Best, and Macleod in turn shared his with Bertram Collip, the biochemist who successfully isolated the molecule. Controversially, a Romanian physiologist, Nicolae Paulescu had published extensively on the use of pancreatic extracts to treat diabetes earlier than Banting and Best and had injected a patient intravenously with his preparation 'pancreine' as early as 1921.

itself actually has many roles throughout the organism in preparing the tissues (not just the liver) for an influx of energy after eating. The specialised 'endocrine' cells that secrete insulin are found within the gut lining in animals such as *Branchiostoma*, interspersed between the absorptive enterocytes. Over time, we see these specific cells evolving as clusters in the bile duct wall (for instance in hagfish) and then separated from the gut altogether in 'islets' of endocrine cells within the pancreas.[18] In this position they are able to sense glucose not in the gut itself, but in the blood stream and regulate its level appropriately by the secretion of insulin. However, they lose the ability to sense the presence of food in the gut. Nevertheless, the endocrine cells within the gut lining ('*entero*-endocrine' cells) have retained this role and secrete other hormones called 'incretins' which act on the islets to then increase the release of insulin from the pancreas after eating—a two-step hormone signal.

A wide array of hormone chemical messengers has evolved for the gut to send messages to other organs or to co-ordinate activities within the gastrointestinal tract itself—over 40 such 'gut hormones' are found in mammals and entero-endocrine cells make up 1 in every 100 enterocytes. For instance, the presence of fat within the upper intestine stimulates the release of a hormone called 'cholecystokinin' which leads to contraction of the gallbladder and the secretion of bile into the intestine. Given that bile is effectively a detergent to emulsify and help absorb fat, this is akin to a timely squirt of washing up liquid into the bowl when cleaning the plates after the Sunday roast. Another example is when nutrients are sensed in the furthermost part of the intestine, risking excretion before adequate absorption has occurred. Hormones are secreted to slow down the motility of the upper gut and thereby improve absorption up stream. Hormones emanating from the stomach and the intestine are also secreted in the brain as hunger or satiety signals and help to control the intake of food—another example of our behaviour being influenced by our guts.

[18] Insulin derives its name from the little islands of hormone secreting cells that are embedded within the pancreatic tissue. These are called 'Islets of Langerhans' after a German pathologist, Paul Langerhans (1847–1888) who discovered them in 1869 but incorrectly thought that their function was in immunity. In the human, pancreatic islets comprise 1–2% of the volume of the gland, but receive 10–15% of the blood flow. Langerhans died tragically young as a result of tuberculosis.

More Midgut

Diverticulum #3.4: Communications between cells

Hormones, as we have seen, are chemical messengers between cells and can act on cells nearby or at a distance. The first such means of communication to develop were between single cells—for instance the slime mold, *Dictyostelium* spends most of its life cycle as a single cell, but under conditions of environmental challenge it secretes a chemical that attracts other cells to form a multicellular slug. Amazingly, the chemical attractant—called cyclical AMP—has kept its signalling role in all animals including ourselves but now acts within the cell to translate the messages from hormones such as insulin, adrenaline and growth hormone. Hormones act within the organism a little like radio messages—being diluted by the circulation, the signal will be loudest to nearby cells and quieter the further away they are. All cells could potentially receive the signal to varying degrees, but selective signals can be sent if only certain cells have the correct receiver (or cell surface receptor) to pick it up.

Nerves also use chemicals to send messages. Called 'neurotransmitters', these are released from the end of the nerve cell and signal only to the cell in close vicinity across a narrow gap called a 'synapse'. Many hormones can act as neurotransmitters and vice versa. For instance, some of the hormones secreted by the stomach and intestine to modify appetite are also found as neurotransmitters between nerve cells in the brain. If hormones are the equivalent of mass transmission by radio, then nerves and their neurotransmitters are the equivalent of getting on the phone and calling just one person.

As to the question of which came first—nerves or hormones—the answer is undoubtedly hormones which are used as signalling molecules between single celled animals. Amazingly, even the hormone insulin is found expressed by an amoeba-like creature called *Tetrahymena pyriformis* although its function in this species is currently unknown.

Before we leave the lungfish, we will take one further look at its intestine. You may remember the infolding of the earthworm intestine called the 'typhlosole', of questionable function. The equivalent in the lungfish is the 'spiral valve' which—as its name suggests—is not straight but completes nine whole turns along the gut, a little like a helter-skelter. It acts primarily to slow down the passage of food and effectively elongate the gut for absorption. Hagfish retain a simple typhlosole, but in the related lamprey this has become the spiral valve. Cartilaginous fish such as sharks also possess spiral valves to compensate for their relatively short intestines. However, they restrict the calibre of the intestine to the extent that large or indigestible items cannot pass through—hence analysis of a shark's stomach contents often provides a long and interesting dietary history!

It is within the typhlosole of hagfish, and the spiral valves of lampreys and lungfish that we see the development of tissue that forms blood cells—both the 'red' blood cells that carry the oxygen around the body, and the 'white' blood cells that play a role in defence. Just as the pancreas develops within the wall of the lungfish stomach only to separate as an organ in its own right during further evolution, so the blood forming tissue in the typhlosole and spiral valve later dissociate as the spleen.

Living on Land

We have now arrived at a crucial point in our journey where we can look to leaving the relative comfort of an aqueous environment to exist on land. Whilst the first fossilised footsteps date back to around 490 million years ago, they are likely to have been left by aquatic invertebrate creatures fleeing a predator across a spit of land, just like modern 'flying fish' that escape by leaping out of water and temporarily entering an aerial realm of existence. At the time that lungfish were able to survive out of water but were still far from being independent of it, invertebrates such as the ancestors of our modern insects were already capable of terrestrial existence, albeit still with aquatic reproduction and larval stages. Four-legged ('tetrapod') vertebrates were a long way behind, for it was not just solid limbs to counteract gravity and air breathing lungs that were required. Whilst these tools are necessary for life on land it turns out that they are not enough. Once again, we will see that the gut itself is key to living on land.

Medical students are fortunate enough to spend some time during their final year outside their University on 'elective'. I used this period to learn a little about tropical medicine in a group of islands that constitute the country of Vanuatu in the south-west Pacific. The island on which I was based was fortunate to host one of the best diving wrecks from World War II—a US cruise ship called the 'President Coolidge' that was converted into a troop ship, accidentally sailed through the minefield protecting the harbour entrance, hit a mine and sank. As well as an extraordinary shipwreck to visit it is also a haven to wildlife, thanks to its unofficial curator of over 35 years, an Australian called Allan Power. He created an artificial coral reef at a point that was used as a decompression stop and a variety of tame marine animals used to frequent it. In particular I remember a large grouper fish (named 'Boris'[19] by Allan) that would regularly come to be fed. He would simply open

[19] Boris was a giant (or Queensland) grouper (*Epinephalus lanceolatus*) that can grow up to 2.5 m in length and weigh as much as 200 kg. We also tamed a large moray eel called 'Nessie' that we used to entice out of its burrow with lumps of meat and hand feed—a practice that I now know is unwise as divers have lost fingers this way…

Fig. 3.4 Acanthostega—a swamp dweller that probably spent most of its life in the water but showed early adaptations to life on land

his huge mouth and suck in a whole wrasse—or whatever fish we had taken down as a treat for him—sideways on. On one occasion he mistook the white mask of a fellow diver as a fish and sucked that off her face just by opening his mouth in front of her. Thankfully she remembered her training and rather than panicking and striking out for the surface she calmly picked up the mask and replaced it after he spat it out.

Imagine if you will, a Boris with legs, adapted to living on land and breathing air. Having somehow caught and incapacitated his prey, he opens his mouth…and nothing happens! Without the support of the water to assist him, he needs moveable and muscular jaws, a tongue, and salivary lubrication in order to ingest his food. Early land colonisers such as the swamp loving giant salamander-like *Acanthostega* of 360 million years ago[20] would doubtless have had to drag their food back into the water simply to be able to ingest it until their foreguts had evolved the mechanisms to deal with it (Fig. 3.4).

The other gut adaptation for life on land involves the reclamation of water and salt. As food passes through the gut it is diluted by secretions containing

[20] Despite its appearance, traditionally used as resembling the first creatures on land, it is now believed that *Acanthostega* could never actually have supported itself out of the water and we may yet find that the first properly terrestrial vertebrates only appeared during Romer's gap (see below).

digestive enzymes, acid and alkali. As much as 9 L of fluid is secreted into the human gut every day and a cow can produce as much as 150 L just of saliva in just 1 day! This fluid will be wasted if it cannot be reabsorbed—along with sodium and other salts that pass along with it. Having focussed so far on the foregut adaptations required for life on land, and before it the specialisations of the midgut, we have paid precious little attention to the 'hindgut'—the part of the gut that we recognise as the large bowel (or colon), rectum and anus. Perhaps this is not surprising as in fishes it is small, comprising an anus or short rectum only. Water and salt are plentiful in the sea but rather limited on land where they cannot afford to be squandered[21] and indeed as we have seen need to be actively secreted. It is the elongation of the hindgut and its evolution as an organ for fluid and salt absorption—the colon—that finally allows life its freedom from the sea. The human colon for instance is capable of remarkable feats of absorption—reclaiming over 90% of the salt and water that enters it.

I suggest we stop to ponder for a while as we have just inadvertently stumbled across something of fundamental importance in our story. As we have seen, life began in the fluid environment of the ocean and required the partitioning of water and salt across the containing membrane (or epithelium). The basic nature of life did not change when faced with the extreme challenge of existing in the alien environment of dry land. Instead, numerous extraordinary adaptations were necessary to preserve the cellular and body composition that evolved in the sea. Whilst we might think that swamps and shallow lakes could provide a 'halfway house', the final transition to living completely on land is actually an 'all or nothing' phase shift. It therefore requires a completely different kind of organism to one that lives in the sea. Evolution—a process of gradual change—would appear to be thwarted when faced with such a herculean requirement for sudden transformation that effectively needs to take place between generations. Its solution as we shall now see was to go one better and produce two completely different types of animal *within the same generation*.

---

[21] Common salt (sodium chloride) has been of fundamental importance to human society throughout history. The earliest population centres developed near sites of salt availability. The 'Natron' valley in Egypt gives sodium its Latin name and chemical symbol (Na); the first European city, Solnitsata ('salt workings') in Bulgaria, developed around a salt mine. The word 'salary' derives from the word salt, and is reflected in the expression 'worth their salt'. In the sixth century, North African merchants considered salt to be literally worth its weight in gold, and in Abyssinia salt slabs were used as coins. Availability of salt has led to wars and altered the course of history—Venice fought with Genoa in 1482 (the war of Ferrara) over the control of a saltworks and the levy of a salt tax by the Pope led to the 'Salt War' of 1540 in Perugia. Population unrest against salt taxes also contributed to the French and American revolutions, and the Indian independence movement in 1930.

The Metamorphosis of Life

The phase transition from water to land, requiring the ability to regulate salt and water flows, is not too dissimilar to that between living in freshwater and salt water. We saw earlier how fishes have adapted to existing in either environment by the use of salt and water pumps in the membranes of their gills, intestines and kidneys. I promised to return to the question of how fishes can live in both. For instance, the Atlantic salmon (*Salmo salar*) hatches from eggs laid in freshwater streams, travels down to the ocean and then in one of nature's extraordinary feats of navigation manages to find its way back to the exact same freshwater spawning ground where it started its life in order to mate, and then die. Such a lifestyle is known as '*anadromous*' meaning 'running upwards'. Eels live life in the opposite direction ('*catadromous*'), spawning in the ocean and living in freshwater as adults.

The ability to differentially handle water and salt between dissimilar environments in such fishes is controlled by a hormone called thyroxine, which regulates the level of expression of the essential membrane pump—the sodium-potassium exchanger (see Diverticulum #2.3 in Chap. 2). The amount of thyroxine in the blood in anadromous fishes correspondingly varies depending on the salt content of the water.

Thyroxine is an unusual but relatively simple molecule that incorporates the element iodine, and related chemicals can be found throughout nature from single-celled algae to sponges and vertebrates. In common with other ancient chemical messengers, its function has changed over time. It probably first arose as an anti-oxidant to reduce the damage from reactive oxygen species, which links it to its role in regulating mitochondrial metabolism which generates them. It is in this capacity as a master controller of metabolic genes that thyroxine is best known to us—individuals with too much of it suffer from weight loss, fast heart rate and sweating. Notably, it also plays a key role in growth and differentiation—children deficient in thyroxine experience impaired brain development and delayed puberty.

We shall now briefly retrace our steps to visit our old friend *Branchiostoma* (the lancelet), in whom we encountered the wheel organ and the pharyngeal slits, for a closer examination of its pharynx reveals yet another specialisation of interest to us. This is a valley in the floor of the mouth called the 'Endostyle' that produces mucus and wafts it backwards on to the pharyngeal slits to trap the filtered prey. It is also within the Endostyle that cells specialise to take up iodine and produce thyroxine. The reason for this dual purpose is that thyroxine is made in an extraordinary way. A mucus-like glycoprotein called thyroglobulin is secreted by the cells of the endostyle and iodine is attached to key amino acids on the large protein chain. This enormous molecule (made up of over 2500 amino acids) is then endocytosed back into the cells where it is digested and the iodinated fragments are released as thyroxine. Just as we saw with other gut-derived organs such as the pancreas, during subsequent evolution the hormone producing cells have budded off the foregut to form a separate gland. This is known as the thyroid gland (hence the name of the hormone) and in humans is found in the front of the neck. Within our thyroid gland, cysts or 'follicles' exist which are circular structures that secrete the mucus-like thyroglobulin into a central 'pond' that is then taken up again into the cells and chemically broken down to produce the thyroxine.

On our way back, let us drop in on the lamprey (*Petromyzon marinus*) whose cousin the hagfish we have previously encountered. Lampreys are jawless fishes (agnathans) that are anadromous like the salmon. In common with many lifeforms they produce a larval stage which looks and behaves completely differently to the adult form. This is a widespread tactic in the animal kingdom which we associate most obviously with insects changing from nymph to dragonfly or caterpillar to butterfly. One benefit to having different body forms for immature and adult animals is that it permits the young to inhabit a separate ecological niche and take a different diet to the adults. As a result, the larger adults do not out-compete their own offspring! Lamprey larvae are specialised to live in freshwater as filter feeders very similar in form to *Branchiostoma*, but as an adult, they migrate into the salty sea and use sharp rasping teeth to feed on other fishes as a parasite. Metamorphosis from a larval stage living in one environment to an adult living in another also allows organisms the opportunity to shift phases abruptly within the space of one lifetime—in this case from fresh to saltwater. In the lamprey, this change takes place entirely and singularly through the developmental actions of thyroxine produced by the endostyle. Interestingly, the endostyle of the larval lamprey can itself no longer be found in the adult form—it has developed into a thyroid gland separate from the gut. Thyroxine is clearly doing much more here than just altering the expression of salt pumps to permit fishes to exist in fresh or saltwater—it is effectively capable of driving the complete physical transformation of the animal.

A Huge Leap for Life

The transformation from tadpole to frog is one that many of us have witnessed in our childhoods and are therefore at risk of taking for granted. It is nevertheless one of the most astounding miracles of nature. During an extraordinarily short period of time, the tadpole completely reorganises itself. The external changes are the most obvious—gills are lost, limbs develop and the tail slowly shrinks away. Those taking place internally are even more dramatic. There are significant alterations in the brain and spinal cord, eyes adapt to be able to see out of water and a new sensory organ—the middle ear—develops. The skin completely restructures and the skeleton changes. Lungs grow, the liver, pancreas and rudimentary kidney reorganise. The intestine undergoes an extraordinary transformation. The herbivorous tadpole has a relatively undeveloped, long spiral intestine (with a typhlosole!) which is required for its diet. Over the course of 5 days this shrinks by 75% and develops villi in the lining that increase the surface area for absorption. A stomach develops along with its secretory glands and the frog is now ready for a carnivorous diet made up largely of insects. Just as we have seen with the lamprey, all of these changes are brought about by just one molecular switch controlled by thyroxine.

Given what we learnt about homeotic genes at the beginning of this chapter, the ability to bring about such a large number of genetic changes through one switching mechanism should not surprise us—although we can still find it awesome! However, there is one outstanding matter to address. I mentioned that natural selection could not adapt to a phase change from water to land between generations, but here it is doing exactly that within the same generation. How can it be that there is a ready-made template for a quite different animal that exists within the tadpole just waiting to be triggered by a surge in Thyroxine? How did natural selection work on this blueprint to fine tune all the necessary adaptations for life on land if there was no opportunity for gradual transition?

To answer this question, it is helpful (as so often) to shift our perspective or viewpoint. We usually envisage the adult as the end form and the larva as the means to that end. From the crucial perspective of reproduction this is clearly the case, as breeding usually takes place in the adult form and these forms are often the most familiar to us. However, it follows that the adult only needs to survive long enough to mate and as a result may be extremely short-lived. For instance, the South Carolina Mayfly (*Dolania Americana*) only exists as an adult fly for about 15 min in order to mate and then die, but spends its previous year of life underwater as a larval nymph. An adult form only briefly inhabiting a new environment does not need a full range of survival equipment. Due to its short life span, the adult mayfly has no need to evolve a means of feeding or of surviving extremes of temperature or of escaping predators. The genetic blueprint for the adult form does not therefore need to be 'complete' in the first instance.

We also need to consider the seasonal changes in environmental conditions that lead to selective advantages in different habitats at different times. Unlike the mayfly, the lungfish did not develop the ability to survive on land in order to mate. Nor did it do so because it saw the opportunities for terrestrial colonisation in its genetically modified offspring many thousands of generations subsequently! The primary purpose of this adaptation was merely to survive the seasonal drying-up of the rivers and lakes until the rains came. Such a period of 'shut down' during drought in such creatures is known as 'aestivation', in comparison to the more familiar 'hibernation' undertaken by some animals during cold periods. Again, only a limited number of adaptations may be required initially in the transformed adult to allow it to survive for such a period in an alien environment. Having made that leap to a different phase (even if only to survive for a short period of time in hardship), it could be envisaged that adaptations to take advantage of the new surroundings might gradually accrue. The animal might as a result evolve to live more and more of its life there, especially given that there would be little competition for food (plants and insects).

Such a notion is backed up by a closer look at the timing of metamorphosis, initiated by the release of the hormone thyroxine. In fishes and amphibians, the burst of thyroxine production to bring about the phase shift is under the control of another chemical messenger called 'CRF', and this is produced by the brain in response to stress. Metamorphosis is thence timed to take place when survival of the animal in its original form is threatened by changes in the surrounding environment such as reduction in oxygen levels, changes in temperature or salinity.

Unlike the insects that adapted to leave the water purely for reproductive purposes, terrestrial tetrapods most likely did so for survival in environmental extremes. Mating in a tenuous state of existence would not be advantageous and land vertebrates were forced to return to water to reproduce, even when spending the majority of their lives ashore. Thus, the amphibians maintained terrestrial and aquatic forms. It was only with the development of the 'amnion'—a membrane enclosing a fluid compartment containing the developing embryo—that vertebrates were able to free themselves from the aqueous realm. The amnion could be surrounded by an impervious shell to prevent desiccation as with bird or reptile eggs, or by retaining it within the body and giving birth to live young—'viviparity'. The creatures that developed this ability—the 'amniotes—are all the reptiles, birds and mammals that permanently inhabit the land, and in which metamorphosis to enable phase change thereby became largely redundant.

The End of the Trail

The stages of terrestrial tetrapod evolution corresponding to the earliest amphibians are currently a critical missing page in fossil history—a period known as 'Romer's gap'[22] between about 360 and 340 million years ago. Our geological record simply skips a track or two between *Acanthostega* and fully land adapted creatures. It was immediately preceded by a mass extinction event of unknown cause which wiped out over 95% of all vertebrate species and was much more significant than that which led to the extinction of the dinosaurs some 300 million years later. Recent fossil finds in Scotland and Newfoundland are beginning at long last to fill in the gap and reveal some intriguing features. For instance, initial interpretations suggests that a strengthened rib cage to support the weight of the creature may in fact have preceded the elongation of weight bearing limbs.

Sadly, it is here, where the trail grows cold but with tantalising glimpses of things to come, that we will end our first journey. It is by no means the end of the story, and I would love to tell you about the wonders of the gut in its myriad forms, from the stomach of the vampire bat to the spiral colon of the lemming, but we will have to leave that for another time as we have other revelations to uncover. Our journey has brought us from the dawn of life over a distance of some 3500 million years, from the first cells to the emergence of life on land. We have watched how containment within a defined space created life but thereby excluded it from the essentials it required; how this was solved by internalising a part of the outside world within the single cell and how this led to predation and the advantage of size. We have observed single cells combining as colonies then developing separate functions to work together as a multicellular organism. We saw the containment paradox at play again with the formation of epithelia and how this was similarly overcome by the necessary formation of a gut that could ingest and digest food where it was surrounded by, but never within the animal. We have considered how the earliest guts may have driven the first 'explosion' of life forms and how the invention of the anus and the 'through gut' may even have brought about the

[22] Alfred Sherwood Romer (1894–1973) studied paleontology, embryology and comparative anatomy of vertebrate classes and defined their relationships in his classic textbook, *Vertebrate Paleontology*. He first identified the missing era in the fossil record (between about 360 and 340 million years ago) that first became known as 'Romer's Gap' in deference to him in 1995.

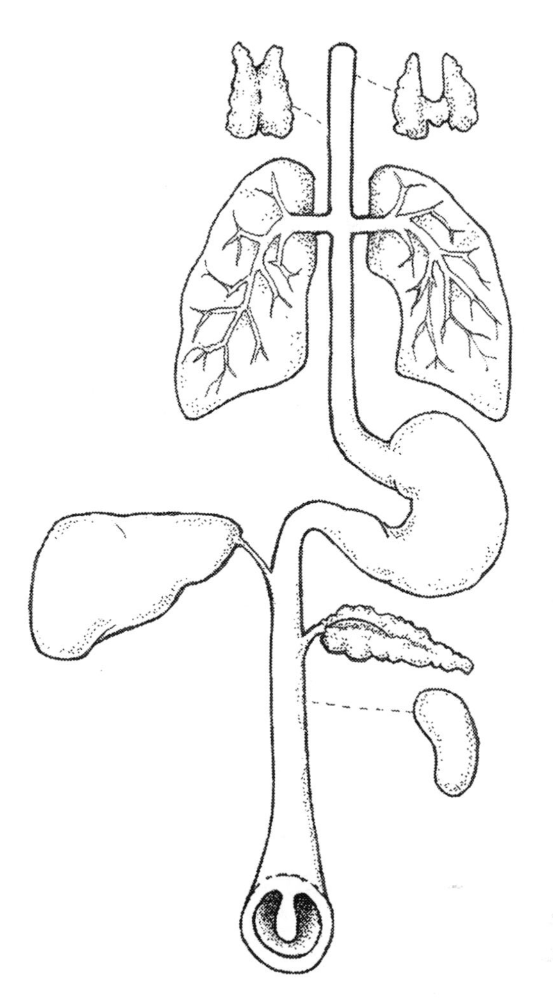

Fig. 3.5 The gastrointestinal tract and the organs that it has spawned. From the top moving downwards—the thyroid gland, secreting the hormone thyroxine; the thymus, producing immune cells (developing from the pharynx); the lungs for gas exchange (from the foregut), the liver for storage and metabolism and the pancreas for digestion (from the mid gut). Below this, the spleen which is thought to have partly evolved from the typhlosole (seen in cross section at the bottom)

Cambrian proliferation. And we have seen how some of the major organs that sustain animal life were derived from the gut, and even driven the process of metamorphosis that allowed it to take the huge step of living on land.

Above all, I hope that I have been able to show you how the gut is the most ancient of organs and how its history is the story of life itself. But there is a parallel narrative that is tightly interwoven throughout the aeons of gut evolution—the epic tale that created our immune system. We will need to retrace our steps almost to the beginning in order to unravel it (Fig. 3.5).

Part II

The Intestinal Palimpsest

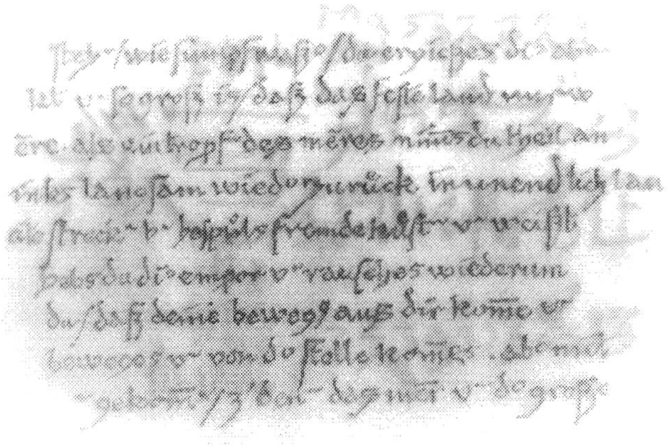

A palimpsest

'Evolution is a light which illuminates all facts, a curve that all lines must follow'
—Pierre Teilhard de Chardin, La Phenomene Humaine, 1955[1]

'Nothing in biology makes sense except in the light of evolution'
—Theodosius Dobhzhansky, Essay title, 1973[2]

[1] Pierre Teilhard de Chardin (1881–1955), a Jesuit priest, palaeontologist and philosopher. 'The Phenomenon of Man' was published posthumously but based on ideas and essays from the 1930s. His relationship with the Catholic Church was strained by his philosophical writings that theologians found hard to accept. Similarly, scientists had difficulties with his theories of evolution of human thought culminating in an 'omega point' of singular collective consciousness which reconciles the scientific and religious strands of his life in his visualisation of God. Outside of 'mainstream' philosophy his ideas have been embraced and quoted by science fiction writers from Philip K Dick to Arthur C Clarke.

[2] Theodosius Dobhzhansky (1900–1975), Ukrainian biologist who (like Charles Darwin) studied beetles before establishing his scientific career in the genetics of the fruit fly (*Drosophila*) both in the USSR and the USA. His seminal work 'Genetics and the Origin of Species' combined the nascent understanding of genetics with Darwinian theory and underpinned the 'Modern Evolutionary Synthesis'.

Introduction

Legend tells of a time nearly 2000 years ago when a Roman High Priest selected the physician to treat his injured gladiators. Applicants were expected to operate on an abdominally eviscerated ape to keep it alive. The only person to accept the challenge was a young man called Aelius Galen who was duly offered the job, and during his 4-year tenure only 5 gladiators died (compared to over 60 in the same period of time prior to his appointment). Galen went on to become the Emperor's physician and the leading medical practitioner of his time. His philosophy and writings would influence Western medicine for over 1000 years. Neither history nor legend has recorded the fate of the ape.

Galen was born and brought up in the ancient Greek city of Pergamon. At the time it was a thriving metropolis and a renowned seat of philosophy and learning. The library of Pergamon was its jewel, second only to that of Alexandria. It is thought that its 200,000 scrolls were largely written on papyrus, an early form of paper manufactured from (and named after) the crushed stems of a plant (*Cyperus papyrus*) that grew in the Nile delta. However, in the second century BC it became an expensive commodity as the enormous demand for papyrus drove the plant to the edge of extinction. Instead, dried animal skins became widely used as a more durable writing medium and their production soon centred on Pergamon. Etymological drift changed the name of the city through the Latin '*Pergamenum*' and French '*Parchemin*' which became synonymous with the product and ultimately gave us the English word—'*Parchment*'. All texts were penned on parchment (or its finer version 'vellum' made from the hides of lambs or calves) until the advent of modern paper in the fifteenth century. And so it was that the two great legacies of Pergamon—Galen's theories of medicine and the very parchment upon which they were written—would endure for more than a millenium.

There is a particular parchment held in a private collection in Baltimore that dates back around 1200 years to the ninth century. At some stage it was housed in the World's oldest surviving library—that of St Catherine's monastery in Sinai. It appears to be a Melkite hymn book[3] written in Syriac, an Aramaic dialect. However, closer inspection reveals that underneath this eleventh century text is much older writing dating back a further 200 years. Using specialist photographic techniques it is possible to read the 'undertext' which appears to be a Syriac translation of 1 of the 11 volumes of a medical book

[3] The Melkites are an ancient branch of the Christian church and derive their name from the Arabic 'Malkee' (ملكي) meaning 'Royal'. Nowadays the term is limited to a specific community of Arabic-speaking Greek Catholics in the Middle East.

called '*On the Mixtures and Powers of Simple Drugs*'. It was written by none other than Aelius Galen. It is the oldest known copy of this important text translated from the original Greek 600 years after the death of its author. Parchment was not cheap to manufacture and so the practice of scraping off the text and re-using it in this way for different content was commonplace at the time. Sometimes many different separate layers of text can be discerned in such documents, which are known as 'Palimpsests'. The '*Syriac Galen Palimpsest*' is just one example of many dating from this period.

In the same way, I envisage the intestine as a palimpsest. The surface is the interface between the animal and its environment—the single layer of cells of the intestinal epithelium one hundredth of a millimetre thick—where containment is necessarily compromised by the need for permeability and large surface area for absorption. Over the aeons of its evolution it has been the meeting place where the perpetual politics of natural selection has played out between organisms—in both battles and alliances. However, due to the way in which nature duplicates processes prior to tinkering with them, the history of immunity is written in the gut in separate strata of time exactly like the different texts of a palimpsest.

As we retrace our steps we will firstly revisit old friends such as the single celled amoeba; the sponges, corals and jellyfish, and the lancelet. We will then see how on these foundations grew the immune systems of the lamprey and lungfish and ultimately those of terrestrial tetrapods like ourselves. And in the human gut we will see the separate under-texts of the intestinal palimpsest that together tells the story of our immune system.

4

The Rules of Social Engagement

Summary In which we uncover the foundations of immunity in the social amoeba. We realise that our immune systems are not merely a means of fighting off infection but evolved for purposes of 'housekeeping' in the multicellular organism and also for avoiding 'cheating' behaviour between different cell types. Just as multicellularity in animals required the development of a digestive system, so it required the evolution of immunity—a rule book of behaviour between cells. Fundamental to all interactions between cells is the recognition of identity—whether belonging to the same species, colony or individual. Just two protein groups—called 'immunoglobulins' and 'leucine-rich repeats'—have permitted the identification of discreet molecular patterns throughout the evolution of animal life from sponges to ourselves. The requirement for defence against micro-organisms evolved predominantly along two pathways—mechanisms to enhance phagocytosis and the secretion of pore-forming proteins that could puncture the cell membrane and kill the cell by breaching containment. Both rely on a dependable means of identifying the microbe. For this purpose, animals developed multiple pattern-recognising proteins specific to unique microbial proteins. However, bacteria reproduce many times faster than a multicellular animal, and through natural selection will rapidly alter or cease production of a protein recognised by animal defences where possible. Animal defences are therefore often targeted against indispensable microbial components recognised by a group of proteins based on leucine-rich repeats called 'toll-like receptors'. Ultimately the need to create more and more fixed pattern recognition molecules hardwired into the DNA of the organism leads to an impasse and a 'dead end' in animal immunity.

J. Woodward, *The Gastro-Archeologist*, https://doi.org/10.1007/978-3-030-62621-1_4

'The rules of social engagement'

'Life did not take over the globe by combat but by networking'

Lynn Margulis

Haven't We Met Somewhere Before?

Many years ago, whilst hiking in the Picos d'Europa in northern Spain I came across a species of snake that I didn't recognise. Clearly a viper, its markings matched none that I had seen before or were in my guide book at the time. Ten years later, the second edition of the guide was published and there looking up at me was a picture of the very creature that I had seen. Between editions, it had become identified as a new species called *Vipera seoanei.*[1] I was reminded of this story when I woke up recently to the news that Europe has another new species of snake. What we previously thought of as the 'grass snake' (one of only three species found in the UK) is actually two species—the 'eastern' grass snake (*Natrix natrix*) and the 'barred grass snake' (*Natrix helvetica*). The newspaper headline erroneously reported a 'new species

[1] Arnold, Burton and Ovenden—Collins Field Guide to Reptiles and Amphibians of Britain and Europe. First edition 1978, second edition 2002 (reprinted 2004). I never leave home without it! 125 species were described in the first edition, 201 in the second. 20 of these w re added as a result of expanding the range that the guide covered, and some were new discoveries such as the Majorcan midwife toad (previously thought to have been extinct), or completely new to science such as the La Gomera giant lizard (how could something so big have escaped detection?!). However the remaining 53 were due to the recognition that species variants were actually individual species in their own rights.

discovered'. Not so. It had always been there under our noses, it was just that we had now discovered that the two types of grass snake were distinguishable. Of course, the snakes themselves had known this all along!

Deciding what defines a species is actually surprisingly difficult.[2] Classically it comprises a group of individuals wherein matings result in fertile offspring. However, as always there are exceptions. Some well-known species can interbreed to produce fertile 'hybrids',[3] and individuals from two separate species may be able to interbreed where they overlap in geographical range but not if taken from the far opposite reaches of their respective territories. Recent DNA sequencing technologies have allowed species to be demarcated more clearly and resulted in an 'explosion' of 'newly identified' species. However, regardless of how tightly a 'species' can be defined, such a community of interbreeding organisms effectively share a gene pool.

Astonishingly, even animals small enough to be single-celled are able to identify another as being of the same or a different species. Sensing the identity—the 'alikeness'—of other cells is the foundation of what we have come to know as 'immunity'. We tend to use this term to mean the way in which our bodies defend themselves against organisms such as bacteria and viruses that cause infections. However, it is becoming increasingly apparent that such a perspective on the 'immune system' is far too narrow, and may not be its original or even its principal purpose. It is probable that immunity actually first came about for completely the opposite reason—as a 'social system' that allowed cells to interact appropriately together (Fig. 4.1). And it almost certainly arose at a time when animals first embraced the team-working of multicellularity.

As we retrace our steps back to our earliest predecessors, we cannot rely on the fossil record but only on existing creatures that resemble those at different stages of our evolutionary path. We will have to keep reminding ourselves along the way that such organisms have had plenty of time to continue to evolve despite their 'primitive' appearance and that time has not stood still for us to be able to use them as a window on our past. Our starting point this time around will be the humble slime mold—a single-celled amoeba with aspirations of bigger things.

[2] The 'species problem' has plagued naturalists for a long time—as Charles Darwin wrote in *On the Origin of Species* in 1859—'No one definition has satisfied all naturalists; yet every naturalist knows vaguely what he means when he speaks of a species. Generally the term includes the unknown element of a distinct act of creation.'

[3] Hooded (*Corvus cornix*) and carrion (*Corvus corone*) crows are an example of two separate species that can mate to produce fertile hybrid offspring. A more interesting example is that of the edible frog (*Pelophylax kl. esculenta*) which results from the mating of a pool frog (*Pelophylax lessonae*) and marsh frog (*Pelophylax ridibundus*). Edible frogs can mate with marsh frogs, pool frogs or other edible frogs and (for complex genetic reasons) the offspring they produce can in turn be pool frogs, marsh frogs or edible frogs! Unlike normal matings, the genetic material of each parent of the edible frog remains separate rather than mixing together. Such species are known as 'kleptons' (from the Greek, 'kleptein' meaning 'to steal') having effectively 'stolen' the genetic identity of one of the parents. This is denoted by the '*kl*' in its nomenclature (as above).

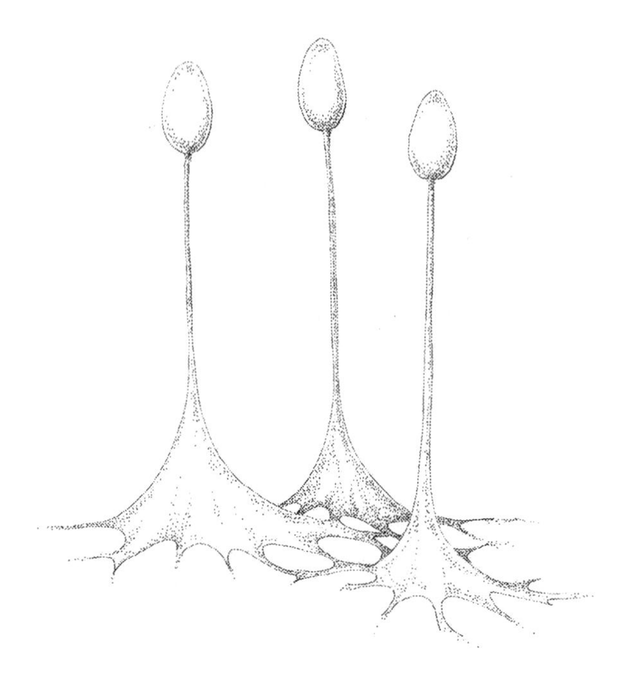

Fig. 4.1 The fruiting body (multicellular phase) of the social amoeba, *Dictyostelium*

Playing by the Rules

We met the social amoeba (*Dictyostelium*) in the last chapter when we came across chemical signalling between cells. *Dictyostelium* feeds on bacteria and other cells by phagocytosis. When food runs out, there is a real danger that the amoebae will simply feed on each other. Instead, in times of famine it employs a different tactic—that of hopeful emigration to a new location (just like the great Irish exodus of the 1840s). It achieves this by secreting the chemical messenger that causes the cells to aggregate into a slug-shaped mass of up to 100,000 separate cells (called a 'grex', from the Latin word meaning 'flock') that then move *en masse*. An extraordinary thing then happens. Cells begin to 'differentiate' to specialise in different functions. Some harden to produce a thin stalk that grows up from the surface and carries aloft an oval shaped 'fruiting body' filled with spores (making it look a little like the CN radio tower in Toronto). Such structures also give mold their 'furry' appearance, although not to be confused with similarly appearing fungi. These can be disseminated widely and release single celled amoebae on alighting in a suitable place.

Such complex social behaviour clearly benefits the species, but at the cost of individual cells that appear to sacrifice themselves for the greater good. About 20% of the cells of the grex die in simply forming the stalk to allow others the chance to survive elsewhere. Whilst it is always tempting to use human 'anthropomorphic' analogies to describe biological processes[4] the altruistic cells of the slime mold have no choice—their individual behaviour is simply dictated by chemical switches. The switch is probably thrown at the time of formation of the grex, meaning that each cell takes an 80% chance of survival at that time. It would certainly be easier for amoebae if they ate each other when they ran out of food. Indeed, it seems highly likely that a single gene regulates this ability to recognise kin and prevent cannibalistic behaviour as it is possible to generate mutants in the laboratory that lack such control and run amok eating their fellow amoebae!

There are many other ways in which amoebae could cheat the rules of social engagement rather than just by eating each other. The majority of the cells that form a grex are found to be genetically identical—most likely as a result of proliferation of an individual cell resulting in its offspring remaining nearby. However, it is perfectly possible for an unrelated amoeba from a nearby patch of cells to join in with forming the grex, and this can lead to trouble. For instance, cells from one 'family' or clonal group could simply hitch a lift on the grex of another but not sacrifice cells in building the stalk. It could then preferentially hijack premium spaces in the 'escape pod' of the fruiting body and be at a selective advantage.

In fact, any type of behaviour that incurs a cost to an organism in order to provide an overall advantage is at risk of being subverted by 'cheaters'. Such cells could be members of other species, deviant members of the same species or even mutated cells within the same animal that grow into cancers. For this reason a rule-book of interactions—between cells and between organisms—needs to be written.

[4]Alfred Tauber in his treatise on the philosophy of the immune system ('Immunity—the evolution of an idea') strongly denounces the use of human metaphors to describe the behaviour of the immune system. The ideas of 'attack' and 'defence' in immunity, as if carried out by premeditated actions of the 'self' may impose semantic constraints on conceptualisation of the processes. Immune responses are more complex than such a bidirectional interaction would imply and are chemically programmed in the same way as the behaviour of the slime mold cells. However, I make no apologies for my liberal use of metaphors as an aid in visualisation of complex processes, as long as they are not taken too literally!

Speaking the Same Language

Amoebae know whether or not they are on the same 'team' through the interaction of proteins embedded in the surface of the cell membrane. All cells exhibit a vast array of such molecules (a little like secret intelligence bunkers bristling with different radio antennae) which allow them to sense, identify and interact with others. The cell surface proteins can be considered as a form of 'language' used by cells to communicate together. Any cell lacking the necessary molecule will be unable to interact with others in much the same way as two people that don't share a common tongue (Fig. 4.1).

Proteins, as we know, are long chains of amino acid molecules stuck together in a long line that fold and twist to produce a three-dimensional shape that enables a specific function. Many proteins can do more than one thing by being made up of separate bits joined together which each have their own function. These different parts are called 'domains'. For instance, one segment could protrude from the cell surface to interact with other proteins in the outside world, whilst another domain fixes the protein in the cell membrane and a third extends inside the cell to transmit a signal to the nucleus. These domains can be shuffled—'cut and pasted'—between different proteins.

In *Dictyostelium* amoebae, a particular protein called 'Tiger' is found extending from the cell surface. If you thought that the 'Tiger' protein of *Dictyostelium* had anything to do with stripes, you would be quite wrong! Biologists love to make up interesting—or even sometimes cryptic—names for their discoveries and in this case the 'ig' in the name is short for 'immunoglobulin' which is an identifiable separate domain contained within the Tiger protein. Proteins containing the 'immunoglobulin' domain are plentiful in nature and comprise a 'superfamily' of similar molecules—we humans have over 750 different types.

Immunoglobulin superfamily proteins are almost always found at the cell surface and make up part of the vocabulary of the language of surface interactions between cells. The reason these particular proteins are so useful is their shape, the most important part of which are two flat sheets like a sandwich. The clever part is that the certain areas of the protein—the edges of the bread of the sandwich if you like—can be altered without the protein changing its overall shape or function yet still making it recognisably different. Such molecules can therefore exhibit enormous variability between members of the same species—a property known as 'polymorphism' (from 'poly' = 'many' and 'morphism' = 'form')—and is the basis whereby individuals of the same species can still express a unique identity. We will encounter immunoglobulin superfamily proteins again throughout our journey, and indeed many people will already be familiar with the name as is it often used to describe the molecules known as 'antibodies' in our own immune systems.

'Tiger' proteins, by virtue of polymorphism, allow amoebae to recognise cells that don't belong to their species (because they lack the protein altogether) or belong to a different colony (by having slightly different 'tiger' proteins). Thus, when you mix cells from two different clonal 'families' of *Dictyostelium* together, if they have sufficiently different 'tiger' proteins they will separate apart again rather than forming a mixed grex. This simple surface interaction between cells prevents any possibility of 'cheating' behaviour between different strains. It is also the fundamental basis of 'immunity'.

Diverticulum #4.1: Diversity and reproduction

The use of polymorphic molecules to identify individuals as belonging to the same species but being discernibly individual within it is of critical importance in social interactions as we have seen with the amoeba, *Dictyostelium*. Sexual reproduction exists for the purpose of spreading genes within the species and preventing the accumulation of harmful mutations. The consequences of not doing so is seen frequently in royal dynasties that encourage inbreeding in order to maintain a 'pure bloodline'. Examples are the haemophilia (a genetic blood clotting disorder) of the descendants of Queen Victoria spread across many of the European royal families, the characteristic protruding lower jaw of the Habsburg dynasty and the physical deformities of King Tutankhamen.

Natural mate selection therefore requires the identification of (limited) *differences* rather than *similarities* between individuals in order to minimise the chances of inbreeding. For instance, two gene clusters in the fungus *Coprinus cinereus* that encode soluble hormone messengers (called 'pheromones' when they affect the behaviour of an individual) and their polymorphic receptors have many thousands of different permutations in order to select a sufficiently different mate.

It turns out that very often the same polymorphic identification markers on cells that we consider to be a part of the 'immune system' (to identify cells that don't belong) also serve the social purpose of partner preference. Mammals use the polymorphic 'MHC' genes (which also code for immunoglobulin superfamily proteins just like 'tiger' proteins) for immune recognition, but they are also found in olfactory receptors and allow some animals to 'smell' the genetic dissimilarity with a prospective mate. This was demonstrated in humans by a remarkable study where female university students were asked to rate the 'pleasantness' of the body odour picked up on the T-shirts of male students worn for two consecutive nights. Those with the most *dissimilar* MHC molecules were rated most highly (or perhaps least lowly...)!

Good Housekeeping

The first specialised cells that could be thought of as having a specific defensive immune role are also found in social amoebae. However, they are rather more like 'housekeepers' than 'policemen'. When the cells of the slug-like grex begin to change into different types, approximately 1 in a 100 becomes a 'sentinel' (or 'S') cell. These cells can move backwards, forwards and sideways within the grex and pick up debris, toxins or potentially harmful bacteria by phagocytosis, like a vacuum cleaner. Within the S cell, many such contaminants can be broken down, digested or otherwise disposed of. However, some organisms are resistant to being killed by the S cells, such as a bacterium called *Legionella*[5] which has learnt to cheat death within the cell and kills it instead. By spreading to adjacent cells such an 'infection' could then lead to the demise of the entire colony. The strategy employed by the S cells to avoid this scenario is to sacrifice themselves together with their load of poisons or harmful bacteria, by falling behind the grex to die in the slime trail and thereby save the other cells.

The ability of amoebae to digest their food depends largely on a single protein called 'Tir A'. In keeping with their role in gobbling up debris and bacteria, S cells have Tir A in abundance. Tir A belongs to another very important superfamily of proteins involved in the social interactions that we recognise as 'immunity'. As with 'Tiger' the name 'Tir A' is derived from a key protein domain, in this case predictably called 'TIR'. In contrast to the immunoglobulin superfamily proteins that are usually found on the cell surface, TIR proteins are found within the cell substance. They usually act as 'go-betweens' to transmit the signals from cell surface receptors to the nucleus. Exactly how TIR proteins came by their name we shall see shortly.

Entrapment

There is an additional housekeeping trick that S cells exhibit in order to rid the grex of harmful organisms—they trap them in a net laced with powerful digestive chemicals. The net itself is made up of DNA molecules which are extremely long and sticky and are extruded by the cells in the vicinity of bacteria. This is the equivalent of the household mop, soaked in antiseptic for cleaning floors and picking up bits of dirt and rubbish. Amazingly, the DNA in the *Dictyostelium* extracellular traps is not that of the amoeba itself but of its passenger mitochondria—clearly considered expendable under the

[5] A particular species of *Legionella—L pneumophila*—is responsible for the human disease 'Legionnaire's disease' that causes an atypical pneumonia with gastrointestinal side effects. It was first identified in July 1976 at a meeting of America veterans in Philadelphia. The society was called the 'American Legion' which gave the disease its name—34 out of the 221 affected individuals died in the outbreak. Legionnaire's disease is associated with water supplies and damp places partly by virtue of the bacteria living within the slime mold amoebae.

circumstances. The Tir A molecule is again required for this extraordinary feat, as are mechanisms for generating reactive oxygen species ('free radicals') within the cell that destroy the entrapped bacteria.

Recent evidence suggests that this use of DNA traps outside the cell is retained throughout nature, even in plant cells. Our own 'Neutrophils' are circulating white blood cells that are very similar to the S cells of the social amoeba—migrating through the blood stream to sites of infection and ingesting bacteria by phagocytosis. They also produce extracellular traps. However unlike the S cells, they are unable to survive the process of casting the nets as they sacrifice their own nuclear DNA as well as that of their mitochondria. The yellow 'pus' that we see in boils and abscesses is largely comprised of this extruded DNA.[6] We will come across other examples of entrapment as a defensive strategy during our journey.

Food…or Friend?

Thus, as we have seen, the earliest form of defence is phagocytosis—eating. This basic mechanism to overcome 'containment' and ingest nutrients diversified from a means of nourishment in single cells into its housekeeping role in multicellular animals, recycling components of dead or dying cells. Phagocytosis only became a form of *defence* with the ingestion of other *living* things. Initially this still worked on a collaborative basis. However bacteria came to live within cells as symbionts (whether through phagocytosis or some other form of ingestion process), their co-existence would appear to have been predominantly peaceful and derived the mitochondria and chloroplasts. It was the process of 'natural selection', working to optimise survival of both predator *and prey* that potentially changed the nature of the early 'immune' system, subverting collaboration into conflict.

Phagocytosis—eating—remains the fundamental mainstay of immune protection throughout the animal kingdom (including ourselves). Cells that look and behave remarkably like amoebae are retained for this purpose and are still described by some as 'Sentinel cells' even in our own immune system. It is perhaps unsurprising (given that it initiated the process of conflict between lifeforms) that phagocytosis should remain part of the solution.

[6] Needless to say, some bacteria such as *Staphylococcus aureus* have evolved ways of getting around entrapment within a DNA net and produce an enzyme called 'DNAse' to break it up and free themselves. Such DNAse has also found clinical uses—notably in thinning the viscid mucus produced in the airways of people suffering from cystic fibrosis, where bacterial lung infections are common and the DNA makes the secretions even harder to clear.

Nevertheless, it is worth a moment's reflection at this point to consider how unlike an amoeba we are, and that we last shared a common ancestor over a billion years ago. Yet some of the fundamentals of our own immune system—including the Immunoglobulin and TIR superfamily proteins—were already in place at that time, and identifiably amoeba-like cells underpin our own immune defences.

Diverticulum #4.2: When phagocytosis fails

There are two main types of immune cells in humans that are capable of phagocytosis—'neutrophils' circulating in the blood and 'macrophages' in tissues (which also circulate in the bloodstream as 'monocytes'). These phagocytes kill ingested bacteria inside the vesicles inside them created by phagocytosis using powerful chemicals that include 'bleach'—the same substance (sodium hypochlorite) that we use to clean toilets and famously kills '99% of known germs' according to one brands' advertising slogan. The bleach is made in the cells by reacting chlorine with reactive oxygen species made during energy metabolism. A particular enzyme is required for this process, and some people are born with a defect in the gene which thereby results in defective phagocytosis. This condition—called 'chronic granulomatous disease'—leads to life-threatening infection with bacteria and fungi causing skin pustules, lung, bone and joint infections. The commonest form is sex-linked, occurring mostly in males and affects about 1 in 200,000 individuals.

Before we leave the social slime mold amoeba behind us, we should ponder another social behaviour that they exhibit. Described as 'farming', some amoebae appear to move on to grex formation and spore dissemination before all the available bacteria have been consumed and they have fallen upon seriously hard times. These colonies allow the bacteria to grow within the fruiting body and spread along with the spores, thus providing a meal for them when they arrive in their new destination. As with everything, this carries a cost. The farmers tend to produce fewer spores than non-farmers, and are only at a competitive advantage if the spores land where there are no bacteria available to feed on. However, farmers can also propagate non-food bacteria within the grex, which bring additional benefits. Some such bacteria produce toxins that decrease the competitiveness of nearby colonies, and others are capable of disposing of environmental poisons. As a result, farmers of beneficial bacteria require fewer S cells and are able to reinvest their resources—in agriculture!

This 'symbiosis' (= 'living-together') of amoebae and bacteria suggests that amoebae can distinguish different kinds of bacteria as well as recognising their own kind. Hence they can tell apart those that pose no threat (or could even be helpful) from those that are harmful. Ignoring or tolerating aliens therefore appears to be a viable strategy for the 'immune' system providing that they do no harm. We have already come across bacteria living in the guts of earthworms in Chap. 3. Indeed, the more we look, the more likely it appears that *all life forms* engage in some kind of peaceful relationships with other organisms. Symbiosis may not always be mutually beneficial to the same degree between host and passenger as frequently portrayed. Although we will come across many examples of mutual benefit from co-existence, there may be no advantage (but equally no harm) to one of the partners. However, symbiosis can also entail harm to one of the partners—'parasitism'. This mode of life is so successful that up to 40% of all animal species are parasites of others. However, on closer inspection it is often found that a benefit of some kind to the host has been overlooked.

Nevertheless, symbiosis provides clear evidence of the initial purpose of the 'immune system' in co-ordinating the peaceful community of life rather than as a means of conflict. It appears that relationships mediated by the 'immune system' are as much about friends as about foes. Therefore, as we continue our journey to uncover its secrets, we should consider the 'United Nations' as a better metaphor for the immune system than a technologically advanced army, although military options can still be called upon for peacekeeping duties!

The Icing on the (Sponge) Cake

Having travelled no further in our journey than the single celled amoeba, we have already encountered the basic rules and mechanisms of immunity and met two of just four major protein families that underpin immune responses throughout evolution. One of these (Tiger's immunoglobulin domain) acts at the cell surface and the other (Tir A) acts within the cell itself. As we move on to the multicellular world of sponges we will meet the other two, which also act on the cell surface and inside the cell respectively. However before introducing them we shall take a quick look at how sponges have sweetened the 'language' of cell surface interactions—using sugar itself.

Sugars exist in many forms—we are familiar with glucose, sucrose and fructose in our foods, as well as other carbohydrates such as the starches found in flour that are made up of long chains of glucose joined together. Proteins can be embellished by the chemical attachment of sugars to make 'glyco-proteins', which dramatically enhances their variability. They can be added as single sugars or as chains, and they can project a considerable distance from the cell surface to create a 'glycocalyx' (derived from the words meaning 'sugar' and 'husk'—and therefore literally a 'sugar coating'). This coating can be as thick as the cell is wide and may even be seen under the light microscope.

The sugar molecules carry a negative electrical charge which attracts water molecules, hence one of the benefits of thick layers of glycoproteins on the cell surface is to produce a hydration layer around the cell that resists desiccation. However, being porous, such a coating also acts as a boundary or 'buffer' zone where interactions with other cells may take place away from the cell surface. Epithelial surfaces such as the lining of the gut or the airway make particular use of the glycocalyx in this way. However, large proteins with numerous sugar attachments can also be secreted from the cell to provide a separate layer of protection. These molecules are known as 'mucins' and in view of the amount of water that they trap, they create the gel-like substance that we describe as mucus, slime or 'snot'. This is also the jelly-like substance that provides the 'middle layer' of sponges and cnidarians (like jelly fish). Cells in the mammalian gut have become specialised to secrete mucus and are called 'goblet cells' due to their shape as they contain a large globule of mucus around which the cell stretches, somewhat reminiscent of a wine glass. They also perform a defensive role as they can be stimulated to release their mucus by signals received from bacteria—the resulting efflux of slime can swamp the bacteria (another example of entrapment) and flush them out of the system. Hence, in some ways the mucus of our guts could be compared to the 'peritrophic membranes' of insects and hagfish.

However, of most interest to us at the moment is that the enormous variety of different types of sugars and the shapes that they make can provide the cell with a specific identity, a bit like a fingerprint. For instance, sponge cells recognise each other by the particular 'flavour' of sugar coating and the mutual attraction of the sugar molecules helps the cells to stick together. If the cells from different sponges are mixed up randomly they segregate and sort themselves on the basis of their different cell surface glycoproteins—just as amoebae do thanks to 'tiger' proteins.

Preparing the Meal

The different shapes produced by glycoproteins make them eminently recognisable to other proteins. Some think that this relationship between sugars and proteins in fact dates back to the earliest evolving life forms, and it is even suggested that this interaction constituted a chemical code that predated that of proteins encoded by RNA. Proteins that recognise sugars are given a specific name—'lectins'—that derives from the Latin word '*legere*' meaning 'to choose' and which also gives us the word 'select'. Lectins are found within the cell, embedded in the cell membrane and can also be secreted to the outside in a soluble form. They are clearly ancient proteins being present in all cellular life-forms, and they mediate a large number of different processes both within and outside the cell.

Over 30 different types of lectin have been found in sponges that can identify separate types of bacteria as well as fungal cells just from their surface sugar patterns. Lectins may have several sugar binding domains, or have tails that can stick several lectin molecules together. In this way, lectins secreted into the medium around cells can bind sugars on different bacteria, thereby sticking them together and 'agglutinating' or clumping them. This effectively immobilises the bacteria and stops them from invading the cell. It also makes them more susceptible to phagocytosis by specialised cells in the sponge similar to the sentinel cells of the amoeba which wander freely within the jelly layer—and are therefore unsurprisingly called 'amoebocytes'.

Amoebocytes have specific receptors on their surface that recognise the tails of the lectin molecules bound to the bacteria and can trigger phagocytosis once attached. In this way, the lectin is effectively forming a bridge between the bacterium and the cell which enables the cell to eat it. This is called 'opsonisation'—derived from the Greek verb '*opsonein*' meaning 'to prepare for eating'! If you remember back to Chap. 1, this is very similar to the way in which cells pick up nutrients for endocytosis into the cell (Fig. 4.2).

This scenario opens up the possibility for all sorts of games between life-forms. For example, some bacteria produce their own lectins that adhere to sugars on the surface of eukaryotic cells allowing them to invade the cell and cause an infection. Others can 'steal' sugar sequences from eukaryotic cells in order to disguise themselves and avoid recognition. One specific complex form of sugar called sialic acid is only found in significant quantities in late evolving bilaterians such as starfish and vertebrates. This is thought to be recognised as a 'do not eat me' signal on such cells, distinguishing them from bacteria. However, nefarious bacteria such as the species causing meningitis in humans (*Neisseria meningitidis*) can pick up and coat themselves with sialic acid molecules to fool the host immune response into not eating *them*. Some bacteria may even have gone as far as stealing the genes that make the sialic acids in the first place. However, just to remind us that immunity is about interactions—both positive and negative—some sponge lectins not only recognise specific bacteria but act to stimulate them to proliferate. Effectively they are 'selected' as symbiotic partners by the sponge—a case of choosing one's friends carefully.

Diverticulum #4.3: A lack of lectins

Humans make a particular lectin called 'mannose-binding lectin' or MBL. It is so-named after mannose, the sugar that it recognises, and in doing so is able to identify a wide range of potential infectious agents from viruses to bacteria and parasites. However, at least 1 in every 20 of us have a genetic defect that results in production of a faulty protein or none at all, leading to 'MBL deficiency'. There has long been controversy over how significant a problem this is for us. Studies of hospitalised patients with infection have suggested that a slightly higher proportion of them have MBL deficiency than the general population, suggesting that this condition does pose a risk. However a long-term follow up study of over 9000 MBL deficient individuals found no increases in infection or mortality and it is likely therefore that this lectin does not play a significant role in our immune systems.

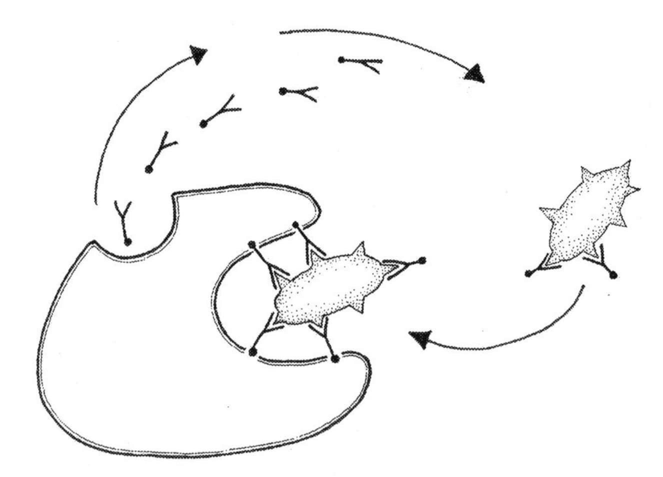

Fig. 4.2 Opsonisation—preparing for eating. A cell secretes molecules such as lectins (in invertebrates) or antibodies (in vertebrates—see next chapter) that recognise and bind to determinants on the surface of its prey. Phagocytic cells carry receptors which bind to the opsonising molecules that then act as a bridge to aid ingestion. The two functions can be separated such that phagocytic cells can focus on ingestion and need only produce one type of receptor (for the opsonin molecule), while other dedicated cell types secrete a range of specialised opsonins to identify different forms of prey by their surface molecules

That's Really Amazing!

Lectins binding to sugar residues are just one example of the way in which cells detect the molecular signatures of other organisms. Such 'pattern recognition receptors' specific bind to certain molecules that may only be produced by a particular species—and hence may give them away as a source of food, a potential partner in symbiosis or a threat. However, the choice of the pattern that is recognised is critically important. For instance, if the surface molecule that signifies a bacterium as being a threat is not essential to its survival or function, then mutants lacking the marker will be rapidly selected and future generations will evade detection. In this way, by rapidly changing their surface molecules, infectious agents such as the Human Immunodeficiency Virus (HIV) and the malaria parasite are able to escape our immune system. This is also the reason why it has proven extremely difficult to date to generate an effective vaccine for either of these diseases.

Other pattern recognition molecules are therefore required that can identify a range of different molecules—not just lectins that recognise the sugar coating, as this can be altered by the bacterium to disguise itself. The 'Toll-like receptors' are one such family of pattern recognition proteins. They are found on the cell surface and are named after a protein found in the fruit fly (*Drosophila*). Supposedly the name comes from an exclamation made by their discoverer, the Nobel laureate Christiane Nüsslein-Volhardt in 1985—'*Das ist ja toll*!' meaning 'that's really amazing!' The specific patterns recognised by different Toll-like receptors include bacterial sugar molecules, specific lipid molecules associated with bacterial membranes, viral and bacterial DNA, and a protein constituent of bacterial flagella—the whip-like projection used for propulsion. Whilst all are chemically very different, they represent an eclectic collection of fundamentally important and indispensable components of micro-organisms, that have been clearly selected as identifying parts of micro-organisms that cannot easily be changed in order to evade detection.

Once again it is the shape of the protein that dictates its function. Toll-like receptors form a characteristic 'horseshoe' shape that sticks out from the cell surface, and is formed by the alignment of multiple stretches of an amino acid called leucine. These segments are therefore called 'leucine-rich repeats'. The benefit of this configuration is that subtle alterations in the intervening sequences can change the shape of the 'binding site' of the receptor, without changing its basic function. Hence by duplicating the basic Toll-like receptor gene and making small changes, different receptors can be made that recognise different molecules. The first Toll-like receptor appears in the sponge and recognises a type of membrane lipid (called 'LPS') found only in bacteria. We humans have 11 different Toll-like receptors, but the prize goes to the purple sea urchin (*Strongylocentrotus purpuratus*) with a collection of over 200, each recognising a different molecular signature.

Let us return to the sponge and its single Toll-like receptor that recognises the bacterial membrane component. Having picked up such a signal, the receptor sends a message from the cell surface to the nucleus. The nucleus responds appropriately to the potential threat by translating DNA sequences into proteins that can be used against it. The 'go-between' molecule that transmits the message through the cell to the nucleus is a 'TIR' protein very similar to the one that mediates phagocytosis in the amoeba (TirA). Even more amazing, this very same TIR protein found in sponges is found throughout the animal world and even mediates immune messages within our own cells.[7]

The TIR domain is therefore fundamental and ancient. It gets its name from the various surface molecules that it interacts with. In this case, the 'T' of 'TIR' comes from the 'T' of 'Toll-like receptor' whose message it transmits. However, it will not have escaped your notice that TIR domains were present in amoebae in the shape of TirA long before the first Toll-like receptors appears in the sponges. This suggests that they existed to provide an information shuttle from different cell surface receptors to the nucleus before evolving an interaction with Toll-like receptors. In fact, 'TIR' proteins are actually so archaic that they are found in all eukaryotic cells. Plants also have their own kind of immune systems which share some molecular similarities with our own and the 'R' in 'TIR' comes from the 'R' of 'Resistance genes'—which code for defensive molecules found only in plants (Fig. 4.3).

The sponge cell therefore detects the bacterium through its specific molecular signature, recognised by a toll-like receptor on the cell surface. This activates a TIR protein inside the cell that binds the DNA in the cell nucleus to switch on the production of a defensive protein that is secreted by the cell to dispatch the bacterium! This three-step process—detection, signal, and response—is a bit like a radar station picking up the presence of invading aircraft and signalling to headquarters where a retaliatory airstrike is launched!

The third step in the process—the secreted defensive protein—is again of a type that is found throughout nature in all animal immune systems. It bears a strong resemblance to proteins produced by our own cells. They are defined by their function as 'pore formers' and take us right back to the fundamental principles of life that we encountered right at the beginning of our journey.

Breaching Containment

No one knows when the first 'pore-forming' proteins came about, but it was undoubtedly a very long time ago as they are made by all living cells including bacteria. Pore formers are molecules that can embed within the cell membrane then join up together in a ring to open up a hole in the middle that

[7] It is called 'MyD88' for reasons that we won't go into here.

Fig. 4.3 The shapes of life. Left: A member of the immunoglobulin superfamily (an antibody molecule). The highly variable sequences are at the two ends of the Y shape. Right: A member of the leucine rich receptor family (Toll-like receptor). The highly variable sequences are at the top of the horseshoe curve. These two protein families provide the variable protein sequences that underlie the immune system by permitting the identification of cells as part of the same organism and of molecules that do not belong to the organism

allows substances in and out of the cell. By breaching the protective hull of the cell, pore forming molecules can make potent weapons. Significant loss of containment almost inevitably results in cell death—like puncturing the skin of a spacecraft or torpedoing a ship below the waterline.

Bacteria make pore-forming toxins which may be used to destroy cells in order to feed on their nutrients. However the presence of such toxins can be linked to the ability of bacterial species to cause disease, although this is often merely a case of self-defence. One example is *Listeria*, the bacterium from unpasteurised milk and cheese that can cause devastating infection in newborn babies. This bug secretes a pore-forming toxin to dig its way out of the phagocytic bubble after it has been eaten by a defensive cell. However, in saving itself it also destroys the host cell and causes disease.

All of the molecules involved in the sponge immune response to bacteria—the Toll-like receptor 'detector', the TIR 'go-between' and the pore-forming 'torpedo' are very similar indeed to molecules that can be found throughout all lifeforms subsequently and are clearly recognisable even in our own cells. Hence, the equipment of immunity may have undergone substantial refinement over the aeons of evolution but the fundamental basis of our own immune system is not really very different to that of the sponge, a modern-day representative of some of the earliest multicellular animals on earth (Fig. 4.4).

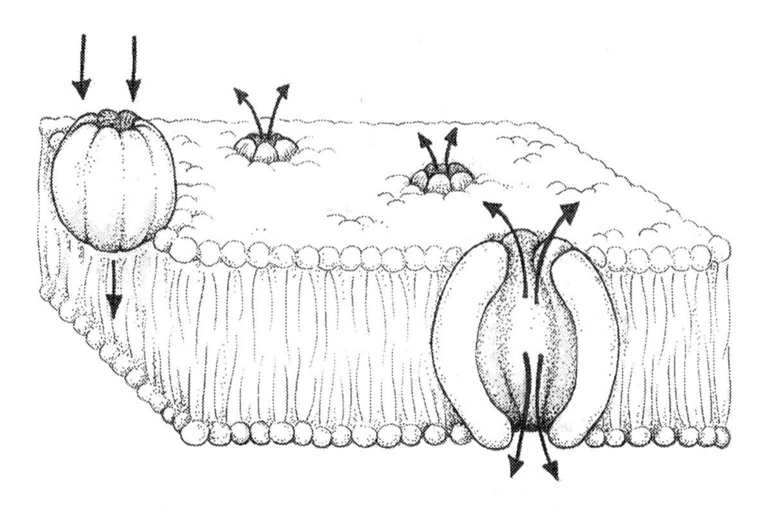

Fig. 4.4 Breaching containment by pore formation. Pore-forming proteins aggregate to produce a hole and puncture the cell membrane. This allows unregulated equilibration of the internal and external environments. Life requires the maintenance of a different chemical composition within the cell, separated from the outside by the membrane and it is therefore extinguished by the mechanism of pore formation. Pore formation (cell puncture) and phagocytosis (cell eating) represent the two main defensive processes employed by the immune system

The Inside Story

Immunology—the study of the immune system—is considered by many medical students to be one of the most complex areas they have to study. Yet, already in our journey, we have met almost all of the key components of immunity, and there is only one major group of molecular 'players' still to introduce to you.

Detailed analysis of the genetic blueprint of sponges shows that in addition to the surface pattern recognition receptors, they also possess a large number of such molecules *within the cell itself.* Being able to recognise invaders once already inside the cell comprises the last line of defence. The majority of such proteins belong to the last of the four groups of immune proteins that I need to introduce to you—the 'Nod-Like Receptor' (NLR) superfamily. These proteins are made up of three different domains that have separate functions, like three separate proteins joined together. The all-important pattern-recognition domain is a 'leucine rich repeat' similar to that found in the cell surface Toll-like receptors. Once the pattern has been recognised by this end of the molecule, the middle part called the 'NOD'[8] region (the 'N' of 'NLR') then sticks to other identical proteins in the cell to form a cluster. The resulting shape of these proteins adhering together looks a bit like a bouquet of flowers. Only then does the third part of the protein—the 'effector' end—do its job by activating pathways and processes within the cell. This third region is ominously called the 'Death' domain for reasons that we will find out about shortly.

Humans have about 23 different NLR proteins. The patterns that they recognise have yet to be uncovered in many cases, however some of them respond to the same molecular signatures as the surface Toll-like receptors. It is important to realise that the shape that is recognised by a protein is 'hard-wired' in the DNA—each is designed to recognise just one specific pattern. Therefore, the repertoire of the organisms' response is limited to the number of different pattern recognition proteins produced. The simple sponge has invested heavily in the use of NLRs with over 130 different types,[9] giving it an enviable range of potential targets and making our collection look rather puny in comparison.

A Noble Death

Within any organism there is a continuous cycle of life and death as cells age or outlive their usefulness and are replaced. However, deliberate cell suicide is one of the earliest forms of defence against infection—we have already come across the seemingly altruistic death of sentinel 'S' cells in the social amoeba as a way of avoiding infection spreading through the entire colony. In a similarly noble form of self-sacrifice, infected cells within multicellular animals trigger a 'self-destruct' mechanism to prevent shedding viruses or bacteria that have multiplied within them and could threaten the whole organism. The death of such cells takes place in a highly controlled manner known as 'apoptosis' which involves them dismantling their internal structures and shrinking down whilst keeping the outside cell membrane intact. The ultimate fate of these apoptotic cells is to be consumed by phagocytes, which thereby 'contain' the infective threat. The name 'apoptosis' (pronounced 'apo-tosis' with a silent

[8] 'NOD' stands for 'Nucleotide binding and Oligomerisation Domain' and is also known as 'NBD'. The nucleotide refers to ATP which provides energy in the form of a phosphate bond (as we saw in Chap. 1) in order to allow the NOD regions of several proteins to join together—'oligomerisation'.

[9] The first sponge to have its full DNA decoded is *Amphimedon queenslandica*, a species relatively recently discovered on the great barrier reef off Australia.

second 'p') derives from the Greek word for the shedding of the leaves of a tree in autumn, an apt and even poetic metaphor.[10]

Apoptosis can also be initiated as a result of other forms of stress occurring as a result of things going wrong—such as gene mutations and defective proteins. Removing such damaged cells is clearly a form of maintaining the integrity of the host and a further example of an 'immune' response playing a more general 'housekeeping' role. In addition, cells can be developmentally programmed to die spontaneously during the growth of an animal. For instance, when the tail of the tadpole shrinks away in amphibian metamorphosis, it is through apoptosis, as it is with our hands and feet which form initially with 'webs' between the digits that undergo apoptosis to free them up before birth.

It should come as no surprise then that the 'death' domains of the 'NLR' molecules within the cell are involved with this 'self-destruct' mechanism. When a molecule from a foreign invader is recognised inside the cell, the NOD regions stick together and the death domains are activated to trigger digestive enzymes that specifically break down the cell's internal structures. The cell dies (along with its invading hoards) to be consumed (tidied up) by housekeeping phagocytes. The NLRs are therefore the last line of defence once the infecting organism has broken into the cell, having penetrated the containing membrane, leaving suicide as the only remaining option.

The 'Self-Destruct' Button

The way in which cells are able to start the suicide sequence themselves is remarkable and shows how we can understand mechanisms so much better by seeing them in their evolutionary context. As we have seen, an infection is not necessary to trigger apoptosis—it can also occur in response to starvation, toxins or physical stress, at the right stage in development, or even if they lose cell surface contact with neighbouring cells. The auto-destruct switch to bring this about is housed within the mitochondria, the energy conversion organelles that are bound by two membrane layers and are present in all eukaryotic cells. Between the two mitochondrial membranes is found a protein that is

[10] In 1972, John Kerr, Alastair Currie and Andrew Wyllie named the process of programmed cell death 'apoptosis' in a seminal article. John Kerr had described the anatomical features of apoptosis by use of electron microscopy at the University of Queensland and subsequently joined Wyllie and Currie at the University of Aberdeen, where the Professor of Greek, James Cormack, is credited with coming up with the name. Andrew Wyllie subsequently moved to Cambridge as head of the Department of Pathology. The Nobel Prize for medicine was awarded in 2002 to Sydney Brenner, Robert Horvitch and John Sulston for identifying the genes in the worm *Caenorhabditis elegans*, that controlled apoptosis, but there was no Nobel for the original discoverers of the process.

recognised as 'alien' by specific NLR-like proteins within the cell—the mitochondrion is after all derived from ingested bacteria in our distant evolutionary past. However, the NLR-like proteins that trigger the self-destruct sequence are not able to interact with this mitochondrial protein as they are kept separated from it by the outer membrane of the mitochondrion. When the mitochondrion is damaged then the 'alien' protein is released into the cell and initiates the apoptotic sequence outlined above as if it were due to an invading organism.

Even more extraordinary is the fact that cells contain their own means of destruction. Rather than relying on damage occurring to the mitochondrion by external factors, cells actually make their own pore-forming toxins to punch holes in the mitochondrial membrane and release the protein, initiating the destruct sequence! These pore-formers are kept in an inactive form in the cell, but can be switched on by proteins that are made during times of stress, or at a genetically predetermined time in the case of developmentally programmed cell death.

We can now tentatively begin to piece together the beginnings of immunity. Microbes that are detected by their unique surface molecules can trigger the secretion of antimicrobial proteins such as pore-formers that destroy them whilst on the outside of the cell. Alternatively, microbes could be ingested by phagocytosis and become contained within an intracellular membrane, for the cell to dispose of them without the microbes ever entering the substance of the cell. Any that escaped into the cell would trigger the suicide sequence as the last line of defence by NLR proteins recognising a specific molecular signature. However, at some stage in the distant past as we have seen, certain bacterial forms evaded destruction inside the cell and were then able to stablish themselves as symbiotic partners such as mitochondria. The permanent presence of an 'alien' within it allowed the cell an opportunity to initiate the self-destruct programme for reasons other than infection simply by the controlled release of a recognisable internal 'alien' pattern molecule. Now that really is *Toll*!

There Are Many Ways to Die...

Simply shrinking away and committing suicide in the face of infection is hardly the image we have of the immune system 'fighting off a bug'. Predictably, the immune cells of animals quickly evolved ways of linking the death of cells to rallying and organising the body's resistance to an infecting organism. One way of doing this is to have molecules capable of recognising the specific patterns of chemicals released by damaged or dying cells themselves, instead of identifying microbial 'threat' signatures'. These 'damage-associated molecular patterns' can trigger responses from immune cells in exactly the same way as the molecular signatures of bugs such as bacteria and viruses. Such patterns signifying cell death may be proteins from the cell nucleus or even DNA itself, which can be recognised by surface Toll-like receptors. It is exactly for this reason that the 'self-destruct' sequence of apoptosis leads to a quiet death whereby the cell membrane is retained intact and the inside of the cell slowly dissolves. When cells die unexpectedly or rapidly, they rupture the containing cell membrane which leads to the release of damage-associated pattern molecules and triggers a significant immune response. This type of cell death—called '*necrosis*'—can occur for instance through extremes of temperature, trauma, poisoning or types of infection.

The machinery of apoptosis can also be used to send out a distress signal to the rest of the organism by activating or priming molecules that work as immune signals and are thereby released when the cell dies unexpectedly. A key protein that is activated in this way is called 'interleukin-1'. 'Interleukins'[11] are chemical messengers of the immune system and signal between different cells and organs in the body. Their name is derived from the words meaning between (*inter-*) and white—signifying their messenger role between immune cells identified as 'white cells' in the blood (as opposed to the 'red cells' that carry oxygen). Interleukin-1 is a very potent 'alarm' signal that co-ordinates a tactical immune response. It binds to cells that express a specific cell surface receptor, and having done so uses the TIR 'go between' molecule inside the cell to transmit the signal to the nucleus. We now come to what the 'I' in 'TIR' stands for—Interleukin-1 receptor. The effects of interleukin 1 are dramatic. In a simplistic analogy it is like a radio signal sent out by a wounded wireless operator to 'call in' an air strike on his position. They include opening up the blood vessels to increase the blood flow to the vicinity, to attract more immune cells out of the circulation to the site of damage, to cause a sensation of pain (to alert our conscious brains to the problem) and by an action on the brain to increase the body temperature and metabolism. The results are the redness, swelling, pain and fever[12] that we call 'inflammation' and usually

[11] The latest interleukin to be added to the list (so far)—IL-40—was described in a publication in May 2017. However, many chemicals that act as immune messengers are not attributed with an Interleukin appellation and so the list is in reality much larger.

[12] These cardinal signs of inflammation—calor (heat), rubor (redness), dolor (pain) and tumor (swelling) are attributed to Aulus Celsus (c25 BC to c50 AD) in his encyclopaedic treatise *De Medicina* ('On Medicine') of 25 AD. The text lays out medical knowledge into three areas—medicine (herbal and drug remedies), surgery and nutrition. This tripartite divide is close to my heart as a gastroenterologist with an interest in nutrition: nutrition appears to have been dropped from the current medical curriculum which

associate with an infection—all caused by our own immune system rather than the bug itself.

The kind of apoptosis that releases Interleukin-1 is called '*pyroptosis*'—literally meaning a 'fiery' death! The reason for its name is simply that this form of cell death is associated with an inflammatory response that can lead to fever in the individual.

Diverticulum #4.4: Keeping the wolf from the door

Systemic lupus erythematous (SLE) is a condition that leads to widespread inflammation affecting the joints, kidneys and skin. It gets its name ('lupus') from the distinctive facial rash that is supposed to resemble a wolf's bite! Blood tests show a widespread immune response against normal components found inside cells—for instance an antibody targeting the cell nucleus is a hallmark feature. Recent studies suggest that SLE arises from an inability to dispose of cells dying by apoptosis, or to clear the debris of neutrophil 'nets'. In other words, a lack of a 'housekeeping' function. As a result, the insides of cells which are usually hidden from the immune system become exposed and falsely recognised as 'alien' features, thereby leading to an immune response—and something of a vicious cycle as this simply results in more cellular debris.

Choose Your Friends Carefully

Because of our preconceived ideas about immunity we have to keep reminding ourselves that it exists to regulate relationships between organisms, not just to pit themselves against each other in endless conflict. As we move on in our journey from the sponges to revisit the Cnidarians (such as jellyfish, sea anemones, corals and *Hydra*) that arose during the Cambrian explosion of about 540 million years ago, we see many such relationships mediated by 'immune' molecules. Within the genome of the sea anemone *Aiptasia* for instance, 13 novel types of lectin molecule are found. The amount of these lectins produced depends on the presence of certain bacteria—however, these are not harmful bugs but rather those that are chosen by the anemone as partners, indicating that these molecules play a significant role in fostering the positive symbiotic relationship between them.

The combination of the host organism and its passenger symbiotic partners has been described as the 'Holobiont',[13] meaning the 'entire organism'. In the

envisages more of a bi-partite divide into just medicine and surgery. Very little is known of Celsus' life or even where he lived, but he is thought not to have been a physician himself but to have collated knowledge on various areas into texts. Another important attribution is the use of the Latin word '*Cancer*' for malignant tumours, translated from the Greek '*Carcinos*', also meaning 'crab' or 'crayfish'.

[13] Again this concept is thanks to the foresight of Lynn Margulis!

same way we should consider ourselves not just as 'human beings' but as colonies of micro-organisms surrounded by a human casing (which some also inhabit!). Immune systems appear to regulate the guest list, keeping them in check and making sure that bad characters are excluded. Unsurprisingly, faults in the immune system can lead to 'immunodeficiency', laying the organism open to infection. However, this is often mediated through the effects of the other parts of the holobiont: symbiotic micro-organisms themselves produce chemicals to exclude or actively attack unwanted guests, and a fault in the 'host' immune system results in a change in their composition that lays the whole holobiont open to attack. For instance, in the laboratory it is possible to generate *Hydra* that lack defensive 'pore-forming' molecules. They actually carry no more bacteria than normal but instead, the composition of the symbiotic population is altered. As a result, such *Hydra* succumb quickly to infections with fungi that would usually be kept at bay by the passengers.

Seen from this new perspective of the immune system, it is clear that different multicellular hosts will produce their own specific immune molecules to select their own partners. As we will see in more detail shortly, early multicellular animals were fundamentally hindered by their slow rate of reproduction and inability to adapt rapidly to altered circumstances—they could only do so by the gradual accumulation of genetic changes selected through consecutive generations. The result is a very cumbersome and unresponsive defensive system. Micro-organisms such as bacteria on the other hand divide—and thereby evolve—very rapidly and can quickly 'invent' ways of circumventing host defences. It therefore makes sense to keep a retinue of passenger microbes that are capable of responding equally rapidly to counter any similar potential threat. The key role of the host immune system is thus to sustain and control its symbiotic passengers which then fend off potential threats—effectively using them to treat 'like' with 'like'. Even though the symbiotic partners are also capable of mutating and potentially of causing harm to the host, it may be a case of 'better the devil you know', as the host and its partners grow up together and usually spend their entire existence together. We shall consider in part 3 what happens when this relationship breaks down.

Diverticulum #4.5: The Human Holobiont

Whilst we have partly debunked the myth of our human cells being outnumbered ten to one by the bacteria that live within our large intestines (in reality we are about 50% human rather than 10%), each of us hosts about 150 different species of bacteria and therefore human genes are thought to make up less than 1% of our 'holo-genome'. The bacteria in our gut take up residence during birth and are initially therefore maternally derived. The particular constituent species are thought to be specific to the individual—like a fingerprint—but can be affected by factors such as age and diet. Antibiotics may temporarily alter the composition of the bacterial flora but it usually returns to its original state afterwards. The importance of the bacterial status-quo is demonstrated by the potential for development of bacterial (*Clostridium difficile*) diarrhoea after the equilibrium has been upset by antibiotics. This can be severe and life threatening, and whilst we can throw yet more antibiotics at these bugs, the best treatment is to introduce another person's faecal bacteria into the gut—a 'faecal transplant'. Thankfully this is usually administered by tube.

The bacteria of the colon (large bowel) ferment non-digested foods—such as plant 'fibre'—and produce beneficial substances that are absorbed and may reduce our risk of diabetes and heart disease. Over enthusiastic adherence to a 'healthy' high fibre diet can sometimes lead to bloating and diarrhoea and is one cause of an 'irritable bowel'. It can usually be treated by simple dietary modification.

However, we are only just beginning to uncover the other effects of the gut bacteria in health and disease—for instance it is becoming apparent that the bacterial composition of the colon can influence our weight, risk of developing diabetes, our mood and experience of pain and perhaps even our allergic responses.

There is already a large industry in supermarket and drug store 'probiotics'. These are preparations that contain live bacteria with the aim of altering our commensal bacterial flora. However, bacteria in such preparations may have difficulty establishing themselves in the colon and many claimed benefits are largely unproven. In the UK they are licensed as 'food' products rather than pharmaceuticals which would require much more stringent scientific proof of effect.

The Universal Adaptor

The major problem faced by invertebrate immune systems is the limited repertoire of patterns that they can recognise as a result of the receptors being 'hardwired' or fixed. The term 'innate' immunity is used to describe this system, being inbuilt and intrinsic. We will discover the alternative of 'adaptive' immunity in the next chapter. In innate immunity, the cell has to produce a different molecule to identify each and every potential pathogen—or friend. To a certain extent, genes can be duplicated and then altered to make up for this (as shown by the large expansion of the NLR family in sponges), but there is a limit to this strategy. The problem is compounded in the case of lectins used to stick to bacteria and prepare them for eating—in addition to the armoury of different pattern recognising molecules, the cell needs to create a complementary array of cell surface receptors to bind them. It is at the next stage of evolution—the cnidarians such as *Hydra* and jellyfish—that we first see how the immune system begins to circumvent this problem by creating a 'Universal adaptor', a bit like the one we all waited patiently for the mobile phone companies to produce so that we did not a need drawer full of different charging cables!

Nature's universal adaptor for lectins is a protein called 'C3' (the C stands for 'Complement'[14]). This comes into play once the lectin has identified and stuck to the surface of a bacterium. The lectin attracts enzymes that cleave the C3 protein in two in the vicinity of the bacterium. One of the two fragments formed then sticks to the bacterial surface and acts as the 'opsonin' to bridge the bacterium to the phagocytic cell rather than the lectin molecule itself. Hence, the phagocyte merely needs to produce receptors for C3 rather than a whole range of different molecules on its cell surface to identify multiple different lectins.

Complementary Applications

Whilst a 'universal adaptor' such as C3 simplifies the range of hardware required for phagocytosis, it can also serve to 'plug in' to different applications as well. As a result, the complement system, with C3 at its core, has evolved over the last 550 million years to become a fundamental 'hub' of different immune processes.

One key application of C3 is the 'Membrane Attack Complex' which is found only in jawed vertebrates. Over time, the gene for C3 duplicated repeatedly but changed slightly on each iteration, generating a whole family of related proteins. Once C3 has bound to the attached lectin on the surface of a bacterium, it attracts its relatives—called C5, C6, C7 and C8. This family gathering effectively builds a scaffold on the cell surface for the last relation, C9. This late arrival is a classic pore forming protein that generates a circle and then punches a hole in the alien membrane to kill the organism. In this way complement can perform the dual functions of enabling phagocytosis and disintegrating targeted microbes.[15]

[14] Paul Ehrlich (1854–1915) won the Nobel Prize (along with Elie Metchnikoff who had previously been an outspoken critic of his theories) in 1908 for his contributions to immunology. He coined the term 'complement' for this specific immune pathway as it 'complemented' cells in their immune actions. He also created the expression 'magic bullet' ('Zauberkugel') to describe the use of specific agents to kill microbes without harming the human host. He discovered (invented) the first such targeted drug called 'Salvarsan', an arsenic-based chemical that worked against syphilis and was the main drug in use for this condition from 1910 until 1940.

[15] What, you may well ask, happened to complement components C1, C2 and C4? These proteins follow what is known as the 'classical' pathway rather than the 'lectin' pathway that I have described above. The

The complement 'system' is of course considerably more complicated than I have described above. Notably each component acts as an enzyme to chop a fragment off the next one in line. This serves to activate the protein and continue a 'cascade' of reactions down the line, whilst the smaller fragments act as chemical messengers to stimulate the process of inflammation in the same way as interleukin-1.

The Clot Thickens

Blood clots form when we cut ourselves in order to prevent the loss of blood—or so we would assume from everyday life. Proteins that stick together to form the meshwork of the clot are always circulating in an inactivated form within the blood stream. They are switched on by recognising and attaching to molecules that are not normally seen inside blood vessels but are exposed when the blood vessel is damaged and blood starts to leak out into the tissues. These proteins work in a very similar way to those of the complement cascade—a protein is activated by splitting and then cleaves the next in turn and so on. It turns out that the proteins involved in blood coagulation closely resemble those of the complement cascade—so closely in fact that it now looks most likely that they actually evolved as a result of duplication of complement proteins and underwent a subsequent change of function.

Whilst at first sight the two systems appear to undertake completely different roles, coagulation and complement remain closely inter-related in ourselves. Protein fragments released during clotting can act to attract immune cells to the clot and initiate the release of inflammatory chemicals. The systems can cross-over such that clotting factors can initiate the complement cascade and in return complement can activate coagulation. One even has to consider that this damage control system to shore up leaks in the circulation evolved as much to prevent things from getting in and causing infection as much as to prevent the blood getting out! In this way blood clots could be considered to be an equivalent in multicellular animals of the DNA nets used by the S cells in amoebae to trap and kill bacteria.

It is in insects and crustaceans (which have an 'open' circulation system in their body cavities rather than our 'closed' circulation where the blood flows in a network of blood vessels), that we see the refinement of coagulation as a fundamental immune response. However, in insects the process of clotting is also linked to the synthesis of melanin, the dark pigment that is made in our

end result is the same, but the 'classical' pathway is triggered by proteins known as antibodies rather than lectins—we will come across antibodies shortly in the next chapter.

skin to protect us from deleterious effects of sunlight. During the formation of melanin, toxic chemicals are generated as a side effect that are used to kill microbes trapped in the clot, which then hardens to close off the defect.

Diverticulum #4.6: Clotting and infection

As a result of its close association with blood coagulation, the immune response to severe infections such as those causing septicaemia (caused by bacteria multiplying in the blood stream) can lead to profound disturbances of clotting.

Widespread formation of blood clots in sepsis causes a condition known as 'Disseminated Intravascular Coagulation' or DIC. Clots form initially in small blood vessels and prevent the flow of oxygenated blood to the tissues. This can in turn lead to complications such as kidney failure, or gangrene of the digits. The extent of blood clotting may be so great that all of the clotting factors are consumed in the process and leaves the remaining circulation unclottable as a result. This paradoxically leads to profuse or spontaneous bleeding elsewhere. It is seen in an extreme form in mengingococcal septicaemia caused by the bacterium which causes meningitis (*Neisseria meningitidis*). One of the first signs of DIC in this condition is the dark red rash which does not blanch on pressure and is caused by bleeding from small vessels in the skin. Severely affected individuals may require amputation of fingers or toes or even entire limbs as a result of blood clots blocking off the circulation.

Stalemate.......

Our journey has now taken us back through single celled creatures to the advent of multicellularity and shown how the 'immune system' came about as a rule book of social interactions between cells of the same and different species. We have seen in the social amoeba how 'eating' in the form of phagocytosis not only led to the problem of infection but also came to underpin the defensive response, and the phagocytic amoeba-like cell has remained as a component of all subsequent animal immune systems. The diversity of surface molecules allowed cells to recognise each other and work out the nature of their relationship, and was significantly enhanced by the ability to generate different forms through 'polymorphism'—highly variable forms of the same protein. This also allowed them to identify important components of potentially dangerous aliens and to initiate a response—secreting chemicals that could kill by puncturing their 'containment' from the environment or by marking them and 'preparing' them for eating. Along the way we have encountered entrapment and clotting as a form of defence. We have also seen how specialised immune cells initially protected the multicellular animal by suicide along with their load of infecting bugs, and how this process became adapted to produce a retaliatory response.

Having travelled no further than *Hydra* and the jellyfish, we have also met most of the cast of the immune system (albeit in rudimentary forms) including molecules that specialise in pattern recognition, as intracellular intermediates shuttling information from the cell surface to the nucleus, as secreted 'weapons' or toxins or as chemical messengers between cells. Throughout our journey we have seen that our prior understanding of the immune system was perhaps rather naive and that the same principles mediate beneficial symbiotic inter-relationships as well as purely defensive or destructive interactions.

However, there is a fundamental problem with the nature of the innate immune system in the form that we have arrived at. All of the pattern recognition molecules are hard wired in the organism's DNA and cannot change except by mutation between generations—which occurs very slowly given the long life-expectancy of multicellular organisms. Bacteria on the other hand reproduce rapidly and can mutate very quickly to circumvent the immune response. Whenever the animal comes up with a new idea to fight them off, it becomes almost immediately obsolete. Whilst the recognition of only essential bacterial components (that cannot be modified without harming the bug) is one way of dealing with this problem, the range of patterns requiring detection just keeps on growing. The only way that the cumbersome immune system can try to stay ahead is to just keep adding more and more pattern recognition molecules—which is really just a form of ongoing stalemate. One is again reminded of the 'Red Queen' from Lewis Carroll's *Alice in Wonderland* who has to run faster and faster just to stay in the same place.

In the last step in this particular direction, we revisit our friend *Branchiostoma*, the lancelet chordate whose ancestors first arose around 520 million years ago and foretells the development of fishes with backbones. Whilst its immune system is clearly up to the job (as these creatures are widespread and have been around for a long time) they are the perfect example of the immune impasse that we have arrived at. The genome of *Branchiostoma* has been sequenced and analysed and comprises vast numbers of genes coding for immune proteins. There are 71 toll-like receptors, 118 NLR molecules, 1215 lectins and over 1500 types of proteins with 'leucine rich repeats' involved in pattern recognition. All in all, the poor lancelet has had to dedicate over 10% of all its genes just to its immune system. Clearly this process of repetitive gene duplication could not continue indefinitely and life needed to find a better way, particularly if it was to explore and colonise new environments where it would encounter new microbial challenges. How the stalemate was broken is now thought to be one of the luckiest flukes in history and led to the evolution of what we term 'adaptive' immunity. Let us find out how this came about!

5

The Orchestra of Life

Summary In which we find that a chance viral infection, harnessed by the vertebrate genome, led to the ability to create myriad pattern recognising molecules by altering the DNA encoding them. This development broke the stalemate of invertebrate immune systems. We discover that the cells bearing these highly variable receptors, called 'lymphocytes' come in a variety of different forms, and we encounter two principle divisions—called B lymphocytes and T lymphocytes. The B cells secrete molecules called antibodies that perform functions related to the primal phagocytosis pathway of immunity whereas the T cells kill directly using pore-forming toxins. We learn how this potent system requires fine tuning and close regulation in order to prevent it from attacking its host and we see how the different cells of the 'adaptive' immune system work in concert to do so. Finally, by looking closely at the evolution and development of the gut we see how the digestive tract and the immune system are intimately associated throughout their evolutionary history.

J. Woodward, *The Gastro-Archeologist*, https://doi.org/10.1007/978-3-030-62621-1_5

'The orchestra of life'

'A symphony must be like the World. It must embrace everything.'
Gustav Mahler 1907[1]

It's Life Jim...but Not as we Know it

The nineteenth century French scientist, Louis Pasteur is remembered for his ground-breaking studies that proved the 'germ theory' of infection and for developing the first vaccinations[2] for anthrax and rabies. However, it is his (almost) forgotten assistant, Charles Chamberland who went on to invent a device that led to the discovery of an entirely new form of life—although the use of the word 'life' to describe it is still questionable.

Chamberland's invention was a porcelain tube that acted as a filter with holes too small for bacteria to pass through. Using just such a device almost

[1] The composer, Gustav Mahler (1860–1911) purportedly made this comment in conversation with Jean Sibelius when comparing their different styles and approaches to symphonic writing. This quote aptly sums up his work which includes all varieties of sounds and musical forms.

[2] The word 'vaccination' originates from the name of the cowpox virus—*Variolae vaccinae*—which bears some resemblance to the more virulent smallpox virus. In the eighteenth century, between 10 and 20% of deaths across Europe were due to smallpox which resulted in 400,000 deaths a year, including 5 reigning European monarchs. Up to 80% of infected individuals died from the disease. Edward Jenner, a Gloucestershire doctor in 1798 inoculated some pus from a cowpox vesicle into a young farm hand called James Phipps who bravely allowed himself to be inoculated with smallpox 2 months later but did not succumb to the disease. The immunity generated to the cowpox was clearly sufficient to protect against the similar smallpox virus. As a result, the use of 'vaccination' became widespread and within 3 years over 100,000 people had undergone cowpox inoculation. The same vaccine—inoculation with the live vaccinia virus—was used until 1986. Smallpox was the first disease to be globally eradicated with the last recorded case diagnosed on 26th October 1977 in a hospital cook in Somalia.

15 years later in Delft in Holland, Martinus Beijerinck[3] was able to show that a plant disease that leads to stunting and mosaic patterns on leaves is caused by a transmissible agent small enough to pass through it. Assuming that the cause was not therefore bacterial, Beijerinck called the infectious substance a 'virus'. Unable to conceive of a living particle so small, he thought that it must be liquid in nature. The name was therefore appropriate as 'virus' is the Latin word for 'poisons' (note that it is plural and therefore when we talk about 'viruses' it is as grammatically incorrect as talking about 'a field of sheeps').[4] Only after the invention of the electron microscope in the 1930's was the particulate nature of the tobacco mosaic virus identified, the first of over 5000 such entities to be described in detail. Being as little as 1/50,000th of a millimetre across, most viruses are far too small to be seen under a standard microscope. The total number of virus particles in the world is estimated at over 10^{31} or in other words, ten thousand billion billion billion, which is more than the number of stars in the Universe. There are likely to be many millions of different types of viruses as all cells are prone to their own specific virus infections but only 219 different viruses have been found so far to affect humans.[5]

Viruses are effectively rogue sequences of nucleic acids—either DNA or RNA—that are incapable of replicating by themselves and therefore fail in one of the basic definitions of life. However, they 'cheat' by infecting cells and hijacking their protein-building machinery to copy themselves and in so doing they often destroy the host cell and cause disease. Their small lengths of nucleic acid encode the proteins—from two in the smallest viral genome to 2500 in the largest—that they require to infect and subvert the host cell machinery.

Whilst we will probably never know for sure, one current thought about the origin of viruses is that they actually started out as cellular life forms that progressively reduced the size of their genetic blueprint to the basic essentials needed for piracy. All they required was the necessary equipment to overcome the containment of cells and get their genes into a cell—and bacteria had developed such tools at an early stage in order to share information by 'horizontal transmission'.

[3] Martinus Beijerinck (1851–1931) is now considered one of the founders of the study of viruses. An eccentric character, he was renowned for being tough on his students and devoted himself entirely to scientific pursuit, considering it incompatible with marriage and family life. He is also known for discovering the way in which plants fix nitrogen by the use of a symbiotic relationship with bacteria.

[4] The word 'virus' was in use to describe snake venom in the Middle Ages, and came to describe an infectious agent of disease in the early eighteenth century. Modern usage has applied it to computer code spreading between and 'infecting' devices, and also to anything that spreads rapidly such as 'viral' video clips on social media. Interestingly, such ideas or representations that spread around people and cultures can change and be subject to selective pressures in much the same way as genes. Richard Dawkins coined the name 'meme' for such a concept in his 1976 book, 'The Selfish Gene' as a shortening of the Greek word 'mimeme' meaning 'imitation' and from where we also get the verb 'to Mime'.

[5] Let's make that 220 now with COVID-19/SARS-2…

Jumping Genes

Viruses may remain dormant within the cell and not replicate until the right circumstances arise. The herpes virus that causes 'cold sores' around the mouth is characteristic of this, being activated at times of cellular damage, for instance after sunlight exposure. Other viruses such as hepatitis B may simply remain as passengers. People who are viral 'carriers' can still be capable of transmitting a disease to others despite showing no signs of it themselves. As we have already seen, defensive pattern recognition molecules (such as NLRs) are capable of identifying alien types of nucleic acids (such as viruses) inside the cell in order to mount an immune response and clear the infection. In order to escape detection, some viruses such as the Human Immunodeficiency Virus (HIV), have gone one step further in order to hide within the cell, by inserting themselves into the very DNA of the host.

Much to our surprise, we now know that a considerable amount of genetic material—up to about 44% of the DNA in humans—is actually made up of small fragments of viruses that incorporate themselves into the host code and randomly move in and out of the genome. They are called 'transposable elements' or 'transposons' for short[6]—but in popular parlance, 'jumping genes'.

By jumping around from place to place in the genetic code they can create havoc. If they randomly insert into the code for a protein for instance, it becomes corrupted and deformed or its translation 'switched off' making it no longer available for translation. In this way, transposons have been implicated in a number of human diseases including cancers.

However, there is one viral gene that infected our DNA about 600 million years ago called the 'Recombination-Activating Gene' (or RAG for short[7]). The protein encoded by this gene is potentially extremely dangerous as it can break DNA and rearrange the coding sequence—effectively a 'cut and paste' tool. Left to its own devices in our genetic blueprint this could be lethal (just randomly the like letters in th rearranging is encesent!), but our ancestors somehow learnt how to tame RAG and control it for their own ends. One way of doing this was to limit its expression so that the RAG protein only appears in a limited number of specialised cells and only at key times in their development. No one knows where the RAG transposon first came from, but this extraordinary and

[6] Barbara McClintock (1902–1992) was awarded the Nobel Prize in Medicine in 1983—the first woman to be awarded the prize without being shared—for her discovery of 'jumping genes' in the maize plant. The commonest transposon in the human genome is a 300 nucleotide sequence called '*Alu*' which is copied up to a million times in our DNA.

[7] RAG has been identified in some organisms that predate the divergence of jawed and jawless fishes—including the lancelet (*Brachiostoma floridae*) and even a sea urchin. However, in jawed vertebrates it has replicated into two genes, RAG1 and RAG2 which act much more efficiently in concert together and its hallmark functions are not apparent in agnathans or protochordates.

apparently fortuitous fluke of nature was to lead to a complete revolution in immune systems that we are only just beginning to understand.

The Immunological 'Big Bang'

We left our journey through the history of immunity at a position of stalemate around 500 million years ago, when ever-increasing numbers of pattern recognition molecules were required just to keep pace with microbial evolution. Given that such molecules are hard-wired into the DNA and can only change by mutation between generations, this is an extremely cumbersome and slow process. What is needed is a way of generating an almost infinite array of pattern detectors that can adapt to the needs of the organism at short notice depending on the threat at hand. Amazingly, vertebrates managed this astonishing feat—more than once—using 'jumping gene' technology.

The principle of the generation of diversity is very simple. Instead of 'one gene—one protein', the gene encoding the pattern recognition molecule has multiple different 'cassettes' that can be shuffled around to create different combinations. This is where the 'Recombination' protein encoded by RAG comes into play as this rearranges the different gene cassettes during the developmental stages of immune cells to generate a wide array of different pattern receptors. An extremely wide range in fact—anywhere from 10^{11} to 10^{18} (a billion billion) different potential shapes depending on the number of cassettes shuffled and the way in which they are joined together. Suddenly, in one 'big bang', we have gone from the 5000 individual genes required for fixed pattern recognition molecules in *Branchiostoma*, to one gene generating countless possibilities (with the help of RAG and a few associated friends). The extraordinary potential unleashed by this revolution is staggering. It means that an organism could be able to mount an immune response to *any* possible threats in its own environment—and also be able to respond to entirely new microbial threats if suddenly transported to alien surroundings. The molecules that are produced in such myriad different forms are called 'antibodies' and are members of the 'immunoglobulin' superfamily that we first met in social amoebae.

However, such great riches do not come without a price. Whilst this may seem like the perfect solution to the problem, there are many obvious downsides. The majority of the billions of rearranged receptors will of course be completely irrelevant and then wasted. Even worse, some may recognise patterns unique to the animal itself and start attacking it—an 'autoimmune' response a bit like 'friendly fire' in military terms. Given the enormous potential and risks of such a revolutionary immune system, evolution has had to invest heavily in the appropriate infrastructure to manage and control its power safely. The complexities of this system are still being uncovered and understanding them is key not only to preventing and overcoming infectious disease, but also in facilitating organ transplantation and treating cancer. Moreover, due to its finely balanced intricacies, malfunctions of the immune system itself can lead to disease.

Diverticulum #5.1: Side Chains and Magic Bullets

Emil Behring[8] and co-workers in Berlin in the late 1880's studied immunity to diphtheria—a serious bacterial infection of the throat and upper airways that killed around half of those (usually children) that it affected. It was already known that the bacterium produced a lethal toxin that mediated its effects. Behring injected sub-lethal doses of diphtheria toxin into guinea pigs and then showed that their serum (the fluid part of the blood without cells) was able to confer protection against diphtheria when injected into other animals that were then challenged with the bacterium. There was no protection however against another toxin (that made by tetanus bacteria) showing that the protective nature of the serum was specific to diphtheria. The word 'antibody' (*antikörper*) was coined by Paul Ehrlich from the same institute in 1891 and his subsequent work led to the large-scale production of serum for 'passive immunisation' against the disease. We still use the same technique to generate antibodies for human use—for instance snake antivenom is produced in horses by injecting them with small amounts of the poison, and at the time of writing a similar technique is being considered for treatment of COVID-19.

Ehrlich postulated that the immunity arose from molecules on the surface of cells that he called 'side chains' which would block the toxin to prevent it from working. He envisaged that cells each produced a wide variety of side chains of different shapes that were ready formed and waiting to engage any potential threats. Having engaged with the toxin through its 'side chain' the cell then produced more of these side chains and released them from the cell surface as 'magic bullets'—hence the immunity could be transferred passively in serum.

His theory was ridiculed on the basis of the low probability of cells bearing so many different receptors that they could match and block any potential or previously unmet molecule. In its place an 'instructive' theory was proposed—that alien molecules 'imprinted' their shapes on the host a little like plasticine moulds. It was only in the 1950's that it became clear that Ehrlich's theory was fundamentally correct in all but one regard. Instead of cells each bearing myriad different receptors, every cell produced its own individual specific type of antibody. Engagement of this antibody with the unique molecule that it recognised would lead to proliferation of the cell. This is the basis of the 'clonal selection theory' of Frank Macfarlane Burnet.[9]

[8] Emil Behring (1854–1917) was the first ever recipient of the Nobel Prize in Medicine or Physiology for his work in 1900. As with many high-profile discoveries, controversy surrounds the award of this prize as Paul Ehrlich had no share in it, nor in the financial rewards that came from the development of the diphtheria vaccine.

[9] Sir Frank Macfarlane Burnet (1899–1985) was a brilliant but often controversial Australian scientist who pioneered immunology research at the Walter and Eliza Hall Institute of Medical Research in Melbourne. His seminal paper 'A modification of Jerne's theory of antibody production using the concept of clonal selection' was published in the Australian Journal of Science in 1957. Despite usually being uniquely accredited to Burnet, the work of many others including the US immunologist, David Talmage, contributed to this theory.

The Orchestra of Life

I chose this as the title of this chapter because the metaphor of music and instruments fits well with the two types of immune system that we have encountered—before and after the immunological revolution brought about by the ability to rearrange DNA to create endless varieties of pattern recognition molecules.

The limitations of genetically 'hard wired' pattern receptor molecules found in invertebrate immune systems are similar to those of instruments tuned to a single note—like a hand bell or a drum. If you want to play a different note, then it requires another instrument, just as if you wish to recognise a different shape you require another gene encoding a separate protein. Whilst it is possible to play a melody with single pitched instruments, it requires multiple different units each tuned to different note—for instance hand bell ringers have to pick up individual bells laid out on a table in front of them. A very simple tune may take several players and require great skill in co-ordination. Taken to its extreme form, a seventeenth century Flemish musical invention—the 'carillon', a large array of bells (at least 23) tuned to different pitches that are often housed in church towers and played using a keyboard. This reminds me of the lancelet (*Branchiostoma*) and its proliferation of individual pattern receptor molecules at the pinnacle—or perhaps the end of the road—for this technology.

Now imagine a violin—where a single instrument can play every single note in the scale and all fractions of notes in between. No longer are multiple different instruments required just to play a simple tune—it can all be performed on just one. In our musical analogy, this is the equivalent of antibodies and related molecules in the vertebrate immune system. However, this invention now opens up dramatic new possibilities. Firstly there are the nuances of phrasing, musical punctuation, timbre and dynamics—which are rather difficult to achieve with hand bells! Then, making just a small 'evolutionary' change to the violin leads to other similar instruments that cover different musical ranges—the viola, the cello and the double bass. Now we can create a string quartet or orchestra and play music in harmony. Evolution of different lines leads to the woodwind section—flutes, oboes, clarinets and bassoons, creating different types of sounds. Each of these instruments also has different versions—for instance the piccolo and alto flute, cor anglais, E flat and bass clarinet and so on—covering different ranges. Then there is the brass section—trumpets, trombones, and tubas. With this wealth of instruments and the sounds they create we can now forge a pallet of musical colours and compose the most magical sound pictures by means of the full symphony orchestra. However, take a close look at the players arrayed on the stage and you will see that right at the back of the ensemble the single note instruments—such as tubular bells and tympani—are still retained as a valued section. In fact, just as it is the percussion section that 'keeps the beat' and holds the band together, so it is the retained ancient invertebrate immune system that co-ordinates our immune orchestra.

Let us now see how the immune system, just like the orchestra, comprises separate sections that play different instruments yet importantly work together (in 'concert') to create a sensitive and dynamic performance. Having taking over 500 million years of evolution from the nascent immune systems of the social amoeba to those of the first vertebrates, we will also see how the main sections of the immune ensemble developed over a remarkably short period of time and how the tunes it plays are still recognisable despite the complex layers of orchestration.

The Same Old Story

In 1977 I was transported as a young lad, awestruck, to a galaxy far, far away (despite never leaving the Odeon in Leicester Square) when George Lucas' film 'Star Wars' hit the screens and redefined cinematic special effects. In 2015, I was sat in the Arts Picturehouse cinema in Cambridge experiencing a similar sense of awe as I watched the (then) latest offering in the franchise, called 'The Force Awakens'. The effects (this time in 3D) were finessed and realistic to a degree that I had never imagined possible and made the original Star Wars film look clunky and old fashioned in comparison. However, the storylines were almost identical!

So it is with antibodies. Updated, embellished and reworked from the original, these remarkable molecules simply retell a familiar tale. You will remember the 'lectins' from chap. 4 as the proteins that recognise the shapes made by sugars. In so doing they agglutinate and immobilise specific bacteria and facilitate their destruction by attracting complement to act as an 'opsonin' for phagocytosis or to initiate the 'membrane attack complex' to punch holes in their membranes. It is the same story with antibodies which do all of the above and act in the exact same way, but in addition to recognising sugars, antibodies are proteins that also recognise other proteins. However, the special effects department has been busy and added on a few extras….

The first thing to note about antibodies—as with any protein—is their shape. Antibodies characteristically form a Y-shape. The two limbs sticking out at the top are identical and can each bind to a specific protein shape recognised by the antibody. As members of the 'immunoglobulin super-family' these ends can be made highly variable just like the edges of the sandwich in the 'Tiger' molecules of the social amoeba that were used for kin recognition. It is the activity of the RAG transposon, chopping and changing the DNA code specifically just for these regions of the protein that provides the enormous range of potential shapes that can be recognised. As a result of chance rearrangement of the DNA, each cell produces its own unique antibody that recognises a specific protein—including those that the organism and its ancestors may never have encountered before.

The other end of the Y—the tail if you like—is designed to interface with other functions, or 'plug ins', much like the complement C3 component that binds lectins. As we saw, one of the benefits of C3 is that it works as a 'universal adaptor' and the cell needs only make a receptor for C3 itself rather than a different one for each lectin. However, the downside of only having the single receptor is the limited repertoire of responses—namely either triggering phagocytosis or killing the organism through complement activation and pore formation. The joy of antibodies is that they can change the tail of the molecule. By virtue of cells having specific receptors for each end, they can engender a different 'flavour' of response to the same foreign protein or organism. For instance, they can be identified by different types of immune cell, or vary in their ability to trigger the complement cascade. The ends of the Y shaped antibody molecule dictate the *class* of immunoglobulin (Ig) and we humans have five different types known by different letters—IgA, IgG, IgD, IgM and IgE. Whilst each antibody-producing cell always makes the same unique protein-recognising end it can alter the class of the tail to change the nature of the response—for instance from IgM to IgG. This ability of cells to change the function of the specific recognition molecule by switching its connector first evolved relatively recently—in the amphibians, only about 350 million years ago.

If you like, these five different classes of antibodies resemble the stringed instrument parts of the immune orchestra—first and second violins, violas, cellos and basses—different instruments but recognisably based on the same structure and played in the same way (Fig. 5.1).

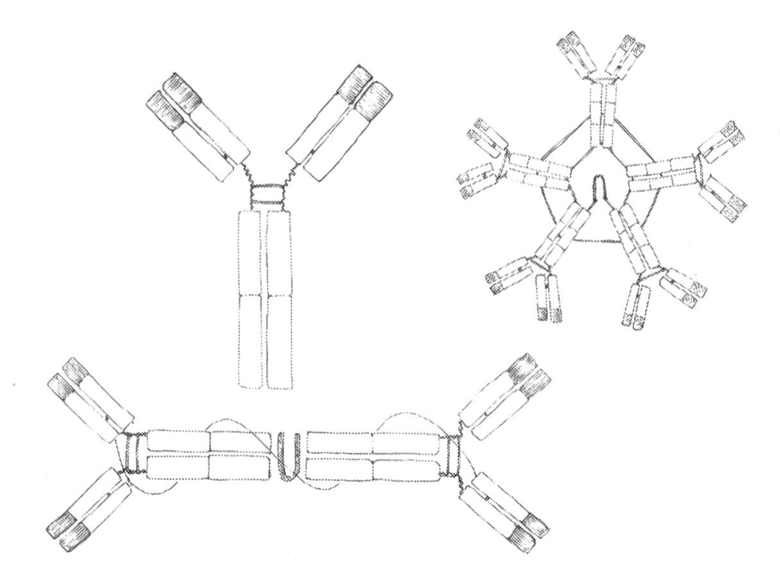

Fig. 5.1 The shapes of different antibody classes. *Top left* immunoglobulin G, *top right* immunoglobulin M, *bottom* IgA. The ends that recognise and bind specific protein shapes are shaded

The Housekeeping Star

In all aspects of the immune system that we have encountered so far, the defensive role against micro-organisms seems to have appeared as a secondary purpose. It is similarly highly likely that antibodies too initially evolved to fulfil roles in maintaining the integrity and social *status quo* of the multicellular organism and its symbiotic partners—the holobiont.

The most ancient of the antibody classes, immunoglobulin M, is found throughout vertebrates as far back as the cartilaginous fishes such as sharks and this dates its first appearance to around 420 million years ago. IgM is actually not just one Y-shaped molecule but 5 separate molecules joined together by their tails into a star shape which means it can bind up to 10 identical foreign proteins. It appropriately gets its letter from the prefix 'macro'—meaning 'large'.[10]

[10] You may be interested in how the other immunoglobulins were given their letters. Mixtures of proteins can be separated by their different electrical charges. When an electrical current is applied to serum the proteins move towards one or other electrode. The different groups of proteins—known as 'globulins'—separated in this way were originally labelled by Greek letters—α,β,γ etc. Antibodies were found in the beta-globulin and gamma-globulin fractions—and these became known as *immuno*-globulins. To avoid confusion (!) they were given the Roman alphabet equivalent of the Greek letters. Hence the immunoglobulin found in the gamma region became known as IgG. It was thought that mouse antibodies would

Animals that have been raised in an environment completely free of any recognisable foreign proteins—either in food or microbes—still produce a range of antibodies. These are known as 'natural antibodies' and are predominantly of the IgM type. These natural antibodies recognise a variety of different structures, some of which are appropriately associated with infective bacteria. However, others are clearly detecting host signatures—such as altered membrane molecules that are found in cells undergoing the form of cell death called apoptosis. These natural IgM 'auto-antibodies' when attached to dying cells can trigger nearby phagocytes to dispose of them—without shedding the molecules hidden inside them that might otherwise incite an unwanted inflammatory response. Even though the presence of 'auto-antibodies' might cause a devastating 'immune storm' whereby the organisms' immune system destroys itself, the opposite is in fact true of the natural IgM auto-antibodies. These housekeeping molecules mop up or conceal any recognisable patterns of the host itself. It is actually only by experimentally depriving animals of these natural auto-antibodies that an overwhelming auto-immune response is 'orchestrated' by the other components of the immune system mistaking the leaked host determinants as traces of alien invasion.

The Self-*like* Gut Contents

Antibodies produced in the gut have their own specific tail and are designated as 'immunoglobulin A'. They are produced in vast quantities—about 5 grams of IgA are pumped into the lumen (hollow centre) of the gut every day in humans. This secreted form of IgA is made up of two molecules joined together by a protein that wraps around it and (rather importantly) prevents it from being digested by enzymes in the gut. IgA is not very good at attracting complement (or acting as an opsonin itself) to prepare bacteria for phagocytosis which is perhaps not surprising given that it is effectively acting outside the body in the space inside the gut without immune cell 'back-up'.

Just as with IgM, we find 'natural' IgA antibodies that can bind (weakly) a whole range of different targets. Similarly, these antibodies are not produced as a result of an infection as they are found in animals that have been raised in

be called IgB (this never transpired) but it meant that the letter B was then not available. Antibodies found in the beta globulin fraction were subtyped as beta-2A and therefore became labelled as IgA. There being no Greek letter equivalent of the letter C, the next immunoglobulin was called IgD. Only IgE's letter came from a property it exhibited—named after the erythema (or 'redness') it can evoke as IgE is involved in allergic responses.

an entirely sterile environment. In effect they are the functional equivalent of the genetically hard-wired 'fixed motif' pattern-recognising molecules that we saw in the invertebrates such as the lancelet and the sea urchin.

One of the ways in which IgA works in the gut is to 'exclude' microbes. Being coated in antibody prevents bacteria from being able to recognise and adhere to the cells lining the gut as a prelude to invasion and infection. However, recognising that the purpose of the immune system is not simply to destroy all alien life forms on sight but to support symbiotic relationships between them, the weakly binding natural IgA antibodies probably protect and nurture the 'friendly' bacteria of the gut. For instance, they encourage binding of the bacteria to the mucus layer which provides nourishment for the bacteria, and also protect them from immune attack. This is very similar to the exclusion of 'self' targets by natural IgM auto-antibodies which prevents activation of an inflammatory immune response. The symbiotic bacteria are in a sense being identified as being a part of the organism even though they comprise different species and are effectively outside the animal by being in the lumen of the gut. The immune system is acting to co-ordinate the holobiont—the animal and all its symbiotic friends—rather than simply the cells containing the DNA of the host. The contents of the gut are therefore not strictly 'self'—but self-like. As we will see later, this relates not just to the friendly bacteria of the gut but also to the food that we eat.

Diverticulum #5.2: IgA Deficiency

An inability to produce IgA is the commonest congenital immune deficiency in humans and occurs in about one in every 600 individuals. It is a benign condition that goes unnoticed by the majority of people who show no symptoms throughout their lives. This is a little hard to explain given that IgA appears to be so fundamentally important that we all flush several grams of it through our guts (and our toilets) every day. However, there does appear to be a small increase in the occurrence and severity of chest, sinus and gastrointestinal infections in IgA deficient individuals. Interestingly—from the perspective of natural antibodies and the housekeeping roles of the immune system—IgA deficiency also leads to an increased risk of auto-immune conditions and allergies. Furthermore, as we would expect from our understanding of the immune system in regulating our symbiotic passengers, the bacterial population of the gut is drastically altered in IgA deficiency, not just in terms of the types of bacteria but also their gene expression. We see this leading in turn to a reduction of fat absorption and reduced body fat in such individuals.

The Players

Orchestral players usually share a belief that their temperament and personality influence the type of instrument that they play. A classic example of this is the poor viola player, always the butt of orchestral jokes![11]

The players in the immune orchestra—the cells of each section—are also very individual. Up until now, the main performers that we have come across have been phagocytes—from amoebae, to the 'amoebocytes' of sponges and jellyfish to our own white cells that still perform this function, the neutrophils and macrophages. The musicians of the vertebrate immune orchestra largely comprise a completely new kind of cell, called the 'lymphocyte'—of which there are many different types. Lymphocytes derive their name from the fluid ('lymph') that is exuded from blood vessels and permeates through the tissues.

The first cells that resemble vertebrate lymphocytes are found in the lancelet (*Branchiostoma*)—not itself a vertebrate, but resembling their likely common ancestor around 500 million years ago. It will come as no surprise to discover that in Branchiostoma, these cells are found within the epithelium of the foregut—in the 'gills' (the pharyngeal slits that serve the purpose of filtering the particulate food). This fits entirely with our understanding of 'containment' with the organism exposing itself to the outside world through its gut,[12] hence the foregut epithelium being the site of first contact with alien molecules, be they indicative of food, friend or foe...

[11] The viola—the alto of the string orchestra—has been made the joke of the orchestra since the eighteenth century. The viola part was often written to simply fill in harmonies rather than ever playing the tune and was therefore the easiest part to play—hence second-rate violinists would be demoted to the viola section. Most of the best viola jokes are unrepeatable, but are generally along the lines of 'what is the difference between the front and back desks of viola players?—about a semitone', and 'how do you know whether a viola player is playing out of tune?—you can see the bow moving'. However, there are well known viola solos such as in the second movement of Beethoven's fifth symphony and concertos have been written for solo viola and orchestra (often played on the viola by violinists such as Nigel Kennedy!)—examples include the viola concerto by William Walton and a work by Hector Berlioz called 'Harold in Italy' that was commissioned by the famous nineteenth century violinist Niccolo Paganini to play on his recently purchased Stradivarius viola. The inevitable outcome—'what is the longest Viola joke you have heard?—Harold in Italy'...

[12] Dermatologists will complain loudly about me discounting their organ here as animals are also exposed to the environment through the skin. However, the skin comprises a sealed, thick, water repellent barrier with it surface area minimised, compared to the leaky, single celled, water-permeable and maximised surface area of the intestinal mucosa which is therefore much more exposed to the environment.

The Antibody Section

Just as with the musical orchestra, there are a limited number of sections of the lymphocyte orchestra for us to familiarise ourselves with, even though there may be a range of different players within them. Instead of string, woodwind and brass sections our ensemble comprises 'B' cell, 'T' cell and 'NK' cell sections.

B lymphocytes are specialised for the production of antibodies. A close look at these cells reveals at least two different types. The least common are those that arise early in our foetal development and are called 'B-1' cells. They are unusual lymphocytes and probably represent a very ancient lineage found in the earliest vertebrate immune systems. B-1 cells are only found in certain locations—in the body cavity around the outside of the gut and within the deeper layers of the intestine itself. Only rarely are they found within the blood circulation. The antibodies produced by B-1 cells are primitive 'natural' antibodies—both IgM and IgA—that undergo very limited gene rearrangement and are therefore more like fixed pattern recognition molecules that lack the usual diverse repertoire of antibodies. They bind weakly and relatively non-specifically to pre-determined structures, usually those associated with internal housekeeping or regulating symbiotic bacteria.

B-1 cells are also capable of phagocytosis—'eating', which suggests that they are indeed evolutionarily ancient, as phagocytosis is the hallmark of invertebrate defences. All of their features—from their location in and around the gut, to their ability to phagocytose as well as the relatively unsophisticated antibodies that they produce—identify B-1 cells as the oldest members of the antibody section of the immune orchestra.

'B-2' cells on the other hand are common in the blood circulation and produce the dizzying array of different antibody specificities which can be fine-tuned to any potential new pattern that presents itself.

The Natural Killer Section

The 'NK' cells are the assassins of the immune orchestra. We do not know quite how ancient these particular types of cell are although their presence is noted at least as far back as the first bony fishes and perhaps as long as 400 million years ago. Natural killer cells do exactly as their name suggests—they sneak up beside animal cells and then quietly stab them in the back and inject them with poison. The method of cell killing is a two-stage process—the first component called 'perforin' is a secreted pore-forming molecule that open up access in the cell membrane and then allows the second part, a toxin called 'granzyme' into the target cell. Amongst its different functions, granzyme disrupts the mitochondrial membranes within the cell and therefore leads to the release of the 'alien' molecule that is recognised and triggers apoptosis (see chap. 4). In this way, virally infected cells can be dispatched without shedding their load of dangerous virus particles (they become contained in a phagocytic bubble and digested when the apoptotic fragments are cleaned up by the housekeeping system).

NK cells are primed to kill in a variety of ways. They have receptors on their surface that recognise the 'tails' of antibodies that have been bound by their variable recognition receptor end to a target. Binding of the antibody tail to the NK cell activates it to kill the bound target with perforin and granzyme in much the same way that antibodies attaching to phagocytes prime them to eat it. NK cells can also be activated by recognising molecules that are expressed on the cell surface of, or secreted by stressed or damaged cells. They therefore function very much as housekeepers to 'tidy up' the mess and in so doing also send chemical signals to phagocytic cells to come and help them hoover up.

However, it is the 'natural' part of the NK cell's name that makes them the most useful, and also the most dangerous of assassins. In fact, they will 'naturally' kill just about anything. Thankfully NK cells have a big 'stop' button that prevents them from destroying every cell in sight. This off switch is a type of protein called 'MHC' that is present on every cell in the body and can be recognised by NK cells. It is a member of the immunoglobulin superfamily, so it is capable of being expressed in many different variable forms and plays a key role in helping cells to recognise each other as members of the same species or organism. However, NK cells are not bothered about the particular identity of the MHC molecule and cannot usually distinguish between cells from different individuals—they only care that an MHC molecule is present. This is relevant because tumour cells and virally infected cells often stop producing MHC molecules on their surface. Without the necessary 'off switch' they give themselves away and are exposed to the risk of assassination by NK cells.

You will remember from chap. 4 the 'Tiger' proteins of the social amoeba which are also members of the Ig superfamily and therefore closely related to the MHC proteins. The primary function of these types of protein in recognising cells has clearly not changed over the aeons of intervening time between 'Tiger' and the MHC. However, the MHC has subsequently evolved a far more significant role—it has become the baton that the conductor uses to direct the entire immune orchestra. So, before we introduce our last orchestral section—the 'T' cells—let us first find out more about our maestro!

Meet the Conductor

MHC stands for 'Major Histocompatibility Complex'. In other words, it is a group of genes that cluster together as a gene 'complex' and encode proteins that are the key determinants of tissue (*'histo'*) matching (*compatibility*). We are all in fact very familiar with the MHC proteins from our everyday knowledge of transplantation, as these are the proteins that vary between individuals and need to be 'matched' when we transplant solid organs (such as kidneys) or bone marrow. If the MHC proteins are not similar, then the transplanted organ will be rejected by the body's immune system. There are two main types of MHC proteins, called Class I and Class II and we each have 6 different class I molecules and 6–8 different class II molecules. However, being Ig superfamily members, they can come in extremely variable forms and there are several hundred configurations of each MHC molecule. Theoretically therefore there are more than enough combinations for all the people on the planet to be completely individual, a billion times over. Thankfully however (for the purposes of organ transplantation at least) genes tend to be passed on together and certain kinds of MHC proteins are more common than others so that such huge potential variability is never realised.

These MHC proteins do much more than simply permit cells to recognise each other. Class I MHC proteins are found on the surface of every cell in the body—except for red blood cells. It is specifically this class I MHC that is recognised by NK cells as their 'off switch'. MHC molecules (being of the same immunoglobulin superfamily) closely resemble antibodies and in exactly the same way, they are capable of binding to specific parts of proteins. In the case of class I MHC molecules, the molecules bound to them are generated from within the cell and 'loaded' onto the MHC before it appears in the cell membrane. This means that a 'library' of different protein fragments that make up the cell itself are displayed with the help of class I MHC on the cell surface. The class I MHC effectively 'opens up' the cell to expose its inner self to the immune system and say 'this is me!—don't attack'. In much the same way that NK cells switch off when they see class I MHC, so the immune system is trained to ignore the collection of self-protein fragments displayed by the class I MHC. However, if something—such as a virus—gets into the inside of the cell then its signature will also be displayed by the MHC on the cell surface and this is recognised as 'non-self' and leads to an immune response being generated against it.

Class II MHC works in a very similar way to Class I MHC but is only found on certain cells. These cells constantly 'sample' their surroundings as a result of internalising parts of it through phagocytosis. Rather than simply revealing parts of their own inside proteins through MHC class I on their surface, they constantly also present parts of the outside world including viruses and bacteria or even cells from different organisms in association with Class II MHC molecules. A variety of immune cells are capable of presenting peptides to the immune system in this way including phagocytes. However one type of cell is particularly specialised for this purpose. Called 'dendritic' cells (from the Greek word 'dendrite' meaning 'pertaining to a tree') they send out long straggly processes that resemble the roots of a tree in order to sample as wide an area as possible.

It is these specialised phagocytic cells—professional 'presenting cells'—that are the conductors of the immune orchestra and effectively use the MHC as their baton. The MHC molecule offers or 'presents' fragments of protein shapes to the lymphocytes to recognise (Fig. 5.2).

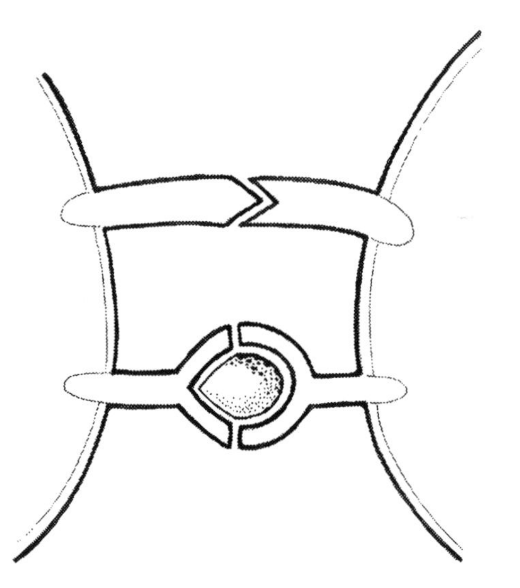

Fig. 5.2 The MHC-T cell receptor 'peptide sandwich'. The presenting cell (such as a dendritic cell) on the right loads up its MHC class II molecules internally with protein fragments (bottom). The T cell on the left recognises the specific fragment through its T cell receptor, but only when presented within the MHC class II molecule. By itself this is not enough to lead to activation of the T cell which requires additional signals through engagement of 'co-receptors' that recognise their counterparts on the presenting cell surface (top). This 'protein sandwich' between immune cell receptors and the MHC molecule forms the basis of adaptive immunity in vertebrates

The T-Cell Section

Finally, we meet the last and loudest section of the orchestra—the 'T' lymphocytes. These are the cells that recognise the protein fragments associated with class I or class II MHC molecules on the surface of cells. They do so by using yet another immunoglobulin superfamily protein that is simply called the 'T cell receptor' which is very similar to the antibodies made by the B cells. Just like antibodies and MHC molecules, the T cell receptor has a highly variable end that can recognise myriad different shapes. Just like B-cells that produce their own specific antibody, each T-cell will produce its own unique T cell receptor which is embedded in the cell membrane and can transmit messages to the cell nucleus upon binding its specific target. T-cell receptors identify protein fragments that are bound to MHC molecules—effectively creating a sandwich with the target molecule in the middle (Fig. 5.2).

There are a number of different players in the T-cell section to introduce to you. The most important member of the orchestra after the conductor—the 'leader' if you like—is a member of this section and is called a 'T-helper' cell. This type of cell is activated by recognising the unique target protein fragment bound to the class II MHC of the dendritic cell and in turn 'helps' the other members of the immune orchestra to play their respective parts.

The other main type of T-cell is called a 'cytotoxic' T-cell. These are cells that kill their targets using a variety of weapons that include perforin and granzyme, the favourite munitions of NK cells. However rather than killing indiscriminately until switched off by Class I MHC molecules (like NK cells) they are directed to their victim by recognising foreign protein fragments loaded onto the exact same Class I MHC molecules.[13] In other words, they represent a more sophisticated or upgraded type of 'killing' cell.

At this stage having met all of the parts of the immune orchestra it is now time that we listened to them in performance to find out how they play together. However, having taken our seats and whilst the orchestra are still

[13] These two types of T-cell can be identified by specific molecules that they express on their cell surface. T-helper cells have a molecule called CD4 and the cytotoxic T cells have a molecule called CD8. The letters CD stand for 'Cluster of Differentiation' and are really just a way of categorising the cell surface molecules that was set up at an international conference on nomenclature in Paris in 1982. Prior to this it was all very confusing as some molecules went by different names, and similar molecules that performed identical tasks in different animals were called different things. The tally of CD molecules currently stands at 371. CD4 and CD8 are closely related immunoglobulin superfamily molecules and bind to the outside of MHC molecules—they make sure that the helper cells bind only to class II MHC and the cytotoxic cells bind to Class I MHC.

The Human Immunodeficiency Virus (HIV) that causes AIDS wreaks its havoc on the immune system by attacking T-helper cells—and it gains entry to the cell by attaching to the CD4 molecule.

warming up and tuning their instruments, I would like to point out some of the older members at the back of the T-cell section that have been around a long time.

T Cell Fraternities

T-cell receptors come in two different forms. Just like the US university fraternities and sororities that are identified by Greek letters (such as 'Phi Beta Kappa' or 'Delta Phi') the two types of T-cell receptor are known as 'Gamma Delta' (γδ) and 'Alpha Beta' (αβ). γδ T cells are present in humans in only small numbers in the peripheral blood (perhaps 1–5% of T cells) with the majority being αβ T cells. For some reason that is as yet unclear, they are much more commonly found in ruminant animals such as cows and sheep where γδ T cells can comprise around 30% of blood T cells.

γδ T cells are the T cell equivalent of B-1 cells in the B cell section. The γδ T cell receptor expresses very limited diversity much like the natural IgM antibodies produced by B-1 cell. It recognises its target molecule in a very similar way to antibodies and does not need to be 'presented' its target by MHC. γδ T cells probably appeared at an earlier stage of evolution than αβ T cells and predominantly fulfil a cytotoxic role rather than a helper role which evolved later with the αβ T cells. The molecules recognised by γδ T cells are often not proteins but lipids found in microbial cell membranes or molecules that are formed during bacterial metabolism. Just like B-1 cells, the γδ T cells are sometimes themselves capable of phagocytosis and presenting target molecules to αβ T cells. And also, just like B-1 cells they appear to play a predominantly 'housekeeping' role in preventing cancers and 'tidying up' damage. γδ T cells are also associated with the gut, being particularly concentrated within the epithelial lining of the intestine where they are seen to move in and out of the gap between the enterocytes in a behaviour known as 'flossing'. Here, and in the skin where they are also prominent, they produce growth factors to maintain the integrity of the epithelium—an appropriate response to damage.

In addition to the above features, B-1 cells and γδ T cells both appear early in the development of an animal and all the evidence suggests that they are the original members of their respective sections that have been retained despite being superseded by more up-to-date models. In fact, a close look at the genes that encode the γδ T cell receptor reveals a direct copy of a B cell antibody gene and suggests that the γδ T cells may be something of a halfway evolutionary house between B cells and αβ T cells.

Diverticulum #5.3: Whole Genome Duplication

It is clear that there is a lot of duplication in the immune orchestra that has generated different sections that are clearly related. During the evolution of animals it is thought that the entire genome duplicated itself 3 times. The first such occurrence was in chordates (such as *Branchiostoma*) before the divergence between jawless vertebrates ('Agnathans'—conodonts, hagfish and lampreys) and all other ('jawed') vertebrates, and the third was at a late stage during the evolution of bony fishes and did not involve our lineage. There is some debate as to whether the second great duplication event happened shortly after the first or immediately prior to the divergence of cartilaginous (sharks) and bony fishes. Nevertheless, the evidence of duplication is easy to see in the immune system. For instance, we have 3 sets each of class I MHC and class II MHC genes (which adds up to six of each as we inherit one set from each parent) as opposed to one 'proto-MHC' in the lancelet that has subsequently been duplicated twice. It is also possible that T-cells arose during this process of duplication from B-cells.

The Music of Immunity

We now get the chance to hear our orchestra's constituent members play in concert. Our piece starts with the conductor—a dendritic professional 'presenting' cell—spreading its roots between cells at an epithelial surface. The dendritic cell ingests an invading microbe (for instance) by phagocytosis and carries it into the body of the cell where it is partially digested and the fragments loaded onto class II MHC on its surface. The oldest section of the orchestra—the fixed pattern recognition receptors—come into play at this stage with a drum roll! If a constituent of the particular bug is recognised by one of these—such as a 'Toll-like receptor'—on the dendritic cell surface then it activates or 'primes' the cell. Once stimulated, the cell then produces many more copies of the class II MHC molecule on its surface and secretes chemical messengers (interleukins) to attract and stimulate other immune cells. It then literally pulls up its roots and migrates along with the flow of draining tissue fluid into thin-walled 'lymph' channels.

Specialised structures sited at the junction of these lymphatic vessels—appropriately called 'lymph nodes'—are pea sized conglomerates of immune cells. These are the 'glands' that we feel enlarged in our neck when we are suffering from a throat infection. The dendritic cell takes up residence in one of these nodes where it can interact with T-cells passing through. Most of the time they will just ignore each other. However, if a T-helper cell has (by chance) made the right shaped receptor to recognise the protein fragments of the microbe presented on the MHC surface molecule of the dendritic cell, then the attraction becomes mutual and the T cell sticks to the dendritic cell. Effectively they become held together by an MHC-alien protein fragment-T cell receptor 'sandwich'. This acts as the 'on switch' for the T cell but by itself it is too weak to fully activate it so that inadvertent stimulation of the T-cell is prevented. However, bringing the T cell into close proximity with the presenting cell allows the engagement of a number of other molecular switches on the surface of both cells that provide a 'verification' signal to the T-cell to complete its activation. This is a bit like making sure that the US president cannot initiate a nuclear war by himself by pressing the big red 'nuke' button on his desk but needs a second in command with a validation code! Once it has been stimulated, the T-helper cell earns its name by secreting a large number of active chemicals—interleukins—that signal to other immune cells to switch them on in turn and generate the immune response. One of these interleukins acts back on the T-helper cell itself to make it replicate and multiply in order to start something of a chain reaction. What started as a T-helper cell solo, crescendos as more and more T-helper cells are cloned, and other sections of the orchestra are brought in as well.

The Two Themes

A traditional symphonic movement has two musical themes or 'subjects' played in turn or together.[14] Similarly, the music of immunity has two main themes—antibody (B cell) and cell-mediated (T cell) immunity. These two themes bear striking similarities to the two ancient forms of overcoming containment in single cells—phagocytosis and pore formation respectively.

Once activated, the T-helper cell disengages from the dendritic cell and helps B cells to make antibodies. Each and every B cell produces its own unique antibody capable of recognising a specific molecular pattern. These cells are usually in a resting state and only produce a small number of specific antibody molecules in IgM form that are embedded in the surface membrane—these effectively act as a 'B-cell receptor' in a similar way to the T cell receptor. Just like the T-helper cell, the B cell requires two signals to activate it—merely detecting the presence of a specific protein that has stuck to the

[14] Possibly as a result of their prodigious output, the structure of classical works became somewhat formulaic with the compositions of Haydn and Mozart and their contemporaries in a pattern described as 'Sonata Form'. This involves an introduction section, an 'exposition' of one or two main themes followed by their 'development' and alteration before 'recapitulation' of the themes in their original form sometimes followed by a 'coda' to end up. Sonata form is a common structure for most pieces from this era—symphonies, concerti and sonatas—but is often used only for the first movement.

surface antibody is not enough. However, once it has 'caught' a molecule on its surface antibody the B cell internalises it, loads it onto a class II MHC protein and presents it back on the cell surface. An activated T-helper cell can then engage with the B-cell in exactly the same way as it did the professional presenting cell—by producing a protein sandwich between the MHC and its T cell receptor. This link now serves to activate the B cell in turn to proliferate and start making large amounts of identical antibody molecules to be secreted rather than just embedded in its cell membrane. In this form of 'Chinese whispers' thanks to T cell help, the B cells can now produce and secrete antibodies against the exact same molecular pattern as was shown to the T-cell by the professional presenting cell.

T-helper cells also provide help to cytotoxic T cells to play the other tune—the cell-mediated theme. Like the T-helper cells, cytotoxic T cells require activation by a professional presenting cell, and sandwich the target protein fragment between the MHC and their T-cell receptor. To be activated, cytotoxic T cells also require a second signal delivered by a T-helper cell. However, this is not provided by direct cell contact between the helper cell and the cytotoxic T cell but through the intermediary of the dendritic cell. In the process of becoming activated, the T-helper cell provides a reciprocal 'licensing' signal to the presenting cell that permits it to switch on the cytotoxic T cell docked elsewhere on its surface. Once again, the unique molecular signature targeted by the immune system has been retained despite cross talk between different cells with separate functions.[15]

Which of these two themes predominates is appropriately decided by the conductor of the orchestra, the dendritic cell itself, depending on the particular chemicals it produced when activated. These are dictated by whether the perceived threat is inside the cell (viral) and requiring a cytotoxic T cell response, or outside the cell (bacterial) and therefore requiring a B cell response. But it is the old fashioned fixed 'molecular pattern recognition receptors' such as lectins and Toll-like receptors that identify the nature of the threat and tell the dendritic cell which interleukins to make. Hence the entire

[15] If you are following closely you will have noted that there is a little detail here that could scupper the whole scheme. As I mentioned before, it is 'self' or internal proteins from within the cell that are presented on class I MHC and external alien proteins that are processed and presented on class II MHC. However, the dendritic cell has to present foreign proteins to the cytotoxic T cells which only recognise them in the context of class I MHC. It turns out that there is a little natural 'fudge' that takes place and professional presenting cells are in fact capable of 'cross-presentation' or 'cross-dressing' of proteins on MHC and are the only cells that can present external proteins on class I MHC...although thymic epithelial cells can cross present internal cellular protein fragments on class II MHC! So much for rules.....

orchestra plays to the drumbeat of its most ancient players in the percussion section.

Practice Makes Perfect

Unstimulated B cells produce IgM on their surface and this is what picks up the foreign protein before it is recycled and presented to the T-helper cell. Following interaction with the T-helper cell, B cells proliferate within the lymph node and the antibodies that they produce are all IgM. There are several downsides to this. Most notably the IgM antibodies bind only weakly to their targets. Nevertheless, IgM is the only antibody produced during the early stages of an infection. The immune system cleverly rectifies both of these issues in a process that is little short of miraculous. Over 3 or 4 days after the start of an infection, defined tiny round structures appear within the lymph node that look a little bit like boiled eggs cut in half. These are called 'Germinal centres'. The outside area—the egg white—is where the B cells are rapidly multiplying. However, one of the consequences of activation by the T-helper cell was to make the B cell produce an enzyme that deliberately inserts errors into its own DNA in a similar way to 'RAG'. Once again this is a really high stakes strategy as mutations in certain genes may lead to uncontrolled proliferation and the development of cancer.

An initial effect of the genetic 'cut and paste' is to alter the DNA that encodes the antibody protein and switch the invariable plug-in end of the IgM to IgG. This class of antibody is smaller as it exists singly rather than being several molecules joined together like IgA and IgM—a bit like a fighter aircraft compared to a bomber! This makes it highly versatile, and IgG makes up over 75% of the circulating antibodies in the blood. However, at the other end of the molecule—the 'business' end—DNA rearrangement is 'tweaking' the recognition portion of the antibody gene to 'fine tune' it.

As we have seen, evolution only occurs as a result of the gradual accumulation of subtle genetic changes between generations that affect the survival of the organism. The same process is happening in the germinal centres in B cells but at a vastly sped-up rate. Mutations are generated in the DNA encoding the variable end of the antibody molecule one million times faster than normal, and the B cells are also dividing rapidly so the intergenerational time is short. As a result, huge numbers of B cells are created, each making a subtly different version of the original antibody. Some will bind more strongly to the invader than the original version, some more weakly and some not at all. The B cells then move to the middle—the egg yolk—of the germinal centre to engage a *second time* with a T-helper cell that has taken up residence there. It is here that they are auditioned to see who is good enough to play in the orchestra. B cells with antibody that binds most strongly to the foreign protein will capture more of it from the surrounding fluid than those with weakly binding antibodies and therefore show the largest amount of the foreign protein on their surface. The strength of the signal delivered by the T-helper cell depends on the number of molecules presented by the B cell. Those with the most will develop into veritable antibody factories making large amounts of high-affinity IgG antibodies whereas those with the least will not survive, being earmarked to die by apoptosis. The runners-up are able to return to the outside area of the germinal centre and undergo further rounds of division to see if they can come up with something better—effectively they are not good enough to perform without more practice!

This microcosm of evolution illustrates a form of Darwinian natural selection, and results in the generation of extremely efficient antibodies. The immune system has effectively overcome the major limiting factor of life—that larger and more complex organisms need to live longer than simple bacteria but therefore evolve defensive mechanisms much more slowly. The 'micro-evolution' occurring in the germinal centre has matched the intergenerational time of bacteria in the equally rapid division of B cells, but then surpassed them a million-fold with the 'hypermutation' brought about by randomly altering the DNA code for the antibody. I am not sure that words adequately convey just how amazing this is!

The T Cell Conservatoire

T cells do not 'hypermutate' to evolve their T-cell receptors in the same way as B cells do with their receptors.[16] However, they do generate large numbers of random forms and need to be educated in a similar way to B cells: some will be completely defective and unable to bind any useful shape whilst others could dangerously recognise host proteins and incite an immune reaction. This is called an 'auto-immune' disease. The school for T cells is a gland called the 'thymus' gland which in humans is found in the upper part of the chest in front of the lungs and above the heart. This how T cells get their name—'T' for thymus.

[16] Actually this is not strictly correct as shark T cell receptors do undergo hypermutation in exactly the same way as B cells. This probably suggests that this was an ancient mechanism that was restricted subsequently to B cells only.

The thymus was first noted by the Roman physician Aelius Galen who thought that it was the 'seat of soul, eagerness and fortitude' or perhaps served as a cushion to protect the internal organs in the chest. In more modern times the thymus has been thought of as a vestigial organ—like the appendix (used to be!). In the early 1900's it was considered to function as a hormone-secreting gland and only as recently as the late 1950's was it realised that the thymus plays a role in immunity when the devastating immune effect of removing the thymus from baby mice was demonstrated. The role of the thymus is most prominent in early life when it is at its largest and it shrinks with age.

The thymus is a very unusual gland that contains epithelial cells, many of which are derived from the endoderm—the same as the gut lining. It also contains dendritic cells, but is predominantly packed with small lymphocytes—immature T cells undergoing development. Similarly to the 'hard-boiled egg' appearance of germinal centres but on a larger scale, the thymus has a dense outer region and a less well populated central zone.

T cell precursors arrive in the thymus from the bone marrow and proliferate in the outside zone whilst rearranging their T-cell receptor DNA (thanks to RAG). This produces vast numbers of early T cells (up to 50 million in the mouse thymus every day) each of which is capable of recognising a different shape through its receptor. However, only about 10% of these survive. These are the cells that have made a 'useful' receptor—namely one that can recognise protein fragments presented by MHC. The thymic epithelial cells are uniquely specialised to cut up their internal proteins differently from normal cells and present fragments that bind only weakly to the MHC. If the T cell is able to bind to the class I MHC in this way then it receives a 'survival' signal to turn into a cytotoxic T cell—and if it binds to class II MHC it can survive to become a T-helper cell. This 'positive selection' passes only those T cells that can bind MHC with 'something' in its binding groove. The possibility of auto-immunity and selecting cells that react against oneself is avoided by the altered protein processing in the epithelial cell which does not present its own internal proteins in the same way as other cells do.

Those T cells that pass the positive selection process by being able to recognise 'something' attached to the MHC then migrate into the centre of the thymus and undergo a process of 'negative' selection whereby any T cell receptors that do actually bind to host proteins are deleted. The epithelial cells in this part of the thymus also do something rather clever. Different cell types in multicellular animals are specialised to produce specific patterns of proteins that define them and their roles. The thymic epithelial cells switch off the controls that make them express only a fraction of the proteins encoded in the genome and they simply express the whole lot. In this way, they can present all of the 'self' proteins that a T cell may encounter anywhere in the body—be they specific to the gut, the skin or the heart—and get rid of any cells that might potentially react to them.

Following positive and negative selection processes, it is thought that as few as 1% of prospective T cells make the grade to graduate from the thymic 'conservatoire' and take a job in the immune orchestra!

Diverticulum #5.4: AIRE and Autoimmunity

A single protein called the 'AutoImmune REgulator' (AIRE) is uniquely expressed in the thymic epithelial cells to cause the wholesale expression of proteins from all cell types and bring about the negative selection of auto-reactive T cells. These include proteins such as insulin and those normally found in other cells of the body such as the thyroid gland. A rare mutation in the gene for AIRE causes multiple autoimmune diseases by ultimately allowing the generation of antibodies against internal organs such as pancreatic islets (thereby causing type 1 diabetes), the thyroid gland, the stomach lining (resulting in pernicious anaemia) and the adrenal glands (leading to Addison's disease). Why the majority of these conditions are directed against hormone producing (endocrine) glands is currently unknown. For the reasons why these patients also experience chronic fungal infections of the skin and mucus membranes—see chap. 9!

The Music Teachers

There are numerous dangers involved in the vertebrate immune response. The chopping and changing of the DNA blueprint (for instance by 'RAG') are high on the list, as is the huge proliferation of B and T cells which needs to be controlled lest they run amok and become cancerous. There are a number of ways in which the immune system manages to exert self-control. The simplest form is to initiate a slow auto-destruct (apoptosis) signal at the time of activation. Each cell therefore has a very limited life span and dies when no longer needed. However, a small handful of B and T cells stay behind and take up long term residence—usually for many years.

These remaining cells are specific to the previously identified protein fragment—the B cell in particular is capable of producing strongly binding IgG antibodies having already been through 'hypermutation'. These 'memory' cells are capable of responding rapidly to a re-infection by the same microorganism, producing a well-honed immune response that would normally take the full machinery of germinal centres and the evolution of antibody responses several days to develop. There is evidence of all sections of the orchestra—even the NK cells—having a form of memory, and this is the principle of vaccination. A small dose of an inactivated bug or a chemical constituent of its surface, is injected to provoke an initial immune response so that the memory cells can respond rapidly and effectively to prevent the infection taking hold after exposure to the living organism.

The members of the immune orchestra also need strict discipline in order to be held back from making inappropriate responses. Members of each section act as 'regulatory' cells, of which there are many different types and they generally act to suppress immune responses by the secretion of specific interleukins. We will revisit the concept of 'tolerance' to foreign substances when we look in more detail at the intestinal immune system, as it is of course rather important that we do not mount an immune response against our food.

The Lost Sounds Orchestra

The modern symphony orchestra has evolved from its early origins in the sixteenth century when music was written for 'consorts'—any collection of instruments to play together, depending on what was available at the time. However, the history of playing music together goes back much further and a recent project called the 'Lost Sounds Orchestra' attempted to reconstruct the sounds made by ancient Greek instruments such as the 'salpinx' (a long straight trumpet) or the 'epigonion' (a plucked string instrument) now known only from preserved works of art.

We can also trace the origins of the immune orchestra back in time through all the vertebrates and plot its evolution as different sections have been added. But the development of the immune orchestra has not progressed simply in one direction from *Branchiostoma*-like predecessors to modern times. A different type of immune system has been discovered that developed by itself quite separately from early beginnings and diverged from the mainstream of evolution in antiquity. It is as if we came across an isolated and undiscovered village high in the Albanian mountains where they still play the salpinx and other lost instruments. This immune 'symphony of lost instruments' is found in the jawless fishes (agnathans) such as the hagfish and lamprey, the descendants of the conodonts.

Lampreys lack MHC molecules and do not have the immunoglobulin superfamily equivalent molecules such as antibodies or T cell receptors. However they do have cells that look and behave in many ways like lymphocytes and they also have highly variable pattern receptor molecules. But, instead of developing the Ig superfamily in this role, they have chosen the alternative pattern recognition structure that we have already come across in Chap. 4—the horseshoe LRR (leucine rich repeat). Just as with antibody and T cell receptor genes, multiply duplicated but different LRR gene 'cassettes' are shuffled by cutting and pasting the DNA to make different surface receptors. In this way, one gene can generate up to 10^{14}—or 10 thousand billion—different shape recognising molecules. These molecules are called 'variable lymphocyte receptors'—or VLRs.

Amazingly, three different types of lymphocyte have been discovered in the lamprey and each produces its own separate VLR—A, B and C and the cells are named after the VLR they carry. Recent studies have revealed that VLRA cells behave very much like T cells, and VLRB cells are like B cells in showing the receptor on their surface and also secreting it like antibody, while the recently discovered VLRC cells are rather similar to γδ T cells! Much is still to be learnt about how these cells work (including how they recognise foreign molecules and which ones they identify). It is particularly notable that VLRA and VLRC cells are found in the gills and the typhlosole (that mysterious longitudinal infolding of the intestine of obscure function) whereas VLRB cells tend to be found in the lamprey blood-stream.

The NK Section Revisited: And Renamed

The differences between the VLR A, B and C cells in the lamprey appear to mimic those between B cells, αβ T cells and γδ T cells. This suggests that these two different systems of lymphocyte—like cells and their highly variable pattern receptors (based either on leucine-rich repeats or immunoglobulin superfamily proteins) have evolved in parallel in the jawless fishes and other vertebrates. It also leads us to speculate that the separation of function of these cell types must have occurred before their specific pattern receptor molecules evolved.

Recent discoveries in mammals have strengthened this case. New kinds of lymphocytes have been described that are commonly found in the gastrointestinal tract and may be evolutionarily precedent to B and T cells as they do not express antibodies or T cell receptors. However, they again mimic the different functions of the B and T cells. They are called 'innate lymphoid cells' or 'ILCs'. As they lack specific receptors to recognise protein patterns, they are instead activated only by chemical interleukin messages secreted by nearby cells. We have already noted the similarity between cytotoxic T cells and NK cells, which are also cytotoxic but do not have a T cell receptor and recognise MHC molecules without a bound peptide. We now identify NK cells as belonging to a group of ILCs called 'type 1' which are thought to predate T cells. There are two other types of ILC—of which type 2 appears to support the function of the antibody production (B cell) pathway of immunity, while type 3 ILCs perform a variety of roles including immune regulation.

This split in the immune system between the two strands of immunity appears to be ancient and we will later trace it origins back to its roots when we examine the immune diseases of the gut (Fig. 5.3).

Coda....The Tale of Fichtelius and Fabricius

We have seen how T cells derive their name from the thymus gland where they develop and hone their repertoire. 'B' cells are similarly named from an organ in which they develop, but bizarrely one that is found in birds, and not in ourselves. It is called the 'bursa of Fabricius'[17] This is a gland that develops from the lining of the lower bowel just above its exit, which in birds is called the cloaca, being a common opening through which both bowel contents and urine are expelled. If this organ is removed in immature birds then they fail to develop antibodies in much the same way that mammals lack T-cells after taking out the thymus gland when young.

Except...that this is not strictly true! In science things usually turn out to be more complicated than we initially thought, which is what makes it so interesting. A close examination of mice that have had their thymus removed after birth, or that congenitally lack a thymus due to a genetic mutation, shows that they do in fact have some T cells after all. And they are found in the intestine.

The T cells found in mice without a thymus are predominantly $\gamma\delta$ T cells and are found interspersed with epithelial enterocytes in the intestinal lining. However, clusters of immature lymphocytes are also found just below the epithelium in small structures called 'cryptopatches' scattered along the intestine. These appear to be a site where precursors from the bone marrow arrive and undergo development before 'seeding' the surface layer. Even though people have been

[17] Hieronymus Fabricius (1537–1619) was an anatomist at the University of Padua and a tutor of William Harvey (who later described the circulation of blood). Fabricius was described as the 'Father of Embryology'. His description of the bursa that carries his name was found after his death in papers on the embryology of chickens and published posthumously in 1621 in '*De Formatione Ovi et Pulli*'. The immunological significance of his discovery was only learnt in 1956 by Bruce Glick and Timothy Chang who experimented with the effects of removing this gland in newly hatched chicks. This ground-breaking research that would revolutionise our understanding of immunology was initially turned down for publication and eventually appeared in the journal—'Poultry Science'!

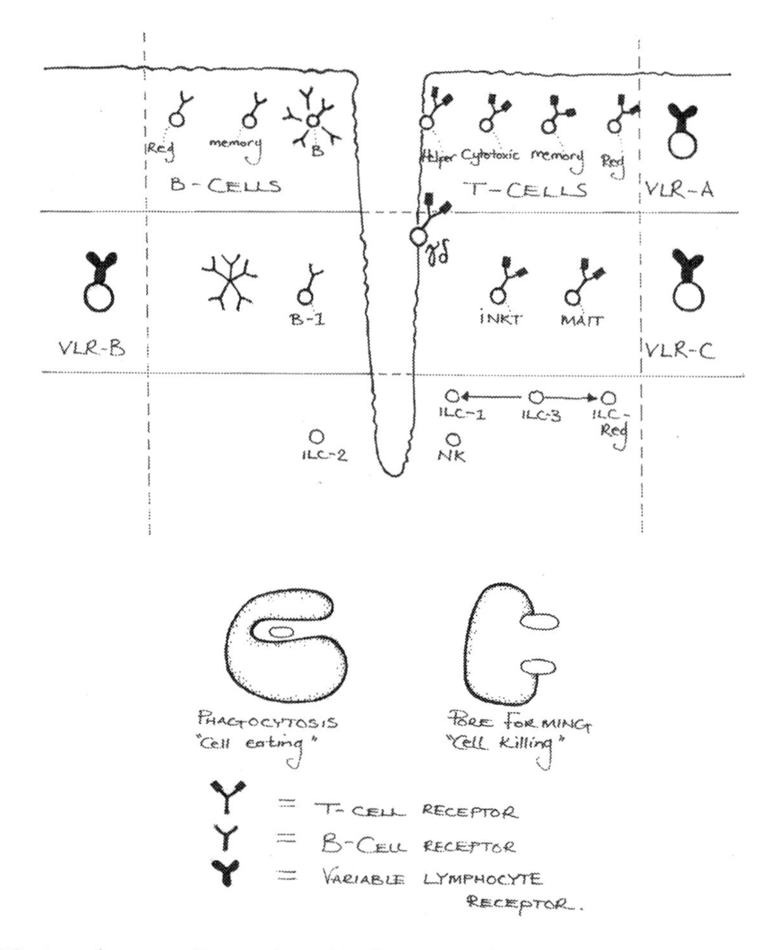

Fig. 5.3 Lymphocyte cell types found in the strata of the gastro-archeological trench, showing the 'great schism'. The lower layers in the trench are the oldest in evolutionary terms. The schism separates the mechanisms based on the two means of overcoming containment—phagocytosis or pore formation. In broad terms B cells are on one side of the schism and T cells on the other. However, helper T cells and gd T cells could be said to lie within the schism as they retain the ability for phagocytosis and bridge the antibody and cell-killing immune mechanisms. Strictly speaking, ILC3s should also be situated closer to the schism as they also retain the ability to phagocytose. The VLR cells of the lamprey are placed approximately near their equivalents in all other vertebrates. *ILC* innate lymphoid cell, *NK* natural killer, *iNKT* invariant natural killer T cell, *MAIT* Mucosal-associated invariant T cell

studying intestines for hundreds of years, these cryptopatches were only discovered at the end of the twentieth century and had clearly been hiding in full sight before their significance was realised—hence their name! Humans do not have structures comparable to cryptopatches but still show some evidence of T cell development within the intestine, independent of the thymus. The potential significance of this in terms of their 'education' is yet to be determined but may relate to the need for the immune system to recognise the holobiont including its intestinal bacteria as well as just the determinants of the host body.

Similar cryptopatches have been found in other animals including amphibians and fishes. Interestingly, in the jawless fishes and *Branchiostoma*, lymphocyte-like cells and cryptopatch-like structures are found within the gills where they are considered to be the precursors of the thymus gland and are called 'thymoids'. Just as the endostyle devolves from the gut lining and becomes the thyroid gland, so the 'thymoids' coalesce and separate from the foregut in mammalian development to become the thymus gland. However, the persistence of cryptopatches in the mouse intestine and the retained ability of animals to develop lymphocytes within the gut lining suggests that this may once have been the primordial site of lymphocyte development. This idea was first suggested by a far-sighted Swedish pathologist by the name of Fichtelius[18] as recently as the 1960's, but clearly by now it should come as no surprise!

Encore!

By its very nature, the immune system is at the cutting edge of natural selection and subject to rapid evolutionary change. On the one hand this causes its origins to be buried deeply in time, whilst on the other it leads to well-defined duplicate layers that we may be able to identify as separate strata in its evolutionary fabric.

Our journey originated with phagocytosis which arose to overcome the necessary containment of life by the cell membrane and we have seen how this has remained the mainstay of invertebrate immunity. Evolutionary tinkering led to improvements such that lectins and then complement and subsequently even antibodies appear in successive strata acting as 'opsonins' to facilitate ingestion. From its ancient beginnings, phagocytosis thereby ultimately begot the 'antibody' or 'humoral' arm of the immune system.

[18] Karl-Erik Fichtelius (1924–2016) was a pathologist in Uppsala in Sweden but gave up this career to become a district physician and a keen canoeist! In his 1968 publication 'The gut epithelium—a first level lymphoid organ' he wrote 'The theory implies that the epithelium of the whole gut….is a first level lymphoid organ and that this epithelium has the same influence on lymphocytes and lymphoid tissue as the bursa Fabricii has in birds'. Fichtelius was a deep thinker with a great sense of humour, and I regret never having had the opportunity to meet him—my PhD thesis was dedicated to him. Ironically, important research that demonstrated the clinical significance of his theory (in coeliac disease) was published in the year that he died. He would also have been interested to know that human tonsils have also recently been shown to have a T-cell progenitor role just like the thymus, firmly demonstrating the link between the foregut and the immune system.

Containment was also fundamentally overcome by the use of pore forming molecules to punch a hole in the membrane. Such pore forming toxins were initially secreted in a non-selective manner, but then aimed at specific targets through the use of lectins and the complement family, and then highly selectively by NK (ILC) cells through the use of perforin and granzyme, and lastly by cytotoxic T cells. This ancient mechanism therefore founded the other main part of the immune system—the 'cell-mediated' arm.

The separation of these two themes of the immune system is a deep rift that dates right back through time and is evidenced in all its evolutionary strata, even in lymphocytes that lack advanced T and B cell receptors (the innate lymphoid cells) and in the alternative parallel development of the variable receptors of the lamprey.

The ability to direct the immune response arose from the variability of protein shapes to recognise the patterns of nature. Initially this served to create an identity for members of the same colony, or animal 'team'. With the advent of multicellularity (for which an immune system was arguably a necessary development) it evolved to maintain the organism and its constituent cell types in check by its regulatory and housekeeping roles. This means of avoiding 'cheating' led to the defensive role of the immune system with which we are most familiar—limiting the damage caused by micro-organisms and other invasive animal species, but also in controlling the development of cancer.

The rapid evolution of microbes due to their short intergenerational time and fast mutation rate created the need for a more responsive immune system than could be generated by 'one gene-one protein' pattern recognition molecules. This requirement presumably also arose from the colonisation of novel environments with unknown threats. Harnessing the RAG transposon allowed for the rearrangement of genes encoding pattern receptors with the immediate result of an ability to respond to almost any molecular shape in nature. However, the potential for self-harm that arose from this revolution required proper checks and balances—the use of MHC to 'present' alien or self-molecules, the need for more than one signal to activate cells and the 'education' of B and T cells in the germinal centre and thymus. This required a new cell type—the lymphocyte—which nevertheless gives away its ancient origins by its need to phagocytose in order to present alien proteins.

In all of this time, we have never strayed far from the gut. Multicellular 'containment' dictates that the gastrointestinal tract forms an interface with the environment, where food-friend-foe recognition is most important. Unsurprisingly, the gut hosts over 75% of all the cells that make up the body's immune system. The earliest evolving immune mechanisms are to be found in the gastrointestinal tract or associated tissues of invertebrates—such as the demonstration of complement proteins in the lining cells of the *Hydra* gut. The finding of early lymphocytes in the typhlosole and gills of lampreys and the developmental origins of the thymus and the bursa of Fabricius in the gut strongly suggest that this was also where vertebrate immunity began. And it is in the gut that we find the greatest depth of complexity and the most layers of the immune system—and appropriately (given that this is where it all started) gastrointestinal immunity is enriched in the oldest, the deepest layers.

For this reason, diseases of the gut—Crohn's disease, ulcerative colitis and coeliac disease—have foxed gastroenterologists and immunologists until recently. Previously, understanding of these conditions was focussed largely on the most superficial and therefore recent layers of the immune system's evolutionary strata. Only by embracing the antiquity of gut immunity and envisaging its role in fostering the 'holobiont' can we even begin to understand the nature of these conditions.

6

Gut Immunity: The Layers of Time

Summary In which we identify the different evolutionary strata within the intestine and see how the immune systems of the gut function. The most ancient and the most modern layers of immunity are most clearly defined, and can be easily distinguished on the basis of their related structures. The oldest ('paleo'-gut immunity) is represented by innate immune mechanisms enriched in the epithelial layer, whereas the most recent ('neo'-gut immunity) demonstrates adaptive immunity in the form of B and T cells and associated structures called lymph nodes and 'Peyer's patches'. Multiple intervening layers of old ('arche'-gut immunity) that bridge the gap between ancient and modern comprise populations of early forms of lymphocyte including innate lymphoid cells, B-1 lymphocytes as well as the $\gamma\delta$ T cells that contribute to the intraepithelial lymphocyte population.

J. Woodward, *The Gastro-Archeologist*, https://doi.org/10.1007/978-3-030-62621-1_6

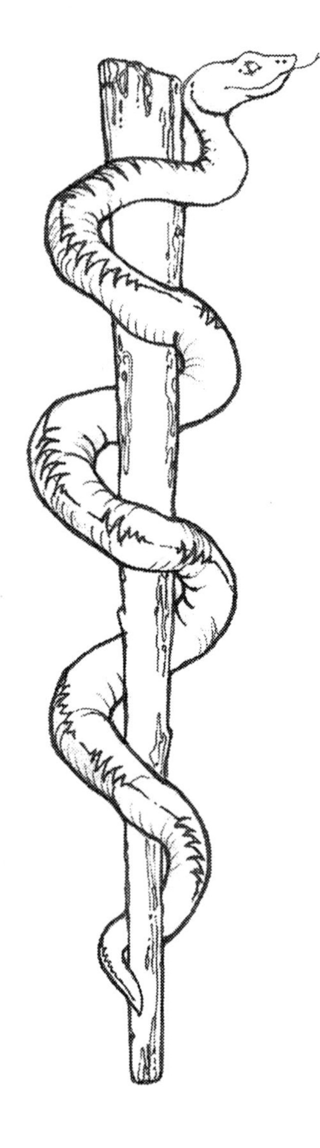

The rod of Asclepius

> *'Onions have layers. Ogres have layers. You get it? We both have layers'*
> Shrek 2001[1]

[1] Shrek (2001) was an animation fantasy film based on the character created by the 83 year old William Steig in 1990. Donkey (voiced brilliantly by Eddie Murphy), is disarmingly honest and as a result of never quite 'getting it' makes the key observations and asks all the right questions that initially sound simple. Science is a bit like that too I think.

Of Worms, Serpents and Dragons

'And the Lord sent fiery serpents among the people, and they bit the people; and much people of Israel died' (Numbers 21 v6). There is a consensus amongst historians of medical science that the 'fiery serpents' referred to in the Bible which were sent down among the Israelites after their destruction of the Canaanites was an infestation with a parasite called the 'guinea worm'. Humans ingest the larvae in contaminated water which then enter the body by crossing the wall of the human stomach and intestine. The female (which can grow to nearly a metre in length) moves through the tissues to the lower leg and creates an intensely painful lesion that is described by sufferers as a hot, burning sensation. Hence the Latin name for the worm, *Dracunculus*, meaning 'little dragon' and the use of the term 'fiery serpents' in the Bible. The skin breaks down to form an ulcer and victims immerse their legs in water to relieve the pain—neatly playing into the hands of the worm that then releases her larvae into the water to complete the life cycle.

Infestation with the Guinea worm was widespread and affected millions until the latter half of the twentieth century but it is now all but eradicated with only 20 or so cases described each year. These worms have been found in Egyptian mummies and the disease was clearly described in a medical treatise written on papyrus dating back over 3500 years. Interestingly, its treatment has not changed since—the end of the worm is found and then twisted slowly around a stick to remove it. This needs to be done very gently in order not to break the worm and may thereby take several weeks to extract. The imagery resulting from the worm wound around the stick has had profound and prophetic repercussions,[2] but most notably it is instantly recognisable as the Rod of Asclepius, the enduring symbol of the medical profession ever since ancient Greek times.[3]

[2] The Bible also has many layers which are compounded by linguistic development or even perhaps deliberate puns in the languages in which it was written. Moses is told to make a 'bronze' serpent and raise it on a stick and by merely looking upon it, the people will become immune to the 'fiery serpents'. This became a religious symbol called the '*Nehushtan*'. The Hebrew words for 'bronze', 'serpent' and 'to prophecy' share a similar root form which may have led to some linguistic confusion in this imagery—although we should be careful of drawing conclusions from the similarity of words alone. There is also some question that Jesus might have been referring to the Nehushtan as a prophecy of his own fate when talking to his disciples shortly before his crucifixion in the garden of Gethsemane.

[3] Asclepius, son of Apollo, is the Greek God of healing. His daughters were Hygieia (Hygiene), Iaso (the goddess of recuperation), Aceso (the goddess of the healing process), Aglaea (the goddess of the glow of good health) and Panacea (the goddess of the universal remedy). The symbol of the rod of Asclepius is often confused with the 'Caduceus', the symbol of the God Hermes. The Caduceus has wings, and *two* serpents wound around it. It has long been associated with commerce rather than medicine and the initial mistake of using it for a medical symbol is thought to have arisen in the USA—although perhaps appropriate given that medicine there has become a form of commercial enterprise!

Along with viruses and bacteria, parasitic worms have lived inside guts throughout intestinal evolution and our gut immune system has learnt to work around them. This is therefore as good a place as any to start our look at the strata of intestinal immunity, as the strategies for coping with worm infestation involve its deepest and oldest layer—the epithelial lining itself.

Diverticulum #6.1: the 'Hygiene Hypothesis'

Towards the end of the twentieth century, it was becoming clear that immunologically mediated diseases (allergies and 'autoimmune' conditions) were becoming more prevalent in affluent societies than the underprivileged. A number of different factors have been implicated including diet, weather, latitude (and hours of sunlight), pollution and medical resources. One theory that emerged was the 'hygiene hypothesis'—that 'westernised' populations live *too* hygienically and thereby lack certain organisms that might actually be doing us some good by altering our immune responses. The theory was first proposed by Strachan in 1989 who observed a greater amount of allergy in first-born children compared to subsequent offspring—who are likely to be exposed to more childhood infections. In support of this, a study showed a higher incidence of type 1 (autoimmune) diabetes in Finland than the adjoining Russian province of Karelia. The genetic composition, culture, climate and environment are effectively the same between these populations which differ only in socioeconomic status. There is now also good evidence from animal experiments to show that bacterial carriage in the gut can assist in downregulating autoimmune responses. For instance, mice that are genetically susceptible to autoimmune diabetes ('NOD' mice) are much more prone to develop the disease when raised under sterile compared to normal laboratory conditions. NOD mice that also have a genetic defect to prevent their surface Toll-like receptors from communicating with the cell nucleus (they lack the TIR domain protein called MyD88) are highly susceptible to diabetes under normal laboratory conditions. Interestingly, the gut bacteria of these two mouse strains differ. Transferring the 'normal' NOD mouse bacteria to the gut of the mutant NOD mouse prevents the development of diabetes. This also suggests that the gut bacterial composition (the microbiome) of the mouse is shaped by interaction with the gut immune system signalling through its toll-like receptors. Intriguingly, human gut bacteria can also be introduced into these different NOD mice—those from Karelian individuals appear to prevent the onset of diabetes, whereas Finnish gut bacteria do not!

With regard to worms, there is no doubt that the presence of such passengers is extremely common in certain parts of the world where diseases such as asthma and inflammatory bowel disease are rare. It is also the case that our immune system tends to be 'skewed' by their presence, as they can secrete certain chemicals to suppress immune responses and evade destruction. However, there is little evidence to suggest that our current standards of living prevent us from picking up parasites—as many as 30% of Europeans carry a species of worm in their intestines without knowing it. Nor is there anything to suggest that these individuals are at any lower risk of chronic inflammatory disease. Furthermore, in clinical trials where brave individuals with asthma or inflammatory bowel disease have ingested pig whipworm eggs, no definite benefit has yet been demonstrated.

The Front Line: the Intestinal Epithelium

The epithelial surface of the gut is its most ancient feature. From the earliest invertebrate origins in sponges and jellyfish it has evolved constantly and whilst remaining familiar in form it has lost and gained functions during the intervening aeons. Originally the site of secretion of digestive chemicals and the absorption and metabolism of nutrients, its digestive and metabolic functions have partly been delegated to offshoots—the pancreas and liver respectively. The lining cells—the 'enterocytes'—have retained their specialist role of absorption but as the interface between the environment of the gut lumen and the interior of the animal they have also taken on a key role in immune regulation. We are only just beginning to realise how on one hand the gut epithelium fundamentally influences the make-up of the human 'holobiont' by selecting symbiotic partners, whilst on the other hand protecting us from invasive disease-causing organisms.

The lining of the human small intestine is characterised by 'villi'—finger like surface projections about 1–1.5mms long—which we first came across in the earthworm intestine (fig. 6.1). They act to increase the surface area for absorption.[4] The human intestine contains approximately ten million villi which help to increase the surface area 30 times over that of a simple tube. As a result, if spread out, the human intestine would cover approximately 32m^2—about the size of half a badminton court (recently revised down from 400m^2—the size of a tennis court). At the base of the villi—the 'crypts'— we find epithelial stem cells that divide repeatedly and their progeny progress slowly up the villus as if they are on a conveyor belt until they fall off the tip and are then shed into the lumen. This journey takes 4–5 days and therefore the entire lining of the intestine is renewed every few days. The human intestine sheds in the region of 10^{11} enterocytes (100 billion cells) every day with a weight of around half a kilogram! The enterocytes differentiate into five separate specialised cell types during their transit along the villus, three of which (the absorptive enterocyte, the hormone-producing entero-endocrine cell and the mucus-secreting goblet cell) we have already encountered. The majority of cells (around 75%) become absorptive enterocytes. Around 10–15% of lining cells in the small intestine (but up to 50% in the large intestine) become 'goblet cells' that produce the mucus and approximately 1% turn into the 'entero-endocrine' cells that produce hormones acting as chemical messengers to signal to other parts of the gut or the body.

[4] Most animals have villi lining their intestines. However the duck-billed platypus from Australia is a notable exception in having only intestinal folds without villi. How villi develop has been studied in chicks and in mice and surprisingly shows two entirely separate mechanisms—suggesting two separate evolutionary processes that have come up with the same best solution. This is an example of 'convergent evolution' where the end result is the same but different evolutionary pathways have developed to get there from separate starting points.

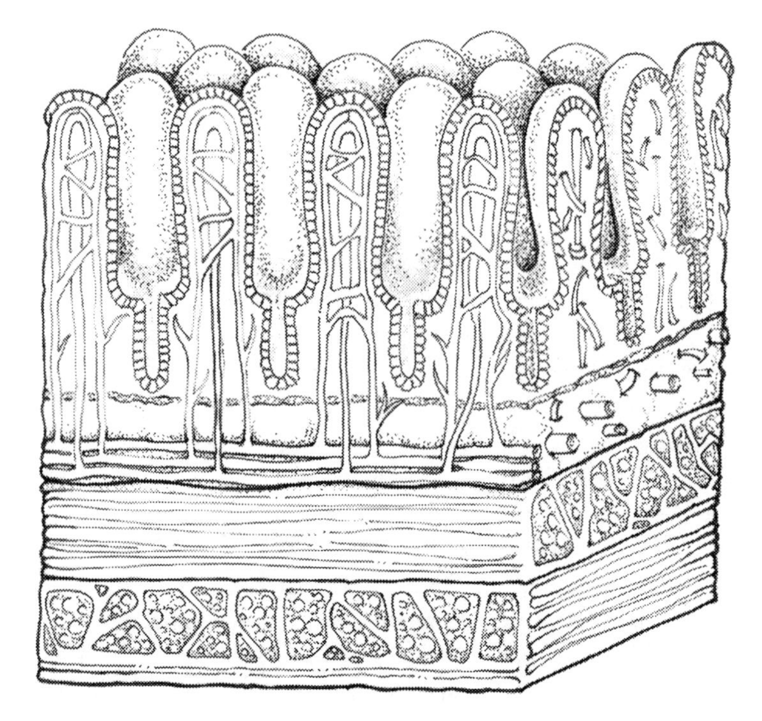

Fig. 6.1 The wall of the intestine. The inside is characterised by the villi—'finger-like projections'—about 1 mm long. The entire lining is covered by the intestinal epithelium (see also Fig. 14) with all its specialised constituent cell types. Below the epithelial surface is the 'lamina propria' and outside this are the two muscle layers—the circular muscle and then the longitudinal muscle. This basic structure of the gut wall is present throughout the animal kingdom—from earthworms to ourselves

A fourth cell type (about 3–5% of enterocytes) are specialised immune cells called 'Paneth' cells that are only found in the small intestine and head in completely the opposite direction, to the base of the crypts where they survive for about a month (we will revisit these shortly). Until the 1970's we thought that these were the only four epithelial cell types present in the gut—but then some very unusual and uncommon cells were discovered, comprising only about 1 in every 250 epithelial cells. They were initially called 'peculiar' cells for very good reason, and went by a number of different names until 2005 since when they have been known as 'tuft' cells.[5] Experiments in mice that have been genetically altered to lack all tuft cells show that they are fundamentally important in the gut's immune response to worms.

[5] A study in 2017 that analysed the products of individual cells from the human gut was able to demonstrate that there are many more cell types than we previously thought. There are no less than 8 different types of entero-endocrine cell that secrete different hormones and two different types of tuft cell, one of which shows significant resemblance to nerve cells.

'Peculiar' Cells

Tuft cells are indeed truly peculiar. They are thin and elongated with a prominent tuft of microvilli—tiny projections looking a bit like Bart Simpson's hair—extending into the gut lumen. There are openings at the base of these projections that connect via tiny tubes to the protein manufacturing machinery of the cell—effectively allowing a highway for large molecules to pass in and out of the cell.

Tuft cells communicate directly with nerve cells. In fact, analysis of the molecules they produce shows many that are similar to those produced by nerve cells. Furthermore, specialised structures resembling the 'synapses' through which nerves communicate are apparent where the nerve ending is adjacent to the tuft cell. Such synapses between nerve cells are extremely narrow gaps across which tiny quantities of chemical signals diffuse in order to trigger a new electrical impulse to travel along the second nerve fibre. Activation of a tuft cell can result in the associated nerve cell firing nerve impulses. The nature of this signal becomes more apparent when one considers that the tuft cell looks very similar to those cells in the nose that sense smell and those on the tongue that can detect different tastes. Astonishingly, tuft cells do themselves produce taste receptors identical to those on the tongue that recognise specifically bitter, sweet and umami (savoury) flavours. A mutation of this taste receptor in the tuft cells prevents mice from expelling worms just as effectively as if they lacked tuft cells altogether.

Thus, it appears that tuft cells act like intestinal 'taste' cells that detect the presence of worms to trigger a response and bring about their expulsion. This response has become known as the 'weep and sweep' as it literally flushes the worms out of the system. The 'sweep' is brought about by the tuft cells stimulating local nerve activation that co-ordinates peristaltic contractions of the muscle. These intrinsic nerves of the gut also bring about the 'weep' by stimulating fluid secretion by the epithelium, and trigger the goblet cells to expel their stored mucus. Just to add to the complexity of the networks that underpin the gut immune system, the nerves that lie just under the epithelium also express Toll-like receptors and are therefore able to sense different bacterial products themselves as a result…

Slime Factories

We have already encountered goblet cells from the earliest days of the gut—you will remember that they are so named for their resemblance to wine glasses as the central globular portion is packed tightly with vesicles filled with mucus. The mucus secreted by the goblet cells produces a layer that separates the gut lining from its contents whilst allowing essential nutrients through. This is similar to the 'peritrophic membrane' of insects and hagfish although in these instances it encases the food rather than lining the gut. None of the digestive enzymes in the gut are capable of chemically breaking down the glycoproteins of the mucus layer, which thereby serves to protect the gut from digesting itself. Mucus is also usually alkaline as a result of the secretion of bicarbonate by neighbouring cells and in the stomach this similarly protects the lining from acid attack. As well as protecting the lining of the gut from chemicals, the mucus (partially) isolates it from the bacteria within the lumen. It is perhaps for this reason that the colon (where most of the gut bacteria live) has a thick coating comprising two recognisable layers, compared to the single layer in the small intestine.

Mucus is made up of large complex glycoprotein molecules that form a gel with water. They also expand dramatically on hydration—the mucus packaged into the goblet cell effectively 'explodes' on release from the cell by increasing its volume over 1000 fold,[6] a little like the opening of an umbrella on the surface epithelium, or a satellite unfurling solar panels when in orbit.

The goblet cell might be envisaged in this way as casting a slimy, sticky net to trap bacteria and worms when triggered to fire. Just like the tuft cells, goblet cells have close connections at their bases to nerve cells—which provide the strongest signal for them to discharge. However, they can also trigger mucus secretion by themselves through fixed pattern recognition molecules detecting bacterial products.

Interestingly the number of goblet and tuft cells within the epithelium also increases during a worm infection. This is brought about by the tuft cells acting through an intermediary cell (a member of the 'innate lymphoid cell' family called ILC2) to signal dividing enterocytes at the villus base to turn into goblet and tuft cells rather than absorptive cells. Given that the entire gut epithelium renews itself every 4–5 days, altering its composition to upgrade the weep and sweep response therefore occurs within a suitably short timescale. It is notable that the tuft cells work in this regard by using an immune cell intermediary to alter the fate of epithelial cells. We see many such instances of interactions of epithelial and immune cells working in both directions and this demonstrates that the epithelium is very much a founder member of the immune orchestra. Along these lines, exciting recent discoveries have shown

[6] Volume expansion of three orders of magnitude—ie one litre per gram of material—is considered one of the definitions of an 'explosion'.

that goblet cells are more than just slime factories—for instance they also secrete a molecule which prevents worms in the gut from eating or reproducing (quite how is as yet unclear!). Even more surprisingly they act as two-way channels for the passage of substances across the epithelium both to and from the gut lumen. It appears that goblet cells can act as a means of excretion of environmental particles from the blood stream (such as the carbon nanoparticles created by diesel fume pollution) into the gut so that they are then lost in the faeces. In the other direction, specialised pathways across goblet cells appear to open and close under the control of nerves to deliver large molecules to the immune cells below the epithelial basement membrane. This breaches the 'containment' of the epithelium and effectively allows the deeper layers of the gut to sample its contents.

The Mucus Matrix

Mucus production by the intestinal epithelium has been a fundamental aspect of gastrointestinal defence since the time of the earliest multicellular organisms. As well as being a protective barrier and a 'trap' for bacteria and other organisms, the mucus layer provides a useful medium for molecules secreted by the intestinal lining cells. In this way they can be concentrated next to the enterocytes rather than diffusing away and being lost in the faeces. The absorptive enterocytes themselves produce a variety of different substances that have activity against micro-organisms.

However some cells of the intestinal epithelium have become specialised in an immune role. Called 'Paneth' cells,[7] they head in the opposite direction to enterocytes to lie in the bottom of the 'crypts' between the villi. Instantly recognisable by the secretory granules with which they are filled, these dedicated cells sense the microbes present in the gut and respond by releasing specific molecules that stick to and 'lace' the mucus.

Diverticulum #6.2

'Wehe dem Kind, das beim Kuss auf die Stirn salzig schmeckt, es ist verhext und muss bald sterben' ('woe to the child who tastes salty from a kiss on the brow, for he is cursed and soon must die').

[7] Named by Joseph Paneth (1857–1890), an Austrian physiologist working in Vienna. He was a friend of Sigmund Freud and also Friedrich Nietzsche.

This old German proverb foretold the discovery (in 1938) of a condition called cystic fibrosis (CF), a rare inherited disease that causes mucus to become too thick. As a result, sufferers are unable to clear mucus from the chest and the thick mucus also obstructs the secretions from the liver, pancreas and intestine. Those afflicted experience frequent chest infections that damage the lungs; pancreatic damage that leads to diabetes and inadequate secretion of digestive enzymes; and cirrhosis of the liver. As many as one in ten children with the condition experience blockage of the intestine shortly after birth, but this can occur at any stage due to the thick mucus blocking the gut.

The condition is due to a mutation of a single gene. Mutations of both copies of the gene are required to develop disease and about 1 out of every 25 Europeans carry one mutation. The gene encodes a transmembrane chloride pump—the 'cystic fibrosis transmembrane regulator' or CFTR. CFTR is found in the epithelial cells of the lungs and throughout the gut and helps to pump chloride ions into the lumen. On passing back into the epithelium, these chloride ions exchange with bicarbonate (also negatively charged) which makes the mucus more fluid. A defect in the CFTR reduces the chloride/bicarbonate exchange and therefore results in thicker mucus. It also prevents the uptake of salt in the sweat glands, hence leading to the 'salty kiss'—and a common test for the condition by measuring the salt content of sweat. Modern treatments for CF have dramatically improved the median life expectancy with this condition—from around 6 months in 1960 to 40 years in 2010.

The proteins secreted by the Paneth cells all carry a distinct air of antiquity about them. One of the oldest is an enzyme called 'lysozyme' which is found in in bacteria and plants as well as in animals and is secreted in tears, saliva and breast milk as well as mucus. Its antibacterial properties were first identified as long ago as 1909 but it was named in 1922 by Alexander Fleming, the discoverer of the antibiotic agent penicillin. This is perhaps no coincidence as the effects of lysozyme on bacteria resemble those of penicillin. Whereas penicillin inhibits the chemical crosslinking of molecules that strengthen the bacterial wall, lysozyme brings about their disruption—resulting in the destruction of the bacterium. Another enzyme secreted by Paneth cells into the mucus is called phospholipase A2 and is also common to most animals. It breaks down the specialised phospholipids of the bacterial membrane to kill the cell and is highly potent—it is also found concentrated in viper snake venom and is responsible for much of the pain associated with their bites.

Paneth cells also secrete lectins that act as pore forming toxins that recognise bacterial surface molecules and aggregate to form a deadly hole in the membrane. Other lectins are produced by Paneth cells including those known as 'intelectins'—a shortening of 'intestinal lectin'—that are very similar to those found in frogs, fishes and as far back as early chordates such as the lancelet. They act to agglutinate bacteria and prevent them from invading the epithelium. Interestingly, none of the lectins have complement binding capability and complement does not appear to be a major player in the gut—it seems that it is a layer of immune history that has never found a role there. In fact, the only significant complement family proteins to be found are those that inhibit complement activation—presumably of components leaking in to the gut from the bloodstream through damaged areas such as ulcers.

Other products of the Paneth cell cocktail include the appropriately-named 'defensins', which function variously as pore-forming toxins and lectin-like agglutinins. One such defensin even links together into fibres and ensnares bacteria much like the sticky DNA nets of amoebae. Defensin molecules are possibly even older than typical pattern recognition lectins. They identify their bacterial targets simply by carrying a strong positive electrical charge which attracts them to the negatively-charged surfaces of bacterial cells. Nevertheless, their release is triggered by Toll-like receptors on the surface of Paneth cells detecting microbial patterns. Close relatives of the defensins are found widely in nature, for instance in the venom produced by a spur on the hindleg of the male duck-billed platypus and in rattle snake venom.

Containing the Threat: the Intestinal 'Barrier'

Absorptive enterocytes themselves provide the last line of defence to invading microbes by maintaining a continuous layer of separation between the gut and its lumen. The cells are all connected together by specialised junctional proteins that act like zippers to produce a barrier to the passage of fluid and salts—as well as microbes—that could otherwise use the gaps between cells as an alleyway to the inside. However, the junctions between the cells can be opened up to become 'leaky' and this is part of the 'weep and sweep' enabling fluid to stream into the lumen as a flush (Fig. 6.2).

Keeping the epithelial barrier intact is clearly critical—any breach in the containment could lead to dramatic infectious repercussions as potentially dangerous organisms and chemicals flood into the animal's interior. Enterocytes however are no different from other cells in having a 'suicide' programme of cell death as a last resort in the case of bacterial incursion. This could be rather counterproductive by opening up a chink in the defensive wall. For this reason, apoptosis is very carefully controlled within the gut epithelium and a number of mechanisms have evolved to quickly close the gap—including the ability of neighbouring cells to detach from the basement membrane and spread laterally in a process called 'restitution'. Lack of contact between epithelial cells also leads to them secreting growth factors and

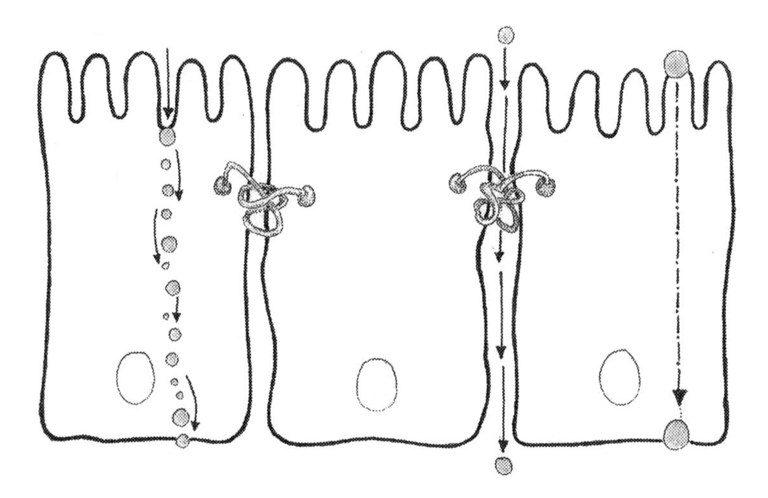

Fig. 6.2 The intestinal barrier. Just as there is a 'containment paradox' for cells bound by their membranes, that still need to allow substances in and out, so it is with the epithelium on the scale of the organism. Adjacent enterocytes are connected by specialised protein links called tight junctions. These proteins are connected to the internal scaffolding of the cell and their leakiness can be regulated from within the cell to permit fluids and small molecules in and out. Substances can also be carried across the cell by 'transcytosis'—endocytosis on one side of the cell with vesicles transported to the other side where their contents are released by exocytosis (left). Large molecules may be similarly transported across the cell by binding to specialised transporter proteins (right)

responding quickly to them. Interestingly, enterocytes in the colon and small intestine differ in their ability to undergo apoptosis. The small intestinal epithelium uses it as a means of shedding invasive bacteria (just like the S cells of the social amoeba), but the colonic cells actually produce high levels of an internal inhibitor of apoptosis in order to prevent cell suicide. This is possibly because of the higher load of bacteria present in the colon than the small intestine and the dire potential consequences of widespread cell loss there. The difference in regulating cell death between the small and large intestine may well explain why cancer (the result of unregulated cell growth with reduced levels of apoptosis) occurs frequently in the colon but very rarely in the small intestine.[8]

[8] Each year in the UK there are only about 300 cases of small intestinal cancer reported which compares to over 40,000 diagnoses of colon cancer a year (making it the fourth commonest cancer in the UK). This is despite the small intestine making up over 75% of the length of the gastrointestinal tract and over 90% of the surface area. Therefore, in terms of surface area, cancers of the colonic epithelium are about 1000 times more common than those of the small intestinal epithelium.

In order to prevent the creation of gaps in the lining epithelium, enterocytes have learnt a few clever tricks. When we look at intestinal epithelium sampled during an infection, the enterocytes often appear to be square in shape rather than tall and thin (columnar). Whereas we used to think that this was the result of damage by microbes, it turns out that this is only half the story—the enterocytes have actually developed a defensive means of extruding some of their cellular substance in order to pump out the unwanted intruders and any harm they have caused. When the infection is cleared, these thinned enterocytes can restore their usual shape within hours!

Enterocytes carry a number of enzymes embedded in their surface membrane, one of which is called 'alkaline phosphatase'. It appears to have a direct action against a specific form of lipid in bacterial membranes. The amount expressed by the cells relates to the bacterial load in the gut and giving additional alkaline phosphatase to animals with gastrointestinal infections helps them to recover more quickly from them. Although this enzyme is found embedded in the membrane of the enterocyte, it is secreted into the gut lumen in the form of tiny membrane blebs that bud off the cell surface. These form at the tips of tiny projections on the cell surface—the microvilli—like little droplets of water on an icicle. Such alkaline phosphatase-rich bubbles exist in high concentration in the mucus where they aggregate bacteria to prevent them from invading cells as well as directly attacking their membranes to kill them.

Enterocytes are also capable of sensing the contents of the gut as they express numerous cell surface molecules—including Toll-like receptors—to identify foreign and potentially threatening substances. They can also transmit their findings by secreting an extensive array of chemical signalling molecules that can act as an alarm signal and attract specialised immune cells to the site of a microbial breach in the intestinal lining. The particular cocktail of interleukins secreted may be specific to the nature of the threat thereby enabling the enterocytes to tailor the response to the invasion—maybe like an emergency '999' call specifying a police, ambulance or fire brigade call out.

Hence the epithelial lining of the gut itself, the oldest part of the intestine, has developed an extensive repertoire of mechanisms for dealing with unwanted passengers. Once again however, we should bear in mind that this is only a part of the functions of the immune system as the same mechanisms foster a balanced environment for the desired symbiotic bacteria. All of the antimicrobial proteins secreted by the enterocytes, goblet cells and Paneth cells appear to encourage the development of a characteristic microbial community in the gut rather than simply sterilising it.

Patrolling the Front Line: The Enigmatic IEL

IEL stands for 'Intra-Epithelial Lymphocyte'—in other words, lymphocytes within the epithelium.[9] They are the immune cells that patrol this single layer of cells (about 1/50th of a millimetre thick) between the basement membrane on which the enterocytes sit and the lumen of the gut. We have vast numbers of them—between 50 and 100 billion, making up over 60% of all the T cells in the body and they are present at a ratio of one to every ten enterocytes in the epithelium. The majority of IELs are T cells and most are of the cytotoxic T cell variety, equipped with the ability to kill cells directly. However, the epithelium is a veritable cultural melting pot of lymphocytes with a vast array of different types present and many IELs are representative of older versions of lymphocytes, co-existing with up to date models. These include Natural Killer cells, some of which also have T cell receptors and are therefore called 'NKT' cells. The latter types of cell often carry a prototype T cell receptor that is 'fixed' rather than variable, like a pattern recognition molecule from the days before the great immunological 'big bang' of recombination. The epithelial T cell population is also enriched in cells that carry the older type of T cell receptor—the γδ receptor—rather than the more modern αβ receptor. For decades we have had little idea of what IELs do and how they do it, but over the last few years it has slowly dawned that these cells comprise their very own specialised self-contained immune system within the tiny space of the thin layer of gut epithelium itself.

Although there is a dizzying array of different types of IEL present within the epithelium, it is possible to categorise them into two main types—'natural' and 'induced'. The 'natural' IELs are those that are found even in the absence of bacteria, for instance in the guts of mice brought up in a germ-free environment and fed with basic nutrients that cannot be recognised by immune systems (which require large molecular pattern signatures). These are generally

[9] Intraepithelial cells were first identified by nineteenth century microscopists and described by Weber in 1847. A number of different names were used for them including 'small round cells' and 'wandering cells'. Initial theories of their purpose and functions were rather weird and wonderful … A process of 'caryo-anabiosis' (literally 'nuclear growth') was suggested in 1912—which envisaged the IELs nourishing the nuclei of enterocytes or even donating their own nuclei to them. This theory existed into the 1950's with scientists attributing a nourishing role to the IELs within the epithelium. A competing theory from the 1920's was that the intestine was the graveyard of lymphocytes and was the body's way of disposing of them through the gut. Experiments were even carried out to show that removing the entire intestine from animals did not increase the amount of circulating lymphocytes, thereby disproving this theory. Interestingly, recent evidence shows that lymphocytes that have failed education in the thymus gland (by producing receptors that identify 'self' molecules) are selectively trafficked to the gut. Therefore this theory of the gut being a site of disposal of lymphocytes may not be such a silly idea after all … .

the more ancient varieties of lymphocyte (NK, NKT and γδ T-cells). For the most part, they recognise molecules within cells that are only released following damage or are expressed on the cell surface of 'stressed' cells. The detection of these signals upregulates natural killer cell functions in the IELs which then quietly despatch the affected enterocyte but additionally secrete growth factors to stimulate nearby cells to divide and close the gap in the defences. In this way, natural IELs can patrol the epithelium, cleaning up damaged, virally infected or malignant cells whilst maintaining the gut barrier.

'Induced' IELs are those that have responded to an infection and remain as memory cells present within the epithelium to mount a rapid response should the infection return. Their receptors are capable of identifying molecules specific to microbial friends, food substances or foes and gradually accrue over time. IELs bear specific surface molecules that interact with epithelial cells to keep them within the epithelium once they have passed across the basement membrane and they tend to stay there for a long time. Therefore, the T cell receptors of the induced IELs represent an archived 'memory' of the specific foods and infections experienced in the past.

The majority of IELs come to the gut from the thymus where they are educated. Interestingly it appears that most of the natural IELs are effectively 'renegades'—they often recognise 'self' molecular patterns which should schedule them for destruction but they are instead exiled to the gut. Similarly, a small proportion of 'natural' IELs may come directly from the bone marrow and bypass the thymus to complete their development within the gut itself in an echo of an original function of the gut lining.

Together the legions of IELs within the gut epithelium provide the whole range of immune interactions—from housekeeping, maintaining the integrity (and 'containment') of the epithelium to recognising and dealing with threats from alien invaders. It even now appears that they influence our whole body metabolism and appetite by regulating the expression of gut hormones produced by the enteroendocrine cells!

Behind the Lines

The gut lining beneath the very surface layer is a site of even more frenetic activity with bidirectional flow to and from the front line. This region—between the epithelium and the muscle layer—is called the 'lamina propria', or 'proper layer' of the gut for reasons that are somewhat lost in obscurity.[10] It

[10] The first recorded use of the term 'lamina propria' dates back to german histopathologists in the 1930's.

is here that the small blood vessels are found. These are the highways that transport oxygen to the gut and carry nutrients from the gut to the liver in the 'portal' system. The lining of these blood vessels has specific 'homing' tags to bind surface molecules on immune cells that have been specially primed to go to the gut. In this way they effectively 'get off at the right station' rather than continuing on their journey through the circulation.

Other vessel-like structures in the lamina propria carry lymph (or 'tissue fluid') away from the gut and drain through 'lymph nodes' in the supporting tissues of the intestine. Large fat molecules that have been absorbed through the intestinal epithelium are also drained through this route—rather than via the blood vessels—and bypass the liver.

The lamina propria is a cosmopolitan hub buzzing with immune cells of all kinds representing every section of the immune orchestra, rather than the very select population of the epithelium. About three quarters of all the antibody-producing B cells in the body are to be found here. The whole of the T cell section is present including helper and cytotoxic T cells as well as the natural killer cells and the other members of the 'innate lymphoid cell' (ILC) section. There are also specialised primitive T cells that express αβ T cell receptors that are fixed or 'invariant' just like the pattern recognising molecules of invertebrates. These recognise a limited range of molecules that are produced only by bacteria as a result of processing vitamins and can therefore provide early warning of their presence.[11] Such cells—called 'MAIT' cells (mucosa-associated invariant T cells)—make up nearly 10% of the cells within the lamina propria and are also found circulating in the blood stream.

The full range of supporting immune machinery is also to be found in the lamina propria including macrophages and dendritic 'presenting' cells. All of this frenetic activity is directly related to the contents of the gut—despite the limitation of the containing epithelial barrier. In order to 'sample' the contents of the lumen, some of the macrophages actually send out thin extensions through the supporting 'basement' membrane of the epithelium and between the lining cells. This has been likened to the periscope of submarines used to visualise what is going on above the surface! Particles are engulfed and travel back through the cell into the underlying lamina propria where they are re-exported—effectively handed on—to adjacent professional dendritic

[11] MAIT cells were first described in 1999. The bacterial products that they recognise are derived from the breakdown of vitamin B2 (or riboflavin). Being a 'vitamin' means that it is not made by human cells and is required in our diet. The molecules recognised by MAIT cells are not presented on the standard MHC molecules but on similar protein called 'MR-1'.

presenting cells. Similarly, goblet cells can provide channels to smuggle large molecules across the epithelial border. Immune traffic in the opposite direction includes the large amount of IgA antibody made by the B cells in the lamina propria. This manages to pass through the basement membrane where it is picked up by specific receptors on the underside of the enterocytes and transported through the substance of the cell to be secreted on the luminal surface. In these ways the containing barrier of the epithelium can be overcome in a very controlled manner, allowing the underlying immune system of the lamina propria to sense and interact with the contents of the gut.

Danger...Minefield!

Yet another type of cell is found within the lamina propria with important immune functions. This is the 'mast cell'.[12] Mast cells are packed full of granules and are often found close to blood vessels. On their initial discovery they were thought to contain nutrients and to have a nutritive function and this explains their name—the German word 'Mast' relates to nourishment or suckling. In keeping with this, we are most familiar with its use pertaining to the breast, hence 'mastectomy' for its surgical removal, or 'mastitis' when infected. However, it turns out that mast cells have no such role and in fact do something completely different. They store up preformed chemicals that have potent local actions when released rapidly from the cell by 'degranulation'. In essence they are like grenades, or landmines. And they are on a 'hair-trigger', primed and ready to explode.

Some of the contents of mast cells and their functions are already well known to us. For instance, histamine released by mast cells in the skin leads to the itchy red blotches ('urticaria')[13] which we associate with allergic skin

[12] It was Paul Ehrlich once again who first described mast cells, thanks to his use of dyes that revealed different populations of blood cells due to the chemicals stored within them. His discovery of 'Mastzellen' were depicted in his doctoral thesis in 1879.

[13] It is well known that stinging nettles (*Urtica dioica)* carry toxins on their hairs which create the typical 'wheal and flare' nettle rash on contact. One of these substances is histamine itself, another is formic acid—the same chemical produced by stinging ants. The genus name of the stinging nettle derives from the Latin word 'urure' meaning 'to sting' but also gives us the word 'urticaria' for the skin condition caused by the release of histamine by mast cell degranulation. The species name 'dioica' shows that there are separate male and female plants (the species is 'dioecious'). Stinging nettles are very common across Europe and have found frequent commercial use—for instance in textiles (almost all first World War German army uniforms were made using nettle fibres in place of cotton) and in food—from nettle soup, to herbal teas, nettle beer and in pastas and polentas. Nettles are added to a Cornish cheese called 'Yarg' and in Dorset there is even an annual world nettle-eating championship for the most raw nettle leaves eaten in a set time (which supposedly originated in 1986 following a dispute between neighbouring farmers...!!).

reactions. Histamine released in the nose and eyes lead to itching and sneezing such as we may experience with hay fever. Many of us will at some stage have taken an 'anti-histamine' tablet to control such symptoms. In its most severe form of generalised mast cell degranulation, the serious condition of 'anaphylaxis' occurs with swelling of the tongue and airways resulting in airway constriction and collapse due to low blood pressure.

Mast cells can be triggered to degranulate by many different kinds of stimuli. These can be chemicals such as snake venom, or drugs related to morphine. Physical triggers such as heat, cold, sunlight or exercise can also activate mast cells—some people even experience urticarial rashes after a cold shower. In addition, mast cells are often closely associated with nerves which can fire them from a distance.

Histamine and other potent chemicals from mast cell granules act on blood vessels to dilate them and make them leaky—this is the reason for the swelling and redness we see with allergic skin rashes. It is also the explanation for the severe drop in blood pressure with anaphylaxis. At a local level this allows chemicals and cells from the blood to access the tissues, resulting in an influx of immune reinforcements. In addition to histamine, mast cells release powerful digestive enzymes capable of breaking down proteins—it is thought that these may have evolved to destroy snake or scorpion venom. These enzymes also loosen up the connective tissue, thereby easing the ability of immune cells and antibodies to permeate freely. An anticoagulant component called heparin prevents blood clots from forming in small blood vessels as a result of immune activation, and interleukin chemical messengers attract or alter the behaviour of immune cells.

Cells similar to mast cells can be found in nature dating back to the earliest chordates. However, it is only over the last 200 million years or so that allergic responses have developed—because they are mediated by interactions of mast cells with antibodies. Mast cells carry a surface receptor for the tail of a particular class of antibody known as IgE. As we have seen, antibody tails act as 'adaptors' to allow the use of different 'apps' or types of response to the specific molecule recognised by the variable receptor at the other end of the antibody. IgE antibodies are thus 'mopped up' by sticking to the mast cell with their tails, leaving their variable binding sites sticking outwards. Any part of a protein floating past that is specifically identified by two different IgE molecules on the cell surface brings them closer together and this triggers the mast cell to fire. As few as 100 such 'crosslinking' events on a mast cell will cause degranulation.[14]

[14] IgE represents only 0.005% of the whole antibody pool in the body. It is generally considered to play a specific role in immunity and allergy due to its interaction with mast cells and therefore to be a modern addition to the immune armoury. Recent evidence however points to a more ancient purpose in a 'housekeeping' role, preventing cancer and tidying up after dying cells—once again demonstrating that defence was not the primary role of the immune system.

The result is a very potent means of priming the immune system. IgE antibodies may be formed following low-level exposure to a potential threat and coat the mast cells where they stay for a long time. These mast cell landmines are thus 'armed' and ready to explode dramatically should the threat return. In the gut this works best with the 'weep and sweep' response to worms where the fluid from the leaky blood vessels flows into the lumen and flushes the invaders away in a wave of diarrhoea.

Smart Structures

The buildings that we construct and live in waste space in the material of the structure itself—the clay brick has changed very little over the last few centuries. Wouldn't it be smarter to use some of the volume of the fabric used in construction for other purposes? The building blocks could perhaps include electronic components and sensors, even heating elements. In the external façade, cavities in the bricks might provide a home to microorganisms that could for instance process air pollution or waste and even generate heat as a by-product.[15] This is the concept of a 'smart brick'.

Of course, any clever idea thought up by humans can usually be found already existing in nature and the gut, which has had such a long time to evolve has well and truly mastered the concept of 'smart' structures. The cells that produce the structure of the intestinal lining are indeed smart and do far more than just build—in fact they orchestrate many of the diverse activities of the intestine.

The scaffolding matrix that underlies the epithelium in the lamina propria comprises fibres of connective tissue that hold it all together. Cells that generate such a structure of interconnected strands are usually called 'fibroblasts', but those lining the crypts of the intestine additionally have features of muscle cells and are therefore called 'myo-(=muscle) fibroblasts'. Intestinal myofibroblasts are ideally placed at the interface between the epithelium and the lamina propria and secrete a whole range of different chemical messengers that act locally on nearby cells on both sides of the basement membrane. Just like a 'smart structure' they do not simply construct the wall of the gut lining but

[15] The exciting but unfortunately named 'LIAR' project ('**LI**ving **AR**chitecture') aims to 'transform our habitats from inert living spaces to programmable sites' and was initiated as a collaborative programme by researchers at the University of Newcastle Upon Tyne in the UK in 2016. It is envisioned as an 'integral component of human dwelling, capable of extracting valuable resources from sunlight, waste water and air and in turn, generating oxygen, proteins and biomass through the manipulation of their interactions'.

interact in both directions—with the outside epithelium and the inside lamina propria, due to their prime position between the two.

These smart cells have the ability to sense their environment—using simple cell surface pattern recognition molecules that we are already familiar with. The Toll-like receptors they carry can identify bacterial components or substances released from inside damaged cells. If these are detected by the myofibroblasts, then it would suggest that there is a leak in the epithelial lining. As a response they secrete chemicals to stimulate the epithelial cells to proliferate in order to seal the hole. Not only that, but going one step further they can identify the particular type of threat that is present—whether it be viral, bacterial or worm—and send messages to the dividing epithelial cells to alter their fate and alter the proportionate composition of absorptive enterocytes or defensive mucus-secreting goblet cells. This is the 'structural' aspect of their role.

On the other side of the basement membrane separating the epithelium from the lamina propria, myofibroblasts also direct immune responses. We have previously encountered the dendritic presenting cells as the conductor of the immune orchestra. However, the choice of music played by the ensemble may actually be dictated by the myofibroblasts which interact with the dendritic cells to alter their behaviour. It appears that even the conductor (the dendritic cell) cannot choose the programme his orchestra plays, despite being the most expert maestro. And in the gut, it is the myofibroblasts, which we mistakenly thought were just the builders of the concert hall, that may choose which pieces from the repertoire are played!

'Patching in': a New Update

Not surprisingly it is the recently-evolved aspects of the immune system of the gut that we know most about. Indeed, most of our understanding of immunology relates to these most superficial layers such that it is only over the last 15 years or so that we have come to realise the importance of the older mechanisms. These recently-added components comprise the classical variable immune receptors of T and B cells and all of the machinery that goes with them. Clearly however the gut immune system was capable of working perfectly well without these 'add-ons' as they only evolved within the last 100 million years or so, and fishes, amphibians and reptiles (not to mention invertebrates) lack all or most of them.

In 1677 a Swiss anatomist by the name of Johann Peyer described scattered oval or round nodules in the wall of the gut that are clearly visible to the naked eye from the internal surface.[16] Humans have about 100 of these lumps in the small intestine, each around 2-3cms across. As this was in an era before any conceptualisation of infectious organisms, let alone the immune system, Peyer thought that his 'patches' were glands for the secretion of digestive juices. In fact, they are part of the lymphatic system that was first described some 2000 years earlier and bear significant resemblance to the lymph nodes that are part of it. We still describe palpable and tender lymph nodes—for instance those in the neck associated with a throat infection—as 'enlarged glands'.

Intestinal Peyer's patches are packed full of lymphocytes and have active germinal centres where B cells are being instructed in the production of high affinity antibodies. They extend through the lamina propria to the mucosal surface where the usual structure of villi is replaced by a smooth 'dome'-like shape. The overlying epithelium is also unusual with the mucus-secreting goblet cells replaced by specialised 'M' cells (M for 'microfold' due to their surface appearance which lacks the small, hairy microvilli of normal epithelial cells). M cells effectively sample the gut contents by phagocytosis of particles and micro-organisms. They then transport them across the epithelium to the underlying Peyer's patch which is rich in professional presenting dendritic cells capable of directing a specific immune response.

Within the Peyer's patch, just as in lymph nodes which we came across in chap. 5, B cells are educated to produce antibodies against surface molecules of the bacteria presented to them by the dendritic cells. However, rather than the IgG type of antibody produced by normal lymph node B cells, in the Peyer's patch they are trained specifically to produce IgA antibodies, the specialised mucosal antibody type that is secreted into the gut. These B cells can remain present in the Peyer's patch but this is of limited value as the educated B cells and their antibodies are really required to be disseminated throughout the length of the intestinal lining. In order to spread far and wide through the gut, they first have to leave it, draining in the 'lymph' or 'tissue fluid' to a standard

[16] Johann Conrad Peyer (1653–1712) was born in Schaffhausen, in a part of Switzerland that extends like a tongue into Germany. He studied medicine in Paris and Basel and then returned to his native town to work as a doctor but also carried out research with two colleagues forming the 'Schaffhausen trio' (not a jazz band!). One of the three, Johann Brunner, discovered glands in the upper part of the small intestine (the duodenum) which are indeed involved in secreting digestive juices and so it is not surprising perhaps that Peyer thought that his patches also performed a digestive function. Peyer published his findings in 1677 in '*Exercitatio anatomico-medica de glandulis intestinorum earumque usu et affectionibus*'. Although it was Peyer whose name has been associated with these structures as a result of his detailed description, they were probably first noted in 1645 by Marco Aurelius Severino (1580–1656), an Italian surgeon, anatomist and philosopher.

lymph node and then into the blood stream. Circulating in the blood stream provides access to the whole of the gut.

However, the intestine only wants back those cells that it has educated in the Peyer's patches as they are producing antibodies specific to the gut contents. It does not just want a random collection of antibody-producing cells from the circulating blood stream. It achieves this in a most ingenious way. Whilst passing through the draining lymph node on its way to the blood stream, the B cell is instructed to produce surface proteins that are able to sense chemicals secreted only by the gut. They also make 'homing' receptors that bind to specific molecules found specifically in the lining of the blood vessels in the intestinal lamina propria. Having stuck to the blood vessel in the gut, by engagement of these receptors, the cells then work their way through the blood vessel and out into the tissue of the lamina propria attracted by sensing the gut specific homing messages. This chemical attraction of cells to bring them together in a location—whilst much more sophisticated—is akin to the aggregation of cells in the social amoeba.

Whilst the Peyer's patches are the forward operating bases of the gut immune system, it is the draining lymph nodes that are the 'headquarters'. Amazingly the homing mechanisms that send the lymphocytes back to the front line in the gut are specific to where the lymphocytes originated such that recirculation to the colon and the small intestine are kept quite separate. The molecular switch that turns on the homing signal in lymphocytes appears to be a derivative of vitamin A called retinoic acid. It is produced by the dendritic cells that migrate along with the lymphocytes back to the draining lymph nodes. However, they are in turn stimulated to produce the retinoic acid by the myofibroblasts of the lamina propria, a clear example of how these cells dictate the repertoire of the maestro dendritic cells themselves.

Diverticulum #6.3: the tonsils, guardians of the gate

At the back of our throats on either side of our tongues and on the pillars of the arch made by the palate are what we know as the tonsils—or more accurately, the 'palatine' tonsils. The tonsils have long been considered a site of recurrent infection and something of a nuisance—their removal by surgical tonsillectomy became almost routine within the last 50 years.

The palatine tonsils make up part of a circle of immune tissue that includes a tonsil on the back of the tongue, bilateral tonsils in the pharynx behind the nose (the 'adenoids') and at the openings of the Eustachian tubes from the middle ear. Known as 'Waldeyer's ring[17]', the tonsils guard the entrance to the lungs and to the gut. This site is also of great evolutionary significance, and all of the historical layers of immunity are to be found blended within their structures. Recent evidence even reveals a thymus-like role as the tonsils have been shown to be a site of T cell development.

Unlike the Peyer's patches, the tonsils are not covered in a glandular absorptive epithelium but in a stratified protective lining more like the skin. It is also similarly specialised to permit the entry of substances and organisms that the underlying immune system can sample in order to prime an immune response. Similarly to the Peyer's patches the tonsils contain germinal centres for the education of B cells in making antibodies—including a particularly ancient class of antibody known as IgD. This has recently been shown to work in a similar way to IgE in degranulating mast cells and may fulfil this role in creatures such as fish that lack IgE.

Whilst the tonsils act to educate and prime the immune system against organisms that could invade the upper airways, tonsillectomy does not appear to have any significant adverse consequences suggesting a degree of redundancy of its role—although throat abscesses are slightly more common in individuals that have undergone this operation. Interestingly, forms of an immune-mediated skin condition called psoriasis (that causes red scaly skin plaques) appear to improve after tonsillectomy.

Layers in Time: Gastro-Archeology in Practice

By looking at the different parts of the gut immune system—the epithelium, the lamina propria and the lymph nodes and Peyer's patches—we should note how we have passed through time from most ancient to most recent. We can now take a step back and look at the strata that we have uncovered just like the archeologist looking in profile at the layers of a trench. In doing so we can share a classification coined by neuroscientists to describe parts of the brain—as 'paleo-' (oldest), 'arche-' (old) and 'neo-' (modern). Whilst these may be used to define the major strata in time, we find that there are multiple layers in each that may blur together. Such a blurring of layers is only to be expected and in no way weakens our analogy of a gastro-archeological dig or a palimpsest. After all, the Victorians continued to use devices that were invented by the Romans despite developing more sophisticated tools themselves, and we

[17] Heinrich Wilhelm Gottfried von Waldeyer-Hartz (1836–1921) who first described the tonsillar ring that bears his name (or part of it!) was a German anatomist who worked in Strasbourg (after it was invaded by the Prussians, the existing French university staff were summarily dismissed) and Berlin. He also worked out the cellular basis of the immune system through analysis of the observations of others, and coined the name 'chromosome'.

can date the origins of our modern cities back to earlier civilisations even though we now have the ability to inhabit new environments (even stations in space).

The 'paleo' gut immune system comprises the lining cells of the gut: the absorptive enterocytes that produce directly-acting antimicrobial chemicals such as alkaline phosphatase; the mucus-producing goblet cells to coat the epithelium and trap microbes, and the specialised Paneth cells that secrete fixed pattern-recognition molecules such as lectins and pore-forming toxins into the mucus.

The 'arche' gut immune system is a broad layer with many substrata that comprises cell types such as the mast cell, the structural myofibroblasts and the early types of lymphocyte. These include the ILCs and natural killer cells as well as the early forms of T cell (bearing invariable or fixed receptors or the $\gamma\delta$ type of T cell receptor). The 'arche' gut immune system is notable within the epithelium (the 'IELs') and present within the lamina propria.

The 'neo' gut immune system is the modern addition of the Peyer's patches and mesenteric lymph nodes—effectively stuck on to the gut. This includes the apparatus for producing honed B cell responses—specialised 'microfold' cells for sampling of the gut contents and germinal centres for producing high affinity IgA antibodies—and the cytotoxic T cell responses that include the induced IELs within the epithelium.

Whilst it is clear that the immune mechanisms of the gut can be separated in time strata in our gastro-archeological dig, it is important for us not to get confused and expect them also to separate spatially within the gut. Nevertheless, this does occur to a certain extent as the oldest 'paleo' immune mechanisms are concentrated in the epithelial layer (for reasons addressed in part 1 of this book) and modern additions of the 'neo' immune system are demonstrated by the Peyer's patches and mesenteric lymph nodes (Fig. 6.3).

The Great Schism

When we take a closer look at the strata of the evolution of the gut immune system we are reminded once again of the large crack that runs like a geological fault right through all the layers of time and appears to keep going beyond the deepest level of our dig. This is the division that separates the antibody-mediated, 'soluble' side of the immune response from the cell-mediated 'cytotoxic' response and is embodied in the modern upper strata by the separation of functions in B and cytotoxic T cells respectively. In the middle layers of the gastroarcheological dig we also see the same separation in functions in cell

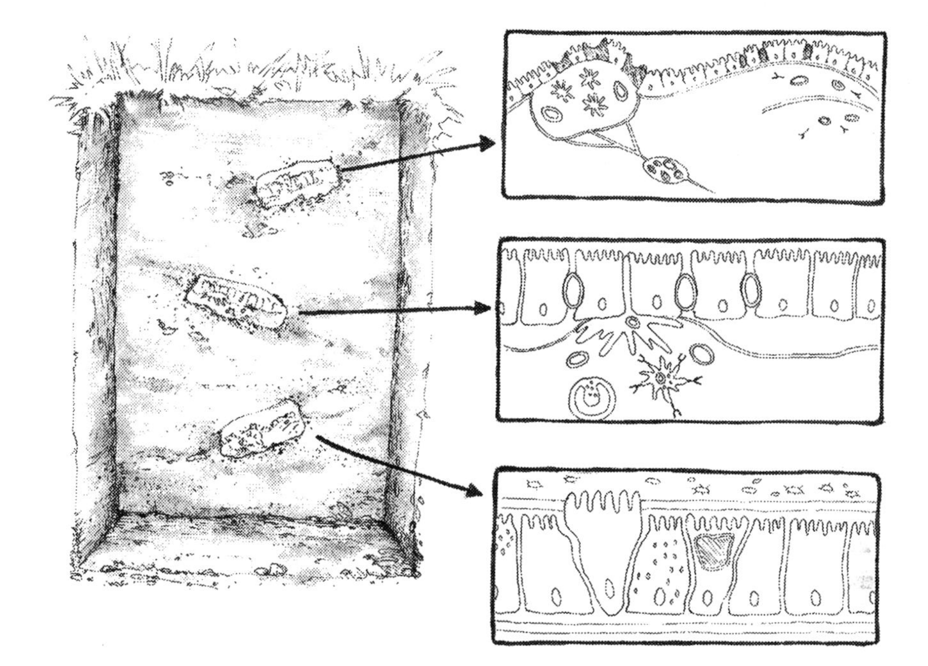

Fig. 6.3 The main strata of the gastro-archeological trench. The lowest (oldest) layer comprises the 'paleo' gut immune system. Here we see the epithelium itself and the defensive functions of the enterocytes. These include the goblet cells which produce the surface mucus, the Paneth cells of the small intestine which secrete 'defensins' to control the microbial population, the tuft cells that act like taste cells to 'taste' invaders, and the absorptive enterocytes themselves that produce enzymes on their surface (including alkaline phosphatase) and express toll-like receptors to detect threats. The middle layer (the 'arche' gut immune system) includes the 'natural' intraepithelial lymphocytes moving between the enterocytes in their housekeeping role, and the macrophages and dendritic cells that push out processes through the epithelium to pick up molecules from the gut lumen. Here we find the intermediate types of lymphocyte including those that lack highly variable receptors (the innate lymphoid cells and natural killer cells) or express invariant T-cell receptors (such as MAIT and NKT cells). We also find effector cells of the lamina propria such as macrophages, mast cells and eosinophils. The upper layer (the 'neo' gut immune system) is characterised by the modern 'add-ons' of lymphocytes with highly variable receptors—B and T cells. Here we find the Peyer's patches that connect to the lymph nodes in the connecting mesentery of the gut and contain germinal centres for the education of antibody-producing B cells that recirculate to the gut from the lymph nodes

types that lack the highly variable receptors of B and T cells—such as NK cells and other ILC subsets. As we have seen, this schism is even apparent in the separate lineage of jawless fishes (modern lampreys and hagfish) that last shared a common ancestor with us over 500 million years ago. Despite not evolving immunoglobulin superfamily variable receptors (antibodies and T cell receptors), their own 'LRR' variable receptors are clearly separated into antibody-like and T-cell like functions. It is at the lowest levels of the gastro-archeological trench that we see the origin of the split, which derives from the two fundamental means of overcoming the 'containment paradox'—phagocytosis and membrane puncturing. Over time there has inevitably been some crossover. For instance, antibodies and complement do not just facilitate phagocytosis but also attract pore forming molecules and cytotoxic cells to their targets. Nevertheless, the separation of the B and cytotoxic T cell roles clearly date back to the two separate solutions of the containment paradox.

The story of the gut and nutrition revealed in part 1 is intimately intertwined with the story of the immune system that we have seen in part 2, right back to the single celled origin of 'eating' by phagocytosis. It is the antiquity of the gut that has led to its enrichment in the 'paleo' and 'arche' immune systems and makes it a palimpsest of the evolution of the immune system. Our unfortunate initial pre-occupation with the modern 'neo' immune system arose simply from the ease of studying the more accessible components of the blood-stream where recent immune innovations are enriched. Only by returning to the gut from a 'gastro-archeological' approach have we begun to unpick the complexities of the immune system and started to unravel the mechanisms of diseases—not only of the gut itself but of the entire immune system. This journey of discovery has really only begun over the last 25 years and there is much still to learn. However, the tantalising glimpses of understanding gleaned by this approach are so exciting that we now find ourselves on the threshold of a new era in medicine.

Part III

The Gastro-Archeologist

The abbey on the island of Iona

'*The triad of consciousness consists of physics and astronomy first, biology second, archeology and history third. These three contribute to uncovering the many falsehoods that surround this world and have almost drowned it for thousands of years*'

Gilgamesh Nabeel

Introduction

Off the west coast of Scotland lies a remote rocky island. Its highest point is less than 100 m above the sea and its low-lying peat bogs pose little barrier to the onslaught of Atlantic storms. It is just 1 of approximately 150 similar islands in the Hebrides. Even in the summer the temperature rarely rises above 15 °C and it experiences over a foot of rainfall each year. Only about 120 hardy islanders have chosen to live there all year round. To get to it from Glasgow requires a 6 h road trip and at least two boat crossings. The ferry to the isle is frequently cancelled in bad weather. Yet, every year about 150,000 people make the journey to this unlikely destination and many return time and again.

It was as a young man that I first experienced the magic of Iona, this jewel of the Inner Hebrides. A south-westerly storm was blowing and raised an impressive swell in the channel. Waves were crashing against the side of the vessel and splashing down to soak those of us huddled on the open deck.[1] As the ramp lowered onto the slipway we had to time our exit carefully. The boat was repeatedly knocked off station and required skilful manoeuvring to bring it round, and there were precious few seconds between waves sweeping over the landing stage to sprint across to *terra firma*. I could swear that one of the boat's company was roped to the harbour railings to avoid being washed away whilst assisting passengers—but this could be an embellishment added by my memory in the drama of the moment. I struggled through horizontal rain against the deafening gale to reach the island's abbey church. The door slammed shut behind me whilst the storm raged furiously outside and as I stood in total tranquillity dripping silently onto the stone floor I understood instantly why this place was so special to so many.

It was on a later visit (in rather better weather) that I sat atop the highest point of the island looking down at the abbey. The evening glow was painting the granite of neighbouring Mull an even deeper shade of pink but with enough sunlight left for the white sand beaches to turn the sea emerald. Behind me were remnants of Bronze Age burial mounds. At the base of the hill, archeologists have unearthed wooden fragments dating back to Saint

[1] The boat was called the Morvern and was introduced by the ferry company Caledonian Macbrayne ('Calmac') onto the Iona ferry route in 1979. There had been considerable resistance to the idea of a car ferry on this route which had previously been served by a flotilla of small 'red boats' and a local councillor famously asked during a public meeting when Iona would be leaving the 'age of the coracle'. 'Morvern' was shaped a bit like a World War 2 landing craft with space for passengers and five small cars in the central well and a front landing ramp. It served the route for 13 years until replaced in 1992 by the larger Loch Buie. It is currently still running as a car ferry in Eire serving the Bere island route in County Cork.

Columba's arrival on the island in 563 AD—credited with the introduction of Christianity to the British isles.[2] The abbey graveyard has been the chosen resting place of early medieval Scottish, Norwegian and Irish Kings including Duncan and Macbeth (of Shakespeare fame)[3] and the island has been a site of religious pilgrimage throughout the centuries.

It is easy to believe that there is an innate spirituality about Iona that has resonated down the millennia. This feeling is reinforced by a similar experience of the view from a little hilltop in Pembrokeshire called Carn Llidi, similarly adorned with bronze age burial sites and looking down on the cathedral of St David's. One can easily get lost in romantic ideals of Celtic mysticism—and not be the first to do so. But, by applying our modern-day thoughts and feelings to the scant evidence of the past we are guilty of interpreting the archeology to fit with our preconceived ideas.

This is the pathway to fantastical beliefs and dangerous justification of ideologies—that of 'pseudo archeology'.[4] In truth, the astonishing beauty of Iona is paralleled by many of its Hebridean neighbours and it is much more likely that our Bronze Age predecessors were more occupied in mere survival or defence from an elevated position rather than day-dreaming. If we are looking

---

[2] Saint Columba landed on Iona in the year 563 AD. Known as Colmcille (literally 'dove of the church') he is one of the three patron saints of Ireland. A well-educated Irish monk, he took voluntary exile across the Irish Sea after an argument over a manuscript that he had copied led to a pitched battle (before copyright law or photocopiers existed!). He travelled widely in Scotland on his mission to convert the indigenous Picts amongst whom he was well regarded as a peace-broker. There is a legend that on one of his journeys he encountered the Loch Ness monster! His successor on Iona, Adamnan, noted that he worked in a hermit cell built on a rocky outcrop on the island. In 1957 Charles Thomas, a Cornish archeologist, identified a possible location of this hut and unearthed fragments of hazel charcoal near a hole possibly used for insertion of a Christian cross. The finds were not carbon dated until 2017 (he had kept them in his garage in Truro!) when they were found to come from a period between 540 and 650 AD which would be contemporaneous with Columba's life on the island.

[3] Macbeth (c 1005–1057) was King of Alba from 1040 when he defeated Duncan I in battle and was in turn overthrown and beheaded by supporters of Malcolm III and subsequently buried on Iona. His life bore little resemblance to the Shakespearean drama and his rule was largely peaceful. He was even considered to be a kind and generous soul—on a visit to Rome in 1050 he distributed coins amongst the poor. It may be that the alternative depiction of him as a tyrant may have something to do with the fact that Malcolm III was a relative of James VI of Scotland and I of England who was on the throne at the time of Shakespeare's writing.

[4] Examples of pseudo-archeology include theories of extra-terrestrial species visiting Earth and endowing early cultures with technological advances (for example Erich von Daniken's '*Chariots of the Gods?*'). More invidiously the German Nazi party tried to use archeological finds to support ideologies of Aryan supremacy. An organisation called the Ahnenerbe (from '*Deutsches Ahnenerbe-studiengesellschaft fur Geistesurgeschichte*' = German Ancestry-Research Society for Ancient Intellectual History) was affiliated to the SS and headed by the police chief, Heinrich Himmler. The Ahnenerbe sponsored expeditions to try to find evidence of prehistoric Nordic influences as far afield as the Bolivian Andes and made a number of absurd and fantastical claims about the origins and unique capabilities of the German people. The roots of such ideology may perhaps be seen in the devastation of German national pride after the First World War. The Ahnenerbe was more recently the inspiration for Indiana Jones' antagonists in the 'Raiders of the Lost Ark' film series!

for historical echoes, then the old radar station and Lewis gun platform on Carn Llidi may be a more realistic construct.

Archeological method has similarly adapted in an attempt to eliminate romantic idealism. The 'new archeology' that materialised in the post-war years replaced the mere collection and categorisation of human artefacts with a 'processualist' approach to understanding more about the societies that left them, based on objectivity and fact. However, it inevitably became entwined with studies of social anthropology, cultural evolution and even politics and ideology. More recent schools of thought have shed doubt on whether we truly have the ability to remove bias and have highlighted how the cultural perceptions and beliefs of the individual archeologist undoubtedly alter our interpretation of the past.

So it is perhaps with our perceptions of 'immunity'. The word 'immune' originated as a legal term in Roman times for protection under the law—effectively freedom from prosecution. Its initial scientific use in the late 1800s similarly implied exclusion, but from the effects of infection. We have traditionally viewed it as the noble struggle of species to survive in a hostile World. Undoubtedly this perspective has been influenced by our recent past when infectious disease was responsible for the majority of human death and misery. Latterly with our improved understanding of the mechanisms at play we have come to envisage the immune system almost militaristically as representing the defences of the host against the battering of hordes of invasive microbes. Such a viewpoint has inevitably coloured our interpretation to the extent that we have even come to label bacteria as 'good' or 'bad', and irrationally almost endowed infectious organisms with a nefarious intent to cause harm.

However, from the new perspective we have gained by following its evolution we have created an entirely different standpoint from which 'immunity' can be interpreted. In the context of the past we now see how 'immunity' originates from the rule book that governs the interactions of different cells in a multicellular organism. Effectively immunity serves as a control of individual cellular advantage for the benefit of the organism—putting the team before the player. The extraordinary leap that took place in the evolution of single-celled organisms into co-operative collections of different cell types relied on an ability to constrain and control their relative growth, to co-ordinate their position and functions and to maintain the integrity of the whole organism. These are the 'housekeeping' roles of the immune system that comprise its fundamental nature and purpose.

However, as we have seen, the advent of multicellularity required more than just a way of bringing about social cohesion, but thanks to the limitations of size (the 'square-cube law'), a functioning gut as well. The immune system and the gastrointestinal tract are therefore evolutionary twins that appeared on life's stage together. Given that they grew up together, it is no surprise that the gut failed to distinguish the organisms that came to live within it from the constituent cells of itself. The rules that bound the organism together would also encompass the microbes inside it—they would be covered by the 'umbrella' of 'immunity' (in the Roman sense of the word as freedom from prose- or perse-cution). As a result, one could envisage symbiosis not as a deliberate state of co-operation between organisms but as an extension of and the inevitable consequence of multicellularity. It follows that 'immunity' is about collaboration and not conflict. Infection is effectively therefore about the 'cheats' in the system, just as tumours are cells trying to buck the 'house rules' of the multicellular animal—the immune system oversees and regulates both.

There is a further complication for symbiotic microbes—that as a result of living *within* the gut of multicellular organisms, they become a potential source of food for the host. For obvious reasons, an ability to avoid being inadvertently eaten has been necessary since the evolution of sexual reproduction in single-celled creatures. The origins of the immune system were therefore not just to distinguish 'friend' from 'foe', but 'friend' from 'food'. Indeed, in the very beginning, at the time of single-celled animals, the 'immune system' was simply a means of ensuring 'immunity' from being eaten by your mate before you had reproduced (after sex, the rules change!). Eating, sexual reproduction and immunity are therefore intertwined throughout evolution. Microbes that evolved evasive mechanisms to avoid becoming supper recklessly caused damage to the host and as a result *friend* that mistakenly became *food* now turned into *foe*. We have come to recognise this as 'disease' and our modern concept of the 'immune system' became limited to the host response to it. We can now see that the much bigger picture of 'immunity' (in the original sense of the word) embraces the peaceful cohabitation of cells making up the organism along with its long-term passengers. When this system breaks down, it also causes 'disease' initiated by the host as we shall now see.

7

What Is Food to One Is Rank Poison to another'

Food Allergy and Intolerance

Summary In which we explore the mechanisms underlying food allergy and intolerance. We start by asking why food allergies are not more common than they are and how the gut maintains a tolerance to ingested proteins. We then begin to dissect the layers of gastro-archeology, by examining the superficial layers of most recently evolved mechanisms. Here we find specialised IgE type antibodies against food that lead to familiar allergic responses through the release of chemicals such as histamine. However, on deeper excavation we find that allergic type responses can be generated even in the absence of T and B cells, whose responses are regulated by innate lymphoid cells. Digging lower and further back in time we see how the epithelium itself and bacteria in the gut may affect allergic responses and how modern food preparation may contribute. We consider desensitisation approaches and the drawbacks of oral vaccination. Finally, we discuss food intolerances as possible 'partial' food allergies that lack all of the layers of the complete response.

From 'De Rerum Natura' (= 'on the nature of things') by the Roman poet and philosopher known as Lucretius (98-55 BC), his only remaining work. From the Epicurean school of thought, Lucretius promulgated materialism in opposition to spiritualism. It contains one of the first suggestions of the existence of food allergy in historical literature.

J. Woodward, *The Gastro-Archeologist*, https://doi.org/10.1007/978-3-030-62621-1_7

Nuts...

'The highest result of education is tolerance'
Helen Keller[1]

The Hair of the Dog

In the first century BC, the king of Pontus, a long-forgotten country in modern day Turkey, was assassinated as a result of eating poisoned food at a banquet. His oldest son, who shared the name of Mithridates, was too young to rule and therefore the widowed Queen Laodice assumed the throne as Regent. She also made no secret of the fact that she favoured her second son for the throne. When he fell ill with a chronic abdominal complaint, the young Mithridates began to suspect that he was being poisoned by his mother, and furthermore, that she may have been behind the plot to kill his father. He escaped the court and lived as a recluse in the Cappadocian wilderness. Whilst there Mithridates deliberately sought to build up his resistance to toxins and poisons of all sorts by taking small amounts of them on a regular basis. He even developed a concoction of herbs, bark, root, flowers and resins of around 50 different plant species which also contained small quantities of poisons and the flesh of vipers. This 'cure-all' came to be known after him as 'Mithridate'. Its recipe was stolen by the Romans and reinvented by Andromachus, Emperor Nero's physician, as '*Theriacum Andromachi*'. 'Theriac' became highly regarded as a medicine, possibly as a result of the

[1] Helen Keller (1880–1968) became blind and deaf following a childhood illness—probably meningitis—at the age of 19 months. With the help of her tutor and subsequent companion of 49 years Anne Sullivan, she learnt the names for objects by feeling them in one hand whilst their names were spelt on the palm of the other. She was able to 'lip read' by touching people's mouths as they spoke and to listen to music through a resonant table top. She became a renowned speaker and writer and an active champion for the rights of sensory disabled people.

number of rare and exotic ingredients which made it accessible to only the wealthiest. The physician Aelius Galen wrote a whole book about it and Roman Emperors supposedly took a daily draft of it. In the middle ages, the Venetians, through their maritime trade, developed a monopoly on the sale of Theriac throughout Europe and it became known as 'Venetian treacle'. The market was only opened up by the publication of its recipe in a French pharmacopeia in 1669—an early example of the free dissemination of medical knowledge! It is said that Oliver Cromwell took copious quantities of theriac to try to prevent the plague but found that as a side effect it cured his acne! Theriac (mithridate) was sold in pharmacies as recently as 1884.

The concept of using small quantities of toxins to try to build up resistance or even cure illness was a pervasive strand throughout medical history from the time of Hippocrates. One example is the 'hair of the dog'—a popular belief that placing a hair of the dog that bit you into the bite wound would prevent the development of rabies. This expression has lived on in the use of alcohol taken the morning after excessive consumption in order to cure a hangover! However, the idea of 'a little of what hurts you does you good' reached its greatest prominence in the art of homeopathy,[2] where poisonous substances are diluted to infinitesimal concentrations and sold as treatments—despite lacking any scientific basis or proof of beneficial effect.

Yet, in what seems evident folly is often found a nugget of wisdom. As we will find out in this chapter, over two thousand years after Mithridates, we are beginning to apply the same principle in the treatment and prevention of conditions where common foods can seemingly act as a poison.

At this point you may rightly ask, what came of young Mithridates whom we left in the wilderness building up his resistance to poison? I am afraid I shall have to leave you in suspense but I promise to tell you later….!

[2]The alternative medicine branch of 'Homeopathy was invented by Samuel Hahnemann (1755—1843) in his book 'Organon of the healing art" in 1810. He was apparently convinced by the ancient idea of '*similia similibus curentor*' (like cures like) by suffering malaria like symptoms when taking the then (and current) cure—cinchona bark, which contains quinine. The basic premise is that substances thought to be associated with a condition are diluted and the homeopath believes that they become more efficacious as a cure the higher the dilution. The art originally also required 'succussing' the solution or tapping the container vigorously on an elastic surface. The 'centesimal' scale represents a 100-fold dilution, hence a 2C dilution is a substance diluted 10,000 (100 x 100) times. It can easily be estimated that most preparations will not contain a single molecule of the supposed active substance—a dilution of 13C would be the equivalent of one drop in all the water on earth, and 40C would provide the equivalent of one molecule in all the molecules present in the entire known universe! Homeopathic remedies were sold at dilutions of up to 200C….! However not all homeopathic remedies were diluted to such an extent—one was sold at a 1/10 concentration and was sufficient to cause toxic side effects. Numerous scientific trials have failed to demonstrate any benefits of homeopathy beyond placebo and it has been recognised as ineffective and potentially harmful (through occasional toxicity, but also harmful misleading of vulnerable patients away from more effective remedies). Astonishingly this branch of 'nonsense' medicine (as it was described by the UK government Chief Scientific Adviser in 2013) was still supported and provided by the UK's National Health Service until as recently as 2017.

Rachel's Story

Rachel was 2 years old when she was first given a peanut butter sandwich by her father. Almost immediately she started crying and rubbing her eyes. Her parents noticed a raised rash that they recognised as 'hives' on her skin which she started scratching. It settled down after about half an hour but they decided to avoid giving her peanuts again. They knew that she had a tendency to allergy—she had developed an itchy rash in the front of her elbows and behind her knees that the doctor had diagnosed as childhood eczema a year earlier.

As Rachel grew up she knew to avoid food containing peanuts. Her school was excellent at keeping nuts out of the school lunch menu and her friends were very understanding around her. She came to know which cafes and restaurants could be trusted to be peanut-free when they said they were.

At the age of 19 Rachel was out shopping with her boyfriend, Greg and stopped at a coffee shop that she had been to many times before. She ate a slice of cake that had no nuts listed in its ingredients. Twenty minutes later she was trying on a dress in a shop and Greg was waiting outside the changing area. A woman came rushing out in a fluster and asked if he knew the young woman inside. He dashed in to find Rachel lying on the floor. Her eyes were open and she looked panicked but was unable to talk, and was gasping for breath. Her face was swollen and red and she was making loud noises when she breathed in or out. Greg shouted to the assistant to call for an ambulance just as Rachel passed out.

When the ambulance crew arrived they gave chest compressions and managed to insert an airway before rushing her to hospital. In the emergency department they were able to establish a heart rhythm and put her on a ventilator machine. Sadly, when they tried to wake her up after 2 days it was clear that she had sustained severe brain damage and was never going to be able to live independently of the machines in the intensive care unit. Rachel's parents and Greg had to make the difficult decision to switch off her life support.

However, that is not the end of her story. Rachel always kept a signed donor card in her purse and had made it clear that she would wish to donate her organs if she were to die. Her family respected her wishes after death and as a result, her own tragedy allowed her to save three other lives. One of these people, whose kidney transplant allowed him to be freed from the need for dialysis, was eating in a restaurant during the week after he left hospital. He felt tingling in his lips and his tongue felt strange. He noticed that he was wheezing, having never experienced asthma in his life and he developed a skin rash. It all passed quickly but was noticeable enough for him to mention to his doctors on his next clinic review. When they tested his blood they found that he had specific IgE antibodies against peanuts and advised him to avoid all nuts for a period of 6 months or so before cautiously reintroducing them under medical supervision. They also gave him an 'Epipen' to inject himself with adrenaline and trained him how to use it in an emergency. The doctors also contacted the teams looking after the other two organ recipients who were then checked and found not to have acquired a food allergy in the same way.

Allergy to Food

Food allergy is commonly reported and appears to be increasing in communities following a 'western' lifestyle. However, not all symptoms related to food are caused by allergy, which is the name given to a particular type of immune response. Allergy is characterised by the formation of a specific class of antibodies—IgE—that identify a specific part of a foreign protein or 'allergen'. As we saw in part 2, IgE antibodies stick to the surface of specialised 'mast' cells and trigger them to shed their granules. This releases chemicals such as histamine that are responsible for many of the symptoms that are due to allergies (and also why they are frequently relieved by the use of 'anti-histamine' drugs). Allergies are therefore caused by the antibody side of the immune response rather than the cell killing side.

Food allergies are most common in childhood. Approximately a third of parents think that their children are allergic to specific foods, but in fact it turns out that only about 1 in 12 1 year olds in the USA are affected. The commonest culprits are cow's milk and hen's eggs, whilst peanut and tree nut allergies affect only about 1 in 100 children. The majority of reactions are not severe and may just cause itchy eyes and a runny nose. More severe reactions lead to difficulty breathing as a result of swelling of the tongue and narrowing of the airways leading to 'wheezing' or asthma. The most severe form called 'anaphylaxis' is where the blood pressure plummets and the circulation fails—which can lead to sudden death, and occurs in about 10 people a year in the UK. Effects on the gut itself can include painful spasms, diarrhoea or vomiting—the 'weep and sweep' response that we saw in chap. 6.

Most children effectively 'grow out' of food allergies—particularly to milk or eggs—by their late teens. However, the same is not true of nut allergies which only about 1 in 5 children lose over time. Adults can also develop food allergies later in life. What is known as 'oral allergy syndrome' causes tingling in the mouth, tongue and lips but without significant swelling or dangerous symptoms. It commonly first occurs during young adulthood and affects as many as 5% of the population, particularly those that already have other allergies for instance to pollen. Seafood and nuts are the commonest culprits of significant food allergies in adults.

The Umbrella of Immunity

It is extraordinary in many ways that stories like Rachel's are not actually more common. The vertebrate immune system, thanks to the 'immunological Big bang' has evolved to recognise and respond to any new protein pattern that it has not seen before. In this way it allows the organism to inhabit different environments and face new challenges without the need for inter-generational evolution of new 'hard-wired' immune molecules. It should by rights therefore identify any new food introduced into the diet as being 'alien' and a potential threat, just as if it were for instance a disease-causing bacterium. This would of course be a Bad Idea for lots of reasons! The immune system very sensibly appears to turn something of a 'blind eye' to anything it needs—notably food or friends. In the introduction to this part we described this as the 'umbrella' of immunity that is extended beyond the confines of the constituent cells of the organism itself to include symbiotic partners as well as food stuffs.

In the gut, digestion *should* make this a little easier. The immune system generally recognises protein fragments about 12–15 amino acids long (within the groove of the MHC molecules that 'present' to the T cell), and the potent digestive enzymes in the upper intestine quickly demolish any protein to much smaller lengths that are not of interest. In theory, therefore there should not be any protein patterns capable of recognition by the gut immune system. Furthermore, the lining of the gut should also act by 'containment' to keep out elements of the outside and isolate proteins away from the inside of the organism, where the immune system might detect them. In reality however, neither of these appear to be the case. The gut is in fact quite 'leaky' and it is possible to detect quite large food proteins in our blood stream after eating. Indeed, most of us even make antibodies to them.[3] How is it possible then that the immune system mounts a response of some kind to food in most of us but only in a few unfortunate people such as Rachel does it cause a problem?

The Tolerant Gut

The intestine is actually buzzing with frenetic immune activity all the time—even in our hygienic modern world, where dangerous food-borne infections are only rarely encountered. This background bustle is the price paid for a

[3] Immunoglobulin G antibodies to food are very commonly found in the blood stream in humans but their clinical relevance is questionable. The presence of IgG antibodies shows that some food proteins can get into the blood stream across the gut where an immune response can be generated and it is likely that their presence merely represents exposure to the protein but does not have any significance in causing symptoms. As is often the case, assumptions are made and put into commercial practice despite the lack of scientific evidence. There are companies that sell IgG antibody blood tests against food, claiming that they can guide elimination diets to treat a variety of different conditions or symptoms. Antibodies to cow's milk protein and wheat are commonly found in most individuals, and these two foods are also commonly implicated in abdominal symptoms for reasons not associated with the immune system. Therefore, any benefit to individuals in excluding these foods is unlikely to be related to the detection of IgG antibodies in their blood. This is perhaps a case of 'cause' and 'association' being conflated at the cost of the unwitting for commercial gain.

responsive defensive system capable of rapid reaction to unwanted invaders. In essence, the gut immune system automatically responds to everything it detects—from food proteins to 'friendly' bacteria as well as disease-causing organisms—but it vigorously and energetically suppresses the majority of this activity. In other words it is an *actively* tolerant environment rather than just simply turning a blind eye to things that are not a threat. When a potentially nasty infection comes along, the mechanisms can be lifted to unleash the underlying defensive response. In this way the gut immune system appears to be permanently primed, poised and ready to pounce. A possible analogy might be an automatic ground-to-air missile defence system that 'locks on' to every aircraft including commercial airliners but is inhibited from firing on them by recognising uniquely identifying radio transmissions. The gut rarely fires the first shot—the immune suppression is usually only lifted as a result of harm to the gut itself.

The cost of this actively suppressed immune response is staggering—several grams of IgA antibodies (the form generally specific to the gut and recognising both friendly and aggressive bacteria) are produced every day and around a trillion lymphocytes (approximately three quarters of all of our lymphocytes) comprise the body's 'border force' in the gut. To put this into context, there are about ten gut lymphocytes for every brain cell that we have. Clearly there must be a good reason for the gut to invest so much of the body's energy and cellular resources in what appears to be a rather pointless and circular exercise. We have already seen how what we call 'immunity' is in reality and above all a 'housekeeping' service. In fact, it is likely that the majority of the hectic activity of the immune cells in the gut in the non-threatened state is directed at promoting positive relationships with the bacteria that live within us, as well as maintaining the epithelial barrier and patching up any leaks. We might see the high cost of maintaining a very active immune system in the gut as the price paid for the necessary breach in the 'containment' of the organism that is required to bring nutrients into the body. The activity of the immune system that we notice—such as when it causes symptoms through combatting potentially harmful infections—is really just the visible small part of the iceberg: for the most part its activity goes effectively unnoticed.

A Tale of Two Immune Systems

The immune 'flavour' of the gut is therefore one of tolerance. It is likely that something new will be accepted as part of the organism itself under the umbrella of immunity rather than automatically treated with suspicion. This is not the case for the immune system elsewhere in the body (with the exception of pregnancy). Take the example of a protein that we have never encountered before. If we eat it, then we 'tolerate' it (even though we may still mount a kind of 'immune' response against it), and nothing bad happens. However, if we first encounter that protein via a different route—through the skin or into the blood stream—then we are likely to mount a severe reaction to it that is more likely to cause harm to ourselves as a result. The really amazing thing is that if the protein is first ingested and *then* we are subsequently exposed to it through the skin or the blood-stream, we have already been 'tolerised' to that particular protein and the immune response is suppressed—not just in the gut but in the rest of the body. In Rachel's case, for some reason this 'oral tolerance' broke down with regard to specific proteins associated with peanuts.

It is therefore as if we have two separate but co-existing immune systems that react differently to challenges even though they share the same mechanisms and types of cell. One immune system in the gut has a wider role that extends its 'umbrella' of protection over symbiotic bacteria and foods, whilst the other ('systemic') immune system in the rest of the body is much more likely to respond aggressively to potential threats. One can perhaps relate this to the efficacy of 'containment' in that the gut is exposed to the environment and leaky, whereas the skin is relatively impervious and presents a more robust barrier and the blood stream itself is entirely enclosed.

Food or Friend?

The gut immune system also behaves differently towards food proteins and bacteria—the tolerance induced to food proteins in the intestine extends to the whole organism, whereas that to bacteria appears limited only to the gut itself. Studying the separate mechanisms of tolerance to food and bacteria is challenged by the fact that feeding alters the bacterial composition of the gut—and both affect the development of the gut immune system itself which only matures after birth! Nevertheless, mice can be reared experimentally on a diet that avoids any protein fragments that are large enough to be recognised by the immune system, or similarly in a 'germ free' environment that prevents them from developing the normal bowel bacteria. Animals that are raised free of food proteins have few T lymphocytes in the small intestine but normal amounts in the colon, whereas those that are raised in a germ-free environment have reduced T cells in the colon but not the small intestine. Such experiments suggest that the immunity of the small intestine (where we digest and absorb our food) and the colon (where we house the majority of our symbiotic bacteria) also behave differently as two separate systems. The small intestine is the predominant site where tolerance of food occurs and the colon is largely responsible for tolerance of bacteria. However, bacteria can also be present in the small intestine (in much smaller numbers) and the tolerance mechanisms of the small intestine differentiate soluble molecules from particulates such as micro-organisms.

Diverticulum #7.1: A Fishy Tale

Whilst idiosyncratic reactions to food were described as long ago as the fourth century BC by Hippocrates, the first clear and specific recorded instances of food allergies date from the 16th and 17th centuries AD. Sir Thomas More in his *History of Richard III* describes an itchy rash developing after eating strawberries, and the Belgian physician Jan-Baptiste van Helmont (who claimed to have chosen to study medicine following a visitation from an angel) described asthma attacks after eating fish in 1662. The term 'allergy' (from the Greek 'allos' = other and 'ergon' = reaction) was coined by von Pirquet and Bela Schick in 1905. The landmark experiment that ignited the scientific study of allergies came in 1921 when Heinz Kustner was working as an associate with Carl Otto Prausnitz in Germany. Kustner had an unusual allergy to fish that only occurred on eating cooked, but not raw fish. Prausnitz took a blood sample from his colleague and injected a small amount of his serum into a patch of skin on his arm. The following day he injected a tiny quantity of an extract of boiled fish into the same spot and developed the typical 'wheal and flare' dermal response of urticaria. Injecting the fish extract into a different patch of skin brought about no reaction. Thus, a component of Kustner's serum—which we now know was IgE antibodies recognising the fish protein to which he was sensitised—was responsible for the reaction.

The 'Prausnitz-Kustner' test remained a standard procedure for investigating food allergies thereafter—injecting a healthy volunteer with the patient's serum followed by a test substance. For obvious reasons this is no longer considered ethical, but 'skinprick tests' have become an essential part of allergy testing. The patient is injected with a small amount of the test substance (along with positive and negative 'controls' of histamine and saline respectively) and will similarly mount a local reaction. There is very little risk of developing a major 'anaphylactic' response to the substance injected.

Dr. Jeckyll and Mr. Hyde

Oral tolerance was only experimentally demonstrated for the first time in humans as recently as 1994, and much remains to be discovered about how it works. The principal mechanism is akin to the way in which the gut immune system detects the protein 'signatures' of harmful invaders. As we saw in chap. 6, the professional dendritic cell picks up the protein fragments as a result of thin root-like processes that it extends through the epithelium to sample the contents of the gut. It may also pick up proteins transported through the passages in the mucus-secreting goblet cells. The dendritic cell then pulls in its extensions and migrates in the tissue fluid to the lymph node where it instructs T-lymphocytes. The only difference is that in the case of tolerance, the dendritic cell secretes chemical messengers that turn the T cell into a regulatory cell that suppresses the immune response rather than a 'helper' cell that stimulates it. The dendritic presenting cells were told how to instruct the T lymphocytes by the scaffolding myofibroblasts underlying the epithelium. One of the key molecules involved in this 'tolerising' pathway is retinoic acid (derived from vitamin A), which also gives the lymphocytes the homing tags that direct them back to the gut from the circulation.

The gut immune system can therefore switch between a predominantly tolerant or the completely opposite aggressive approach towards foreign proteins—just like Dr. Jekyll and Mr. Hyde.[4] It seems that in individuals such as Rachel who mount an inappropriate immune response to food, there is a defect in the switch mechanism that allows certain food proteins, but not all, to be recognised as potential threats whereupon they trigger an allergic reaction. To understand why oral tolerance breaks down, only in some people, and increasingly in 'westernised' societies we need to start digging through the layers of gastro-archeology. Inevitably the keys to unlocking the secrets of tolerance and its pitfalls will be found in the deepest layers (the 'paleo' and 'arche' gut immune systems) as an ability to accept food rather than react to it would seem to be a fundamental property of the gut immune system. Time to start digging our trench!

Brushing Away the Topsoil

A few years ago, our understanding of food allergy went no deeper than the surface layer—from after the immunological 'Big Bang' that allowed antibodies to recognise an almost infinite array of new molecules. We knew that it was a specific class of antibody—IgE—that was involved. At its variable end IgE identifies the culprit food protein, whilst at the other end, it has a 'plug-in'

[4] *'The strange case of Dr Jeckyll and Mr Hyde'* was a novella written by Robert Louis Stevenson in 1886 whilst convalescing from illness in Bournemouth. In the story, the benign Dr. Henry Jeckyll becomes increasingly unable to prevent his transformation into his murderous *alter ego*, Mr. Edward Hyde, despite concocting a potion that initially works to good effect, and takes his own life. The character is thought to be based on Stevenson's friend Eugene Chantrelle, a French teacher in Edinburgh who murdered his own wife by poisoning her.

module that allows it to stick to mast cells. These are the cells that then release their granules full of stored chemicals to initiate an inflammatory response—increasing the blood flow to the gut, attracting immune cells from the blood circulation, stimulating nerve cells and increasing the leakiness of the epithelial lining. The response of food allergy in the gut is therefore like the 'weep and sweep' response to worms.

Being 'allergic' to something requires an initial 'sensitisation' whereby the offending molecule is identified and IgE antibodies are produced that are capable of binding to it. Mast cell numbers are increased by sensitisation and their surface receptors become loaded with the IgE antibodies with their variable peptide-binding ends pointing outwards. In this way they are primed and ready to react as soon as exposure to the protein re-occurs. In the case of food allergies, the initial sensitisation may occur at any surface—skin, lungs or gut lining. When the food is then eaten, there are primed mast cells already present in the gut wall and coated with IgE antibodies ready to react. However, any food protein that gets into the blood stream through the gut can then trigger mast cells at other sites, hence food allergy often leads to symptoms outside the gut such as itching skin, 'wheezing' from airways constriction or in extreme cases such as Rachel's, anaphylactic collapse (Fig. 7.1).

Because the natural response of the gut is one of 'tolerance' there is growing evidence that whilst it often occurs in the gut, the site of initial sensitisation may actually be away from the intestine in many cases. For instance, there is

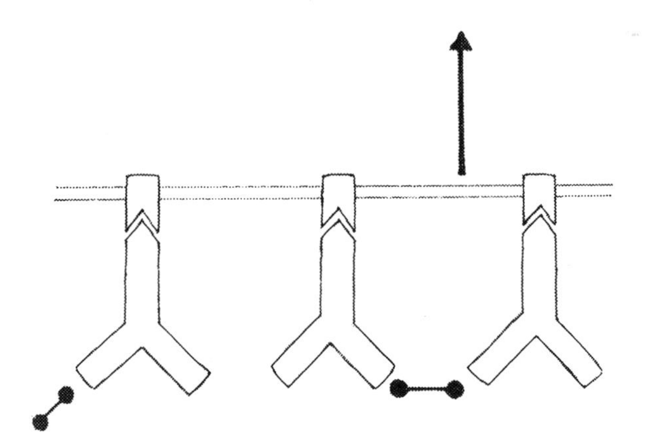

Fig. 7.1 Activation of mast cells by immunoglobulin E. The cell membrane of the mast cell carries receptors that bind immunoglobulin E (IgE) antibodies which are formed during initial sensitisation reactions—and can persist attached to mast cells for months or years. The mast cell is activated when adjacent IgE molecules are connected by the protein that they recognise as shown on the right. Allergens are therefore commonly found to contain repetitive sequences that can bind more than one antibody

a very strong link between the childhood skin rash, eczema, and the subsequent development of food allergies. It would appear that the chronic skin damage allows the sensitising protein access through the skin in order to prime the allergy. Once again, the principle of 'containment' is at play—a specific protein called 'filaggrin' appears to be important in maintaining the barrier function of the skin. Those individuals with a genetic mutation in filaggrin that prevents it from working properly have a high risk of developing eczema in the skin—and their risk of food allergy is increased by as much as 30-fold.

The increasing use of skin preparations—including those containing peanut oil and ironically used to treat eczema—can thereby lead to the priming sensitisation reaction leading to food allergy when the substance is subsequently eaten. This maybe the route that ultimately led to Rachel's tragic situation during her shopping trip. Along these lines, we might also consider that modern soaps and detergents could affect the integrity of skin barrier, and perhaps even the leakiness of the gut, if we fail to rinse off the washing up liquid having cleaned the dishes!

Thus, in the very uppermost layer of the gut's immune history we have already begun to construct a plausible mechanism of how allergies to food might come about. Indeed, we were once satisfied with this superficial explanation as being the entire story. However, it leaves many questions unanswered—why, for instance does the gut immune system mount an allergic response rather than its usual tolerant one? Why does the reaction involve the antibody arm of the immune system rather than the cell-killing response? Why is it that the antibody response is of the unusual IgE type rather than the other classes such as IgG? Why are only some foods implicated and only some individuals affected?

It is time to get the trowel out and dig a little deeper.

A Surprise Finding Unearthed

Actually, immunologists have many tools for answering these questions, of which a trowel is not one! The equivalent tool in this case is the use of gene-editing technology. This allows the possibility of silencing (or 'knocking out') one or more genes from a laboratory animal. Originally this could only be done for the whole organism with the result in many cases that the animal failed to survive early developmental stages. Newer techniques allow the capability of knocking genes out of only some specific cell types within the animal, or doing so conditionally so that the gene can be silenced at a particular time.

This sometimes enables the later functions of genes to be identified even if they are lethal when mutated in the early embryo.

The top layer of our gastro-archeological dig—the antibody producing (B-cell) and cell killing (T cell) immune responses—can be neatly removed by deleting the gene for RAG. You will remember that RAG is the protein that rearranges the DNA that codes for antibodies and T-cell receptors in order to generate the enormous variability in shape detection in the vertebrate adaptive immune system. Knocking out RAG is not a lethal mutation as RAG's role is restricted specifically to the lymphocytes and for the one purpose of generating diversity in their receptors. In 'RAG knockout' mice, the clock is therefore effectively turned back to a time before the immunological 'Big-Bang'—when *Branchiostoma*-like creatures were at the pinnacle of evolution and entirely dependent on hardwired pattern-recognition detectors.

The big surprise—given that we thought that allergy was dependent on the triggering of mast cells by IgE antibodies—is that it is actually possible to induce allergy-like responses in the RAG knockout mouse, which has no antibodies of any kind. Of course, such responses are not directed to the particular protein or sensitising substance as this could only be done through the use of the specific antibody receptor. However, the physical effects are the same, and suggest that all the mechanisms required to mount the response were present before antibodies came along and fine-tuned it to specific molecular signatures.

The Next Layer

The cells responsible for co-ordinating the allergic response lie in the next layer of our trench. These are the cells that resemble lymphocytes but lack the highly variable cell surface molecules of B and T cells. They are the recently discovered innate lymphoid cells or 'ILCs'. Despite lacking surface antibodies and T cell receptors, they behave in many ways like their more sophisticated modern counterparts, even to the extent of being divided by the 'great schism' between the two main arms of the immune system. In the case of allergies, it is the specific innate lymphoid subset called type 2 ILCs that underpin the response. We came across these cells in chap. 6 as they are involved intimately with the unusual 'tuft' cells of the intestinal lining in protecting against worms.

In the gut, ILCs lie underneath the epithelium in the lamina propria. They act as a form of communications hub that integrates signals from different sources. When activated (by chemical messages from nearby cells) they in turn secrete interleukins to signal to other cell types in order to bring about the allergic response. Importantly they stimulate mast cells to proliferate and to shed their granules. They also signal to other immune cells in the circulation to attract them to the site. Chemicals from ILC2s similarly initiate the whole process of antibody production by B cells—through activating and instructing the professional dendritic presenting cells to orchestrate the process. ILC2s additionally have surface molecules that can connect directly with T cells to activate them, and they even present foreign proteins to lymphocytes on MHC molecules, behaving much like dendritic cells. In this way the response can be directed to individual protein shapes.

We have seen how the allergic process can still occur (to a degree) with complete loss of B and T cells, achieved by deleting the RAG gene. However, knocking out a gene crucial to the development of type 2 ILCs almost completely abolishes the possibility of such a reaction. These cells are therefore clearly fundamental to the response.

If ILC2s are the driving force behind the B and T cells that leads to IgE antibody formation against the proteins responsible for sensitisation and allergy, then what is it that triggers the ILCs? In the case of food allergy, where does the food itself fit in to this picture?

The Depths of Time

Deeper in our trench we reach the really ancient layers of the gut—the epithelial cells of the gut lining itself. It transpires that these cells can themselves direct the behaviour of the gut immune system. We have already seen how the tuft cells effectively taste the presence of worms and secrete a chemical messenger in response, to trigger the immune response. Similarly, the absorptive cells of the gut lining can respond to threats and also send out signalling molecules in order to direct a specific type of immune reaction. The chemicals they produce that lead to food allergic responses are called alarmins, which are usually released from the cell when its membrane is breached as a result of damage. The alarmin molecules therefore act as a signal that something is attacking the epithelial cells and could enter the organism through the gut lining. However, there is now increasing evidence to suggest that some can also be secreted in response to a variety of non-lethal stressors such as stretch or the mere presence of chemical toxins. A particularly important such alarmin—called HMGB1[5]—is usually present in the cell's nucleus but when it is

[5] HMGB1 stands for 'High Mobility Group Box -1' protein. I do not feel the need to explain why as it is not a particularly interesting explanation! However, the key chemical signals involved in signalling from

released it signals to nearby cells through their surface receptors. One of these is actually a Toll-like receptor—one of the earliest pattern recognition molecules which we first came across in the sponge. The epithelial lining cell has therefore effectively produced a molecular mimic of a protein pattern associated with bacteria, in order to switch on an immune response that is normally associated with them, even in the absence of any such organisms. HMGB1 can also be picked up on another receptor, aptly called 'RAGE'—the reason for which we will come across shortly. The knock-on effects of HMGB1 docking with its cell surface receptors includes the secretion of other active messaging molecules, some of which act directly to activate both ILC2s and mast cells themselves.

Hence, we have uncovered a pathway that links damage to the epithelial lining cells with the allergic response through the ILC2s, leading to antibody production by B cells and mast cell degranulation to cause the symptoms. This is easily understood when we consider the immune response to worm infection, where the invading organisms can clearly cause injury to the gut. However, food allergy—which closely resembles the immune response to worms—is more difficult to grasp as the relatively small number of foods associated with allergy are not commonly thought of as being toxic or damaging, except perhaps by those who are already sensitised to them. Given that initial harm appears to be critical in triggering the whole process, we need to take a closer look at the foods that cause allergy, to see if we can find any evidence to suggest that they are in fact not as harmless as we think....

A 'False Alarm'

It is the protein part of the food that is recognised by the immune system, and specific proteins in culprit foods have been identified and studied. In peanut allergy, for instance, there are 11 proteins[6] that have been identified as capable of inducing allergy. They are named after the species (*Arachis hypogaea*) as Ara h 1 to Ara h 11 of which Ara h2 and Ara h6 are the most potent. Most plant proteins that cause allergy are in related families, and most function as storage proteins in seeds, to provide the nutrients for the seed to germinate. Animal

the epithelial cells are called interleukin 33 (which is related to interleukin 1), thymic stromal lymphopoeitin (or TSLP) and interleukin 25 (secreted by the tuft cells). Interleukin 33 in particular is relevant as it is produced as a result of HMGB1 activating the RAGE receptors, and is shortened by the action of proteases to its active form.

[6] Actually (as purists will be keen to point out), there are 12. They are still numbered Ara h 1–11 as the 12th is called Ara h 3b

proteins that cause allergies are rather more diverse in their function, but many of those found in seafood, for instance, are contractile proteins associated with muscles.

All of the food-derived allergic proteins that have been identified to date (numbering around 700) belong to a small number (less than 2%) of different types (or 'families') of proteins, yet there are few similarities between them. Most are resistant to digestion in the gut—which means that they remain large enough to be identified by the immune system. Many—particularly the seafood muscle proteins—have repetitive sequences that mean that they can bind to more than one IgE antibody, and this 'crosslinking' of the antibodies on the surface of the mast cells is what leads to their degranulation. There is little evidence of any allergy-causing foods directly damaging the gut itself. However, some proteins that cause allergy (particularly in the case of the house dust mite which causes asthma) belong to a class of enzymes that digest other proteins and are called 'proteases'. Given that worms secrete such proteases in order to break down the barrier of the gut wall and invade, it is possible that similar proteins could directly cause damage to the intestine, which might sound the alarm to the immune system. The damage may not be to the cells themselves but to the junctions between them, to make the gut 'leaky'. This might allow larger proteins to cross the barrier, and as a result bypass the usual 'tolerance' mechanisms.

Similarly, on closer inspection, some allergy-inducing plant storage proteins have a role in protecting against fungal or bacterial infection (have you ever wondered why seeds don't usually rot in the ground?). They do this by being able to adhere to the lipid membrane of the microbe contaminant, to disrupt it. Whilst there is no evidence to date of any such harm being caused to the gut lining cells by these proteins, it could provide a plausible explanation for triggering an alarm signal if this were to be the case.

However, given that only some people are allergic to these food proteins whilst everyone else can consume them without showing any signs of toxicity (and their immune response is tolerant), it must be that any direct damage caused by them must be relatively trivial. Nevertheless, it may be enough to trigger an exaggerated response in certain individuals, leading to sensitisation.

Why Do Food Allergies Only Affect Certain People?

Human beings are astonishingly diverse. Every protein that we produce will be recognisable for what it is, but most will have subtle variations between individuals such that none are exactly identical. Natural selection will usually have phased out any seriously deleterious mutations and therefore most alterations will have relatively trivial effects. Many of the genetic variations will be in the 'non-coding' DNA that is involved in switching genes on or off, rather than in the protein itself. Whereas the easiest genetic diseases to understand are those that involve mutation of a single gene, most are caused by minor changes in a large number of different genes. In this way, genetic causation of disease can be seen not just as a single on/off light switch, whereby you either have the one faulty gene or not, but rather as a picture made up of different pixels (each being a different gene involved), where the overall image is recognisable but may vary in its colours, brightness or minor features. Similarly, diseases for which a large number of different genes are involved may vary in the way they present, their severity or the types of symptoms, but still be (perhaps questionably) the same disease.

One way to try to pick out the different genes involved in such 'polygenic' diseases is to carry out a 'genome-wide association study' or GWAS. This technique examines changes in single letters in the code that have been identified at specific points in the genome. Given that DNA is usually transferred in chunks when organisms reproduce sexually, the presence or absence of well-defined single letter changes can be used as a tag to mark a specific region of the DNA. Scanning the DNA for these tags in (many) individuals with the disease compared to those without effectively points towards parts of the genome that are more likely to be involved in the disease process. It does not tell us anything about particular mutations, but acts like a large 'you are here' arrow, which allows us then to look at the nearby genes (or for the most part, non-coding DNA) to try to work out which might be involved.

GWAS studies carried out in patients with food allergy highlight a relatively small number of pointers in the DNA that may help us to understand the underlying disease process. One of these is specific only to peanut allergy, and tags a region of DNA close to that encoding MHC class 2 molecules. These are the molecules that present foreign proteins to the immune system. It would have been really useful if studies had then shown that only certain MHC proteins were capable of binding with the sensitising peanut peptide sequences, as this would have meant that we would easily be able to identify those individuals at risk, or not, by whether or not they had that particular kind of MHC (spoiler alert—see the next chapter on coeliac disease!). However, unfortunately the area of DNA that is tagged appears to be involved with altering the *level* rather than the *type* of MHC protein expression.

However, other food allergy 'hits' on the GWAS are of great interest. One is close to the gene encoding the filaggrin protein that (as we saw earlier) maintains the skin barrier and mutations of which are associated with eczema. The association of eczema with food allergy is well known and thought to represent a route by which proteins could cause sensitisation in children, by entering the body through a different route from the gut. Interestingly, filaggrin is also found in the oesophagus, which has a multi-layered epithelium like the skin rather than the single layered secretory lining of the intestine. GWAS also predicted the involvement of another protein involved in maintaining the containment barrier of the skin, which is similarly expressed in the oesophagus. It is possible therefore that sensitisation to food proteins in children with eczema occurs not through damaged skin, but through the faulty barrier of the oesophagus.

The other significant genetic association found for food allergy was with a specific interleukin—interleukin 4. This chemical messenger is produced by the ILC2s in response to alarm signals from the epithelium, and it appears to be one of the key switches in the allergic response. It activates T cells to become 'helper' cells to shift the immune reaction to the 'antibody' rather than the 'cell killing' response, and induces the switch in the antibody docking end from IgM to the IgE that sticks to mast cells. Mice which are bred to lack either interleukin 4 or its receptor do not develop food allergies.

It seems therefore that in order to develop food allergy, an individual has to have the right mix of genetic mutations in key genes, and then sensitisation only occurs to certain food proteins with specific properties that may result in low-level activation of an 'alarm' signal from the gut epithelium.

This of course merely begs the next question.

Why Are Food Allergies Becoming more Common?

The increase in food allergy over the last 25 years or so cannot be due to changes in the genetic makeup of the human population, which would of course take far longer. Studies reveal an increasing incidence of food allergy in societies as they become 'westernised'—a cultural change that involves dramatic alterations in diet and food preparation, environmental pollutants and lower exposure to rural microbes. There is evidence to suggest that all of these factors might influence the development of food allergy, but the extent to which each one is responsible is currently unclear.

Food processing can frequently change the conformation and chemical nature of proteins. Heating, for instance, usually denatures proteins to change their folded structure—this is why egg white changes from transparent to opaque when you fry an egg. This might make them change their shape so that they are made more or less recognisable by antibodies. For instance, Kustner's fish allergy (see diverticulum #7.1 above) only occurred with boiled and not raw fish. Pasteurising milk results in the milk proteins aggregating together, and this change to a particulate nature seems to make the protein more accessible to the immune system—or serves perhaps to crosslink IgE antibodies by repetition of its protein sequences. The peanut protein Ara h1 is more likely to cause allergy when peanuts are 'dry roasted' than if they are cooked in other ways.

Heating, baking, or dry roasting appears to be particularly good at leading to sensitisation changes in proteins. One of the causes for this is that it leads to the addition of sugar molecules to the protein as a result of what chemists call the 'Maillard' reaction. The resulting molecules are called 'advanced glycation end-products'—or AGE for short. The amount of such AGEs is increased in milk 86-fold by microwaving for 3 minutes, or nearly 700-fold by turning it into milk powder. Infant milk formulas have 60–70 times the amount of AGE as breast milk. AGEs are also produced by bacterial metabolism, and it is postulated that the high sugar consumption of western diets can increase the generation of AGEs in the gut. Consumption of one sugar—fructose—has increased six-fold (from 7 to 40 kgs per person per year in the USA) over the last 50 years.

We now discover how RAGE, one of the alarmin receptors (for HMGB1 as above) gets its name—as the 'receptor for advanced glycation end-products'. Hence, chemical alterations induced by modern food processing might act to 'short circuit' the allergy response by triggering a 'false alarm' as a result of mimicking the epithelial stress response. Instead of epithelial cells initiating the immune response by secreting alarmins, the AGEs themselves in the food act on the alarmin receptors to bring about the same effects as if there had been damage to the epithelial lining. Such a hypothesis is tempting and may help to explain many facets of the recent increases in food allergy—including the lower rates in rural areas (where less than half the sugar is consumed as in urban populations), and from Mediterranean and Asian cooking which generates fewer AGEs and includes foods that reduce AGE production (for instance fruits and soy). However, at the moment there is only circumstantial evidence that AGEs are involved in human food allergy responses.

Diverticulum #7.2: The Fourteen Foods

By a European regulation introduced in December 2014, caterers and food wholesalers are now required by law to be able to provide detailed information about whether a product contains any of the 14 foods that are identified as the most likely to cause allergy. The list is as follows:

Cereals containing gluten (see next chapter)—wheat, barley or rye.

Crustaceans—such as crab or lobster, shrimps or prawns.

Eggs.

Fish.

Peanuts.

Milk.

Soya.

Tree nuts—such as cashew, almond or hazelnuts.

Celery.

Mustard.

Sesame.

Sulphur dioxide—used in dried fruits, beer and wine, some meat and vegetable products (this is clearly not a protein and the reason for its effects are still unclear).

Lupin—used as a source of flour for bread making.

Molluscs—such as oysters and mussels.

The tragic death of a teenager from sesame allergy after eating it in an unlabelled product from a large UK outlet in 2016 exposed a loophole that permitted stores to omit labelling on products made freshly on site. The resultant tightening of legislation in this area—called Natasha's Law—was passed into UK law in 2019.

First Contact

The predominance of food allergy in childhood shows that events in early life are critical to its development. The first exposure to food allergy-causing proteins may even be before birth, as they have been found in samples of amniotic fluid surrounding the foetus. Babies will also be exposed to food proteins in their mother's breast milk.

The human gut immune system is undeveloped at birth and requires both food and bacteria in order to mature. The way in which the immune system and its microbial passengers develop together leads to a unique relationship that endures through life. The environment of the gut is extremely tolerant in early life, in order to develop this affiliation with its microbiome, and is assisted in this by chemicals found in breast milk that encourage lymphocytes to become 'regulatory' T cells rather than 'helper' cells. There are also large amounts of antibodies (IgA) produced in breast milk that serve to protect the baby by providing an umbrella of passive immunity in the gut, when the child is unable to produce its own. These antibodies also tend to shape the microbiome of the infant gut which becomes very similar to that of its mother, also as a result of picking up bacteria in the birth canal and from close association with its mother during feeding.

It was once thought that the presence of potential allergens so early in life and before the maturation of the immune system could lead to the development of food allergy. Indeed, it was once recommended not to introduce potential allergy-causing foods until 3 years of age, and pregnant mothers were also discouraged from eating them. However, a Danish study showed that eating peanuts during pregnancy actually *decreased* the risk of the offspring developing peanut allergy. Similarly, a large project called 'LEAP' (Learning Early About Protein Allergy) showed that only about 1 in 30 children at risk of peanut allergy (for instance with eczema, other allergic type conditions or strong family history) who had peanuts introduced into their diet as early as 4 months went on to develop symptoms with peanuts, compared to about 1 in 6 of those where the introduction of peanuts was delayed.

There may therefore be a narrow window of tolerance in the developing gut immune system early in life which effectively sets the scene for future immune responses. As we have seen, it may also be that the importance of early oral exposure is that it pre-empts first contact by other routes—for instance through the skin—which might lead to sensitisation rather than tolerance. However, the maturation of the gut immune system requires not just food but bacteria, and therefore there is a possibility that the way in which the microbiome develops could also shape our future responses to food.

Gut Bacteria and Food Allergy

There is now abundant evidence of the involvement of the gastrointestinal bacterial flora with food allergy. Mice that are raised without bacteria in their gastrointestinal tract ('germ free') preferentially produce an allergic rather than a tolerant response to food proteins, therefore it appears that bacteria are somehow protective in this regard. However, it is not only the presence of bacteria, but specific types of bacteria, which are important in shaping the response. In one famous experiment, mice were generated that had a genetic defect in the receptor for interleukin 4. This is the molecular messenger that lies at the heart of the allergic response. The effect of this 'activating' mutation is to permanently switch on the receptor, as if high levels of interleukin 4 were always present. This makes it easy to sensitise these mice to proteins and cause an experimental food allergy. The investigators found that these mutant mice had completely different bacteria in their gut to those mice without the mutation. Amazingly, by transferring these bacteria to the normal mice, they could also transfer the allergy to a specific protein. This could then be switched off again by transferring regulatory T cells into the mouse, and even more amazingly doing so also restored many of the changes in the bacterial composition of their guts!

It appears, therefore, that these effects are not just due to the bacteria themselves, but that there is a two-way interaction between the gut immune system and the bacteria. The mechanisms involve some of the deepest layers of the immune system, such as the cell surface Toll-like receptors that recognise bacterial proteins and transmit the signal to the nucleus by their TIR protein family intermediaries (which we first met in chap. 5). We have seen how these Toll-like receptors detect the molecular signatures of threatening bacteria in order to mount a defensive response. However, for symbiotic bacteria covered by the immune umbrella of the host animal, we now see the opposite. The proteins secreted by these bacteria are detected by Toll-like receptors which tell the gut immune system to become tolerant instead of mounting a defensive response. It appears that within the repertoire of fixed pattern-recognising molecules such as the Toll-like receptors there are 'activating' and 'suppressing' pathways. As a result of triggering the suppressing pathways, gut lymphocytes signal to the lining epithelial cells of the gut and to the goblet cells to alter the secretion of their ancient defensive proteins (defensins) in order to downscale their response. Hence the bacteria are signalling through the epithelium to the gut lymphocytes to effectively stop their attack, and the lymphocytes alter the behaviour of the epithelial cells to make the inside of the gut a more homely environment for the bacteria. In this 'tolerogenic' environment, allergic sensitisation to food proteins is also reduced.

The above pathway depends on the identification of bacterial molecular signatures (by 'fixed' pattern recognition molecules) in the immune system. It now appears that there are other ways in which the gut bacteria produce a tolerant environment, and this depends on the food that we eat. A diet high in fibre (not just the stringy bits on beans and bananas, but actually any complex plant molecules that escape digestion in our upper intestines) effectively provides nourishment for our gut bacteria. As a by-product of metabolising our leftovers, they produce chemicals such as butyric acid, a very short-chain fatty acid molecule. Butyric acid appears to work directly on the structural cells—the epithelial and scaffolding myofibroblast cells—to produce retinoic acid from vitamin A. As we have already seen, the downstream effects of this are to instruct the immune system to generate regulatory T cells and a tolerant response. This is another example of the bacteria influencing the immune system of the gut to make it a friendly place.

The best bacteria for producing butyric acid are the bifidobacteria, well-known from commercial 'probiotic' drinks that can be bought in supermarkets. These types of bacteria are also the most commonly-found bacteria in breast-fed infant guts. Furthermore, in experiments where bifidobacteria are given to allergic mice, their food allergy response is diminished. A common source of bifidobacteria is in unpasteurised dairy products, and this might also explain why rural communities tend to be protected from allergy compared to urban societies—but only when exposed in early childhood (Fig. 7.2).

Mithridates...and a Cure for Food Allergy

You will remember that, at the beginning of this chapter, we left Mithridates in Asia Minor (over 2000 years ago), in a bit of a paranoid state trying to build up his resistance to all the potential poisons that his enemies might use against him. After 4 years he emerged from hiding, retook the throne from his mother and imprisoned her and his younger brother. Both were to die in prison, possibly as a result of Mithridates' experiments with poisons. Mithridates became King Mithridates VI of Pontus, known as Mithridates the Great. He substantially enlarged the empire that he inherited and challenged the power of Rome, in alliance with Greece, in the 'Mithridatic' wars. However, despite enormous success he was ultimately defeated by the Roman general Pompey in 66 BC, and his people rebelled against him. With nowhere else to flee he attempted to take his own life...with poison. According to legend both his daughters died by the poison first, but it did not work on Mithridates as he was entirely immune to its effects. He had to ask his servant, Bituitus to run him through with a sword.

Mithridates' successful mitigation of the effects of poisons by gradually building up small amounts is exactly the same principle that is now used in the immunotherapy of allergies. For food allergy, the route of administration of small amounts of the protein can be by mouth, under the tongue or across the skin. Oral immunotherapy—feeding small amounts of the protein to allergic individuals—involves initially determining the degree of sensitivity and the largest amount that can be ingested without experiencing symptoms. The dose is then gradually increased until the individual is able to manage enough to make accidental exposure safe. This is called 'desensitisation'. It appears to work by gradually increasing the number of regulatory' T cells rather than helper cells that are made by exposure to the protein. There also appears to be a gradual

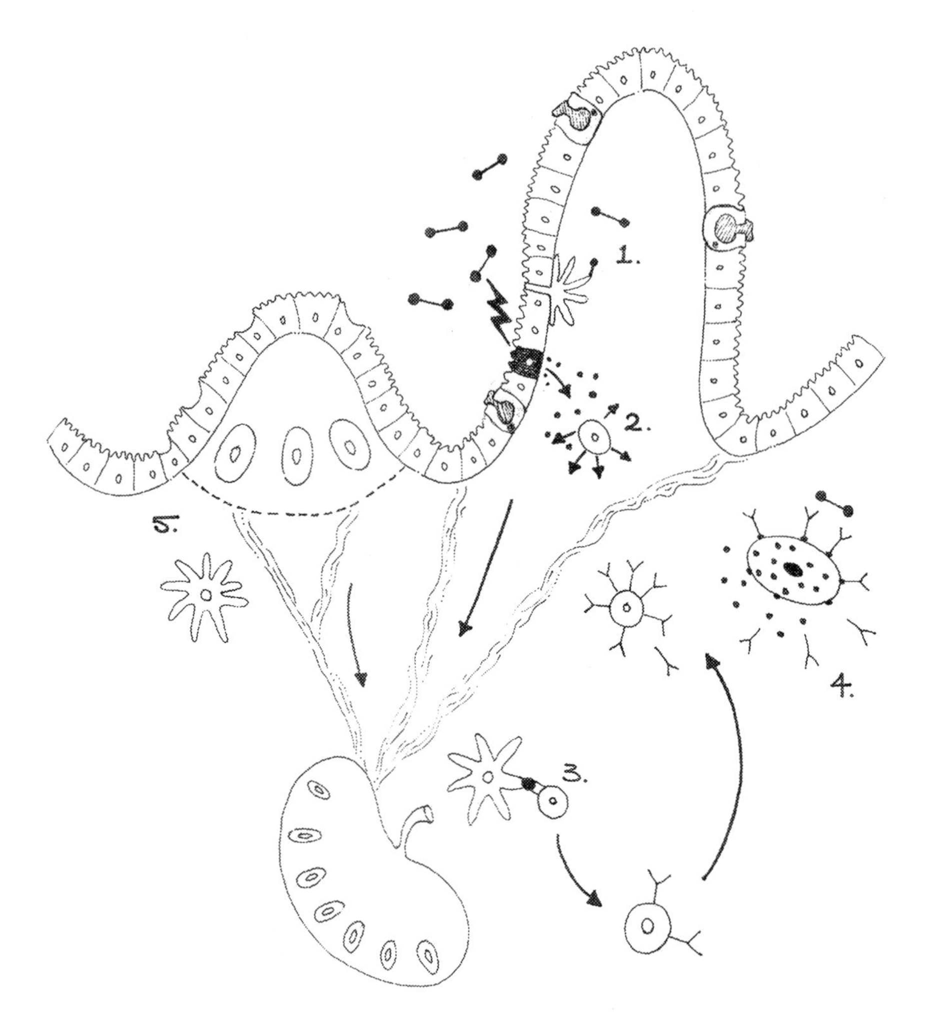

Fig. 7.2 Current understanding of food allergic responses. (1) Food allergens cross the epithelium through a 'leaky' epithelium, through macrophages extending processes across the cell layer, or through channels opened up in goblet cells. Dendritic cells bearing the food proteins migrate to the mesenteric lymph nodes. (2) ILC2 cells in the lamina propria are activated by specific chemical messages ('alarmins') secreted by the epithelium. These may be in response to actual damage caused by the food allergen, or simulated damage due to activation of the RAGE receptor by molecules formed during food processing. ILC2s secrete a large number of different chemical messengers (interleukins) to activate the immune response and make it deviate towards an allergic antibody response. The most important of these messengers produced by ILC2s is interleukin-4. (3) T helper cells are activated by recognition of the food protein on the dendritic cell in the mesenteric lymph nodes and educate B cells in the germinal centres to produce immunoglobulin E type antibodies. B cells migrate in the blood stream and home back to the gut where they produce antibodies to coat the mast cells. (4) Mast cells which are already coated with the IgE antibodies produced by sensitisation (steps 1–3) degranulate when exposed to more food allergen proteins that cross the

switch away from IgE antibody production to a class of IgG (called IgG4) antibodies, specific to the food protein. These IgG4 antibodies seem to interfere with the binding of the IgE antibodies. Whilst the majority of patients can be successfully desensitised using oral immunotherapy, allergic reactions during food reintroduction are common. In order to maintain tolerance after desensitisation small amounts of the offending protein still need to be taken, and it really only protects against low-level accidental exposure. The goal of sustained unresponsiveness to effectively return patients to a state of non-allergy with occasional high-level exposure is much more challenging to achieve.

A number of different techniques are therefore under investigation to try to improve oral immunotherapy. Starting off desensitisation using a sub-lingual approach—placing a small amount of protein under the tongue—appears to be safer than oral immunotherapy as none of the offending protein is absorbed from this route into the circulation. Applying the protein to the skin and covering it with a patch also has potential benefits—side-effects inevitably include an itchy rash at the site, but there is no need for gradual dose escalation as with oral immunotherapy. There is also a suggestion from experiments with mice that the regulatory T cells produced in the skin can migrate to the gut and directly inhibit mast cell activation, and that it might be possible to bring about sustained unresponsiveness as a result.

Finally, it appears that the use of 'adjuvant'[7] treatments with oral immunotherapy may bring about significantly better results. The immune system tends to work through duplication of signals to reinforce a response. This is effectively a form of ratification to ensure that a reaction is not inadvertently initiated—like the need for co-authorisation by a second high-ranking general or politician to launch a counter-attack, when radar detects signals of an enemy invasion. Hence by providing more than one signal to the immune system, we may be able to replicate this process.

[7] The word 'adjuvant' derives from the latin verb *adiuvare*, meaning 'to help'. It is of different derivation from 'adjunct' which comes from the verb *adjungere*, meaning 'to add on' or 'join'.

Fig. 7.2 (continued) epithelium on re-exposure to the food. Mast cell degranulation is also stimulated by interleukins produced by ILC2s. The result is a rapid 'allergic' response occurring after eating the culprit food whereby the contents of the mast cell granules (such as histamine) lead to increased fluid secretion in the gut, and nerve activation causing pain and vomiting. Mast cell chemicals reaching the circulation can cause distant effects on other organs such as swelling and narrowing of the airways (asthma) and itchy skin rashes (urticaria). (5) The normal 'tolerance' regulatory response resulting from education of dendritic cells and B cells in the Peyer's patches is suppressed by the chemicals secreted in the above steps

Adjuvant treatments under investigation include the use of an infused artificial antibody against specific chemical messengers (interleukins), or one that blocks IgE or its receptors on mast cells.

There are also early signs that probiotics may help significantly in improving the response to oral immunotherapy—particularly a bacterium called *Lactobacillus rhamnosus*, which produces butyric acid from undigested carbohydrates. At the time of writing, the available evidence suggests that oral immunotherapy can moderately desensitise individuals with peanut allergy, but actually tends to increase the risk of anaphylaxis and severe reactions over time. There is therefore a substantial need to improve such treatments in order to achieve sustained unresponsiveness in individuals, and it may well be that mimicking the immune system's multiple signals will provide the necessary breakthroughs.

Diverticulum #7.3 Oral Vaccines

Vaccines are a means of establishing protective immunity to an organism to prevent future infection with it. The safest way to do this is to use some of the distinctive molecules on the microbe's surface that the immune system can recognise (rather than the whole organism, which could cause the disease). Other forms of vaccination use killed or inactivated bugs that can still trigger an immune response but without the risk of infection, or by the generation of less potent or 'attenuated' microbes that have been bred artificially to lack certain factors that can cause disease. This is effectively the same as giving someone a mild dose of the infection.

Given the effective containment provided by skin, most infections either affect or access the organism through mucosal (i.e. moist) epithelia such as mouth, lungs or gut. However, the overwhelmingly tolerant environment of the gut immune system makes it extremely difficult to develop effective vaccines for oral delivery, hence most require an injection! In order to develop a vaccine, it is necessary to fool the immune system into thinking that there is a real infection occurring. Usually this occurs as a result of signals resulting from actual damage caused to cells as a result of the infection. As we have seen, this multiple-signal requirement acts as a failsafe device to prevent an inadvertent immune response. The vaccine designer therefore has to try to reproduce the effects of an infection without causing real harm, but the use of adjuvants in order to replicate the 'damage' or 'threat' response can sometimes result in minor symptoms such as fever. So far, the only adjuvant licensed for use in an oral vaccine is an inactive form of a cholera toxin, which helps invoke a protective immune response against the cholera bacterium (*Vibrio cholerae*). The only two other oral vaccines so far available use weakened live strains of the organism—rotavirus and typhoid (*Salmonella typhi*) respectively.

One promising candidate as an adjuvant for future oral vaccines is flagellin—the protein found in the bacterial propulsive flagella—which is recognised by a specific Toll-like receptor. It may also be that active immunity can be triggered (effectively short circuited) by artificially administering the specific chemical messengers (interleukins) that signal threat in the immune system. Along with advances in the delivery system—for instance using microparticles that can be taken up by the gut immune tissue—such developments are beginning to promise the development of a new era of oral vaccination.

Partial Food Allergies?

The rationale for a gastro-archeological approach to gut conditions is that it is all too easy to focus on the surface layers that we now know represent the culmination of the effects of all the deeper layers of the gastro-archeological 'trench'. Therefore, whilst it is unlikely that we would see an effect occurring in only the upper strata without all the supporting layers below it, it is quite feasible that we could invoke a type of immune response where the upper, more modern layers are not engaged. We have already seen, for instance, that it is possible to generate mice with a 'RAG' knockout that are incapable of producing B or T cells (hence no antibodies). However, they can nevertheless produce many of the effects of an allergic response through the ILCs and the effects of the chemicals secreted by the gut cells themselves.

There are indeed a number of conditions that we are now beginning to recognise as having all of the hallmarks of an immune response, but without the IgE antibodies and without mast cell activation. Our fixation on just the modern superficial layers of the gastro-archeological dig has meant that our understanding of the mechanisms of these conditions is still incomplete.

One of these conditions is 'Food-protein enterocolitis syndrome' or 'FPIES' for short and results in acute vomiting and diarrhoea after ingestion of the culprit food, but without the classical IgE antibodies being produced in the majority of cases. A chronic form associated with signs of damage of the intestinal lining is also rarely described. Although described in adults, this condition is more commonly found in children.

A condition that I come across frequently in my adult practice is called 'eosinophilic oesophagitis'—which really needs an easier name to say and spell as it takes up quite a long time in my clinic explaining this to patients! Let us just call it EoE (from its American English spelling) for now. This is a condition where the gullet (oesophagus, or esophagus in American English) is infiltrated with immune cells called 'eosinophils'. These are a specific kind of blood cell that are attracted out of the circulation into the tissues during an immune response, to help bring about inflammation. They are present in all allergic-type responses including food allergies and asthma. There is now good evidence to suggest that EoE is a form of food allergy. In EoE, all of the chemical messengers that are secreted to bring about the condition are found to be secreted in the oesophagus, yet these patients do not have IgE antibodies against specific foods. In a sense this is similar to the mice that have been genetically altered to be unable to mount an antibody response.

Unlike with typical food allergies, symptoms do not occur rapidly in EoE after eating specific foods. Instead, the effect is gradual, taking days or even weeks and causes swelling and narrowing of the gullet that can result in a feeling of food getting 'stuck' after swallowing or even blocking the gullet altogether. Given the slow onset and improvement of the condition with introduction and withdrawal of different foods and the lack of specific IgE antibodies, it can be difficult to identify a particular culprit. However, by dietary avoidance of one or all of several identified high-risk foods (wheat, dairy, seafood, eggs, nut and soy), it is usual for the condition to improve after a period of some weeks. How and why these foods cause this allergic response without their food proteins being specifically identified (by antibodies and T cell receptors) is unclear. It seems likely though that they somehow subvert the older layers of the allergic response—for instance by triggering a threat-like response from the epithelial cells.

What Is the Difference Between Food 'Intolerance' and 'Allergy'?

I am frequently referred patients to my clinic with symptoms that they have attributed to different foods. In many cases the individual is convinced that a specific food is the cause, but are unable to identify the culprit. This can be because they are trying to apply a 'food allergy' approach—thinking that only a single specific food will be responsible for the symptoms. In general, if a patient is allergic to a specific food then they will be fully aware of which food they are allergic to without needing to seek my help to identify it! Those that are not clear as to the cause are usually food 'intolerant' rather than 'allergic'. The symptoms that are described are usually gastrointestinal—bloating, a change in bowel habit, nausea—or unspecific such as lethargy, headaches or feeling generally unwell. The time of onset and duration of 'intolerant' responses may be highly variable, whereas allergic responses are usually rapid and dramatic. Intolerances can also involve multiple different foods, unlike allergies which are usually very specific.

Trying to get to the root of the problem in these settings can be extremely challenging. Firstly, there is a natural tendency for us to blame any symptoms we experience on something that we have eaten, understandably so if the symptoms are gastrointestinal. However, it may be that it is an underlying problem with the gut itself rather than an individual food—for instance, we know that the motility and sensitivity of the bowels can be considerably altered by circumstance, such as an exam or job interview. Ongoing stress or even the effects of different medications can alter the background function of the gut, leading to symptoms after eating (almost anything). Very often, a patient in such a situation gradually cuts out different foods that they think might be responsible, and arrives in my clinic on a highly restricted diet, yet still experiencing the symptoms.

Some of the effects of foods are predictable. For instance, whilst a diet high in fibre carries significant health benefits, it can lead to bloating and loose motions due to the indigestible carbohydrates that pass through the intestine and into the colon, where bacterial fermentation produces gas. Others might have direct effects due to the chemicals that they contain, such as the stimulant effect of caffeine on the bowels, which can work on the colon within 4 minutes of a cup of coffee and have the same effect on the gut as a 1000 calorie meal.

However, even when such potential factors are excluded, there still remain many people with often vague and unexplained symptoms that they are certain relate to what they have eaten. Given that modern medicine has largely failed to help provide an explanation or cure for such symptoms, it is not surprising that patients often turn to unorthodox approaches. There is a growing industry of 'fringe' practitioners, tests and treatments. Some are loosely based on scientific knowledge. Many are (at best) prematurely launched onto a commercial footing without adequate evidence to support them, while others are clearly deliberate attempts to defraud vulnerable individuals.

Using the 'gastro-archeological' model, it may be possible to consider potential explanations for such scenarios of food intolerance. Just because no food-specific IgE antibodies are found does not necessarily mean that the immune system is not involved. Rather than taking a 'side-step' in the same layer (for instance by looking for other classes of food-specific antibody such as IgG) we should dig deeper, just as we did to explain the nature of EoE. However, we should not also forget that there are two strands of the immune system. In this chapter we have just followed the antibody ('soluble') aspect of the gut immune system to describe food allergy. When we follow the cytotoxic ('cellular') arm we uncover further interesting possibilities to help explain food intolerances. Let us gather up our tools and start another trench on the other side of the 'great schism'.

8

'The Evil Bread'

Coeliac disease and Gluten Intolerance

Summary In which we identify coeliac disease as an immune response to a class of protein called 'gluten' which is found in the cereals, wheat, barley and rye. The cytotoxic cell-killing side of the immune response is implicated in coeliac disease and leads to damage of the intestinal lining. It is therefore very different to food allergies which are caused by the antibody type of immune response. As we again dissect out the immunological layers involved, we find that the most recent adaptive immune mechanisms are critical in the disease process. As we dig deeper and further back in time we find that there are important initiating factors arising from older immune processes involving the epithelium itself. We find that much of the damage that occurs is due to subversion of the normal activity of the housekeeping intraepithelial lymphocytes within the surface epithelium. This is triggered by a stress response signalled by the enterocytes and possibly due to a toxic response of the gluten or other proteins ingested along with it in wheat. Finally we consider that a newly recognised condition of 'non-coeliac gluten or wheat sensitivity' may also be driven by the deeper layers of the gut immune system but without the most recent additions of the adaptive immune response.

J. Woodward, *The Gastro-Archeologist*, https://doi.org/10.1007/978-3-030-62621-1_8

The 'evil bread'

'Whoever has silver, whoever has jewels, whoever has cattle, whoever has sheep shall take a seat at the gate of whoever has grain, and pass his time there'

From the 'Myth of Cattle and Grain', Sumerian text, third millennium BCE.[1]

Aisha's story

Aisha had two hobbies which she found balanced each other rather well. She loved baking and would fill the house with exquisite aromas whilst trying out new recipes for different breads, pastries and cakes—which were rapidly devoured by her appreciative teenage sons and her husband, after he returned from work in the city. However, when not in the kitchen she could be found in the gym or running in the fields around her house. She would time herself running over the same course and quickly reached a standard pace that suited her.

During a family summer holiday in Portugal, Aisha experienced a bout of gastroenteritis. Her bowel habit never really settled after returning home and she noticed that she was having to open her bowels 2–3 times a day, more than before, with much looser bowel motions. She also found that she was unable to match her pre-holiday running speed, which she attributed initially to having had too much of a good time whilst being away! Strangely enough, despite eating out most nights on holiday her weight had reduced and continued to fall after her return home. Aisha started to become concerned when she failed to improve after about 6 weeks following her return, especially as people were noticing a change in her complexion and her usual 'bubbly' personality. Her family doctor was unable to find any significant abnormality when he saw her, but sent off samples of her stool for parasites and bacterial infections, and blood samples for a variety of routine tests. There were no abnormal organisms found in the stool. However, when the blood test results came back, Aisha was shown to be anaemic (a low red blood cell

[1] The 'Myth of Cattle and Grain' is a Sumerian creation myth dating from around 5000 years ago. It describes how *An,* the Creator, generates the cattle-goddess Lahar and the grain goddess Ashnan. Thence follows a debate over the relative merits of livestock and agriculture. Ashnan wins the argument, as depicted in this quote.

count) and was also deficient in iron and folic acid, two essential substances that the body needs to make blood cells. There were also minor changes in her liver blood tests, suggesting mild inflammation in the liver.

The doctor referred Aisha to the hospital for further investigations. An endoscopic examination was carried out where an endoscope (a tube with a video camera on the end) was passed into her mouth and steered through her oesophagus and stomach into the intestine. Small samples of the lining of her small intestine were taken to examine under the microscope for changes. Two weeks later, the doctor telephoned her to tell her that the samples had shown that she had a condition called coeliac disease, and that this explained her blood test results and her symptoms. The treatment involved a diet where she avoided a substance called gluten. This is present in wheat (hence all the bread, pastries and cakes that she so enjoyed baking) but also in barley and rye, and therefore necessitated a significant change in her diet and lifestyle. However, within 3 weeks of cutting the gluten out of her diet, she felt like a 'new person' and already noticed an improvement in her bowels. 3 months later, she was beating her pre-holiday running times and felt even better than before her holiday.

Following advice from her doctor, she arranged for blood testing for her two sons, one of whom was also found to have coeliac disease. Aisha did not give up on the bread making, but instead became expert at baking gluten-free bread (and cupcakes) that tasted every bit as good as their gluten—containing equivalents (actually rather better in my opinion!)

Ironically, Aisha had no idea of the origins of her name. 'Aisha' is derived from the Arabic word meaning both 'life' and 'woman'—the two being inextricably linked through ancient associations with fertility. One of the symbols that designated the deities (usually female) of fertility in Middle Eastern religions was a 'spike' of wheat. Hence the same word, 'Aish', has also come to be used as a word for 'bread' in Egyptian Arabic....

'A kind of Chronic Indigestion Which is Met with in Persons of All Ages'

This was how Dr. Samuel Gee, a paediatrician at St Bartholomew's Hospital in London described what he called the 'Coeliac Affection' in a report in October 1887.[2] Whilst this has been attributed as the first proper description of coeliac disease, he acknowledged that it was probably initially identified in the first century AD by a Greek physician called Aretaeus who hailed from a region of modern-day Turkey called Cappadocia.[3] Gee derived the name 'coeliac' from the

[2] Samuel Gee (1839–1911) was a physician who worked in both Great Ormond Street Hospital and St Bartholomew's Hospital in London. The description of coeliac disease is how he is best remembered, although he was also the first to describe a condition of cyclical vomiting syndrome. There are numerous anecdotes of his excellent teaching and his legendary punctuality. In one story he was catching a cab to the hospital for a clinic but was running late. The door of the cab accidentally opened *en route* and caught on a tree, which ripped it off its hinges—but he insisted on continuing the journey minus the door lest he be late.

[3] Aretaeus was one of the great Greek physicians of the first millennium CE—probably living in the second century. He came from Cappadocia, a region now in Turkey, studied in Alexandria and practised in

Greek word used by Aretaeus, '*Koiliakos*', meaning 'pertaining to the abdomen'—effectively 'tummy trouble'. An attempt to call the condition 'Gee-Herter' disease (co-recognising an American physician who described it some 20 years later) thankfully fell by the wayside. 'Coeliac' is spelt 'celiac' by American authors, and has been adopted as such in some European countries.

For a condition that now affects about 1 in every 100 people in Western societies, it is extraordinary how so little was understood about coeliac disease until so recently. However, that it was in some way related to diet was apparent from early on. Gee himself noted 'if a patient can be cured at all it must be through means of diet'. He apparently treated one child with a diet of Dutch mussels ('a quart of the best' every day!) and eight children were similarly cured with a diet of bananas in the USA in 1924.[4] It was not until 1953, by the work of a Dutch paediatrician called Willem Dicke,[5] that it became known that it was not what was *in the diet* but rather what was *excluded from the diet* that mattered. Dicke realised that certain types of cereal flour were toxic and could cause relapses of coeliac disease. The culprits were identified as wheat, barley and rye.

At about the same time, the next piece of the puzzle of coeliac disease was put into place by Dr. John Paulley, a gastroenterologist working in Ipswich hospital in the UK. He examined the intestines of patients undergoing laparotomy for the condition of 'idiopathic steatorrhoea' (literally meaning 'fatty stools of

---

Rome, but little else is known of him except for his writings. Whilst following the traditions of Hippocrates he was a free thinker who was not tied to dogma. As well as describing coeliac disease for the first time he also produced the first accurate description of asthma. It is most likely due to his extremely vivid accounts of sufferers of certain diseases that we are able to credit him with their 'discovery'.

[4] Sidney Haas (1870–1964) was an American paediatrician who made a claim to fame in 1924 when he published his recommendations for diet in children with coeliac disease that included up to 8 bananas a day. He made the observation that people in the cities in Puerto Rico tended to develop coeliac disease (or 'sprue' as it was then called) and ate bread, whereas in the countryside where the diet was much more banana-heavy the condition was rare. In one of those all-too-common 50:50 moments in life he chose completely the wrong way, believing that it was the health benefits of the bananas rather than the detrimental effects of bread that lead to coeliac disease…

[5] Willem Karel Dicke (1905–1962) was a brilliant paediatrician who became director of the Juliana Children's Hospital in the Hague at the age of just 31 years. His recommendations regarding the diet in coeliac disease were the subject of his thesis but also benefited from a long period of observation over many years in the interwar period. He only published after the war and continued to investigate the nature of the toxic component of wheat. His interest was initially stimulated by the report of Haas (see above), in which one patient experienced symptomatic relapses after eating wheat. A rather charming story has become part of medical mythology . Dicke lived through the Dutch winter famine of 1944–45 when cereal crops were in short supply. Children with coeliac disease paradoxically did better than their non-coeliac peers. At the end of the war, the Allied air force mounted 'Operation Manna' to drop food supplies into Holland. Supposedly on reintroducing bread into their diet, the coeliac children became worse again. The story goes that Dicke developed his theory on the toxicity of wheat as a result of these events. I would love this legend to be true—not least because a relative of mine was a pilot in 'Operation Manna' and flights took off from my village airbase. However, whilst Dicke will certainly have witnessed with interest, his ideas were firmly established at the time, and the winter famine and its relief were not the key factors that led to his discovery.

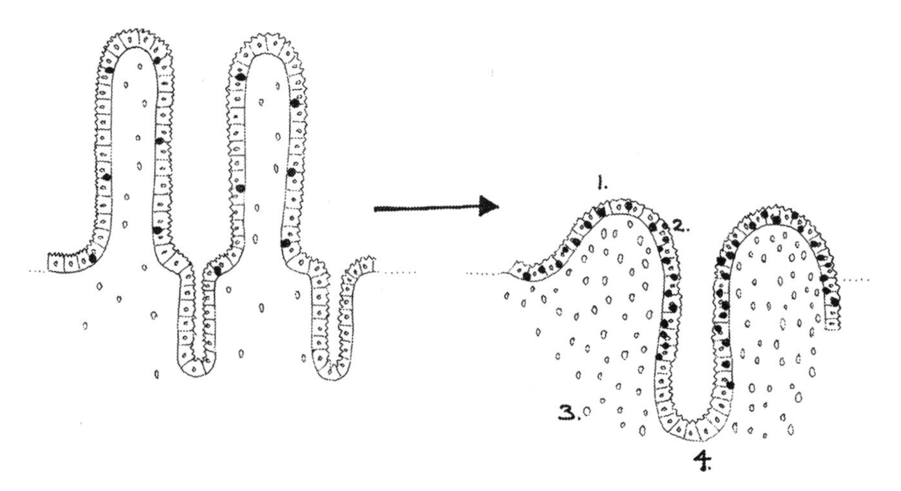

Fig. 8.1 The damage to the intestinal lining caused by coeliac disease. There are four key differences between the normal intestinal lining on the left and the intestine affected by coeliac disease on the right: (1) 'Villous atrophy' whereby the normally tall and thin villi become shortened or barely distinguishable above the intestinal surface. They are also expanded by increased numbers of lymphocytes in the lamina propria below. (2) Increased numbers of intraepithelial lymphocytes within the epithelium. (3) Infiltration of the lamina propria with activated lymphocytes—predominantly antibody-producing B cells. (4) 'Crypt hyperplasia'—deeper crypts with evidence of increased epithelial cell proliferation as a result of growth factors secreted by the inflammatory response (notably the intraepithelial lymphocytes)

unknown cause'). On close examination of the intestinal lining under a microscope, he described the changes that are now associated with coeliac disease—a loss of the surface 'villi' or 'finger—like projections' that increase the surface area of the small intestinal lining, and an increase in numbers of lymphocytes in the epithelium and below it (Fig. 8.1). Within the next 5 years, a device was invented (the 'Crosby[6] Capsule') that could be swallowed, take a 'pinch' biopsy of the intestinal lining and retrieve it for doctors to examine in the laboratory.

This ability to sample the surface of the small intestine without undergoing an operation led quickly thereafter to the understanding that coeliac disease was associated with inflammation of the small intestine and loss of villi—so-called 'villous atrophy'. This helped to explain why the condition causes

[6]Various devices for sampling the lining tissue of the intestine were invented in the mid 1950's in Argentina, London and the USA. The one that became most widely used was the Crosby Capsule, named after Dr. William Crosby (1914–2005). This device was a capsule that the patient swallowed. After a period of time they would be given an X-ray to make sure that it was in the small intestine. It was connected to tubing that allowed the operator to apply suction—this would draw the intestinal lining into a hole in the end of capsule, and a sliding mechanism across the hole then neatly excised it. The capsule could then be dragged out of the mouth to retrieve the sample. Crosby was a military doctor who served in Korea and had wide interests in haematology and oncology, as well as a passion for languages. He studied Russian whilst serving in Korea, and published an acclaimed translation of Baudelaire's poetry.

diarrhoea and deficiencies of specific vitamins and minerals, as the inflamed and damaged intestinal lining is compromised in its ability to absorb nutrients.

How cereals bring about this damage and cause coeliac disease, but only in certain people, is still being worked out as we will see shortly from our gastro-archeological trench. However, before we start digging, it is no coincidence that I started this chapter with a quote from an ancient myth. For some reason, myths abound in relation to coeliac disease and there are some that we should debunk from the outset!

The Myth of Coeliac Origins

When I was a medical student, I was told that coeliac disease was more commonly found the closer one travelled to the West Coast of Ireland. The extraordinary non-scientific justification of this story was that the Irish used potato rather than wheat as their subsistence crop (until the well-known blight of 1845), and were therefore intolerant to the cereal through lack of exposure to it! In truth, coeliac disease is actually now known to be as common in Tehran (for instance) as it is in Galway—affecting about 1 in every 100 people in both. One of the populations with the highest proportion of coeliac disease is thought to be the Berber people of North Africa—the blond haired and blue-eyed descendants of the Phoenicians of ancient times.

It appears that the development and spread of coeliac disease coincided with the migration of humans from the so-called 'fertile crescent' of the Middle East, where cereal crops were first cultivated as long as 30,000 years ago. Thus, susceptible populations moved westwards across North Africa, north-west into Europe and Scandinavia, and south-east into northern India and modern-day Pakistan. Secondary migrations from Europe through exploration and colonisation led to the condition being prevalent in Caucasian populations of North and South America, South Africa and Australasia. Indigenous peoples of the Americas, Australasia, Africa and Asia are relatively protected from coeliac disease, as are the Indians of Dravidian origin from the south-east of the sub-continent and Sri Lanka. However, we are just beginning to see some individuals from these groups with the disease. Whilst this may be due to the spread of the westernised diet, the dissemination of Caucasian genes conferring susceptibility may also be implicated. For instance, among my coeliac patients are two Ethiopians, both of whom have European grandfathers as a consequence of the Italian invasion of the country (then 'Abyssinia') in 1936. As a physician specialising in coeliac disease, it is often useful to remember the history and geography one learnt at school!

The Myth of Age at Diagnosis of Coeliac Disease

Coeliac disease used to be thought of as a condition that one was born with but that only caused symptoms at around the time of weaning onto cereal grain-containing foods. However, it is now very apparent that the diagnosis of coeliac disease actually occurs in most people in adulthood—and at just about any age (the oldest person diagnosed with it in my practice was 95 years old at the time). This of course assumes that the condition is detected at the same time that it develops, which is not always the case. For instance, people often identify symptoms dating back many years that resolve when following a gluten free diet. This suggests that the condition has actually been present for this period of time. On the other hand, I have come across many that have been found not to have coeliac disease when investigated, only to develop it at some stage in the future suggesting that it has arrived at that time. We will return later to factors that might trigger the condition to arise in those who are genetically susceptible to it.

The Myth of Symptoms Associated with Coeliac Disease

Originally, coeliac disease was considered to be a life-threatening condition as children diagnosed with it early in life suffered from diarrhoea and weight loss—'failure to thrive'. In those severely affected a fatal outcome was usual. It is now clear that these cases are extremely rare and that the vast majority of people with coeliac disease—in fact as many as 9 out of every 10—do not have any symptoms of diarrhoea or weight loss. Instead, a diverse range of complaints is now found in people with the condition. Many present to their doctor with fatigue or a low blood cell count (anaemia). This may be secondary to the inability of the intestine to absorb sufficient nutrients such as iron or folic acid, which are needed to make blood cells. Abdominal symptoms such as bloating or cramping pains are commonly found, and some patients even describe constipation rather than diarrhoea. These are often misdiagnosed as being due to 'irritable bowel syndrome'.

Many people are found to have coeliac disease but describe no symptoms whatsoever, being investigated for the condition as a result of an affected family member. In fact, it is now thought that as many as a half to two thirds of all people with coeliac disease go through their entire lives without ever being diagnosed with the condition. Whilst it might seem at first sight that such people are fortunate, the opposite may actually be the case. Fatigue and underperformance are difficult to quantify. My concern is that such people might never know how well they might have felt if only they had been diagnosed with the condition and received treatment. Furthermore, when patients with coeliac disease present later in life to the doctor, it is often as a result of disability due to skeletal deformities caused by weak bones—a condition known as osteoporosis. Longstanding, undiagnosed coeliac disease can lead to this by causing malabsorption of the calcium and vitamin D needed to maintain bone strength. Another condition that occurs in older people who have undoubtedly had coeliac disease for a long time without a diagnosis is a specific and exceptionally rare type of cancer affecting the immune cells in the intestine called a 'T-cell lymphoma'. In this case it probably arises as a result of chronic, constant stimulation of the immune system in the gut.

The Myth of Coeliac Disease as a Food 'Allergy'

Coeliac disease—as we will see shortly—is quite different from the food allergies that we have looked at in the previous chapter. Whilst it is an immune-mediated condition and indeed one that is directed against a specific food component (gluten, which we will look at more closely shortly), there are significant differences between coeliac disease and food allergy. One of the most significant is that coeliac disease causes chronic damage to the lining of the small intestine, which usually develops gradually (over days or weeks) when eating gluten and resolves over weeks to months when it is taken out of the diet. Allergies, on the other hand, produce very rapid responses (within seconds) that resolve equally quickly when the offending substance is no longer present, and do not leave lasting damage in the gut. As a result, there are very few people with true food allergy who remain ignorant as to the culprit that brings it on. Contrast this with coeliac disease, where it took nearly 2000 years from the time of Aretaeus to that of Willem Dicke to work out that it was caused by some substance found in cereal grains, even though a dietary cause had long been suspected.

As we start to delve below the surface in our understanding of the mechanism of coeliac disease, we will come to realise that the coeliac disease and food allergies lie on opposite sides of the great schism that has divided the immune system throughout time. Food allergies largely result from the 'antibody' arm of the immune system, mediated by B' cells in the most superficial strata; coeliac disease is from the other side of the great divide, the cell-mediated or 'cytotoxic' aspect represented by 'T' cells and their predecessors.

Starting the Coeliac Trench: the Surface Layers

It is time for us to look at how coeliac disease comes about, and to do so we need to start digging our gastro-archeological trench, just as we did for food allergies in the last chapter. Once again, we will start with the very top-most layers involving the most recently-evolved immune mechanisms. Until very recently it was thought that coeliac disease could be completely understood without having to dig any further at all.

The reason for this is the key observation that only people with one or two specific forms of class II MHC molecules (labelled as either 'DQ2' or 'DQ8') can develop the condition. You will remember (from Chap. 5) that the MHC proteins are highly variable cell surface molecules that provide us with our individual identities. These are the 'tissue types' that we try to match between donor and recipient in bone marrow transplantation, for instance. We also learnt in Chap. 5 how the specific class II MHC molecules serve the purpose of presenting fragments of proteins to the cells of the immune system to recognise and choose to ignore (if from oneself), or to respond to (if not). Therefore, finding that only very specific MHC molecules are involved in coeliac disease suggested that a protein fragment of some kind was being presented to T 'helper' cells to trigger the response. Class II MHC molecules other than 'DQ2' or 'DQ8' presumably lacked the ability to bind and present the protein fragment to T cells. People bearing such non-compatible class II MHC molecules could not then develop coeliac disease, as the initiation step of the immune response would not be supported.

This important discovery of limited class II MHC association with coeliac disease provided the ability to unlock the mechanism of the condition. It would seem that all that was left was to find the protein fragment that was recognised by the specific MHC molecules and we would have uncovered the whole story, without scraping far below the surface of our dig!

The tale gets much more complicated than this as we shall see, but is worth pointing out that the class II MHC association does provide us with a very useful test for *not* having coeliac disease. If you lack the specific DQ2 or DQ8 types of the molecule then it is highly unlikely that you are capable of developing the condition. Unfortunately, it is not helpful if you *do* have the necessary DQ2 or DQ8 tissue type, as these are found in as many as 25% of individuals! Hence, assuming that about 1 in 100 people have Coeliac Disease, and a quarter of the population has the correct class II MHC, then only about 4 in every 100 of all the people with DQ2 or DQ8 will have the condition, and 96 will not. This is not, therefore, a useful test for coeliac disease, but it is a very useful test for not being able to have it!

To find the protein that is presented to T-cells to initiate the immune response in coeliac disease we clearly need to understand something about cereal protein chemistry, which is rather an alien world for gastroenterologists!

About Cereals

Wheat is the most abundant food crop in the world, and trade in wheat exceeds that in all other crops combined. It is the leading vegetable source of protein (which comprises about 15% of its dry weight—more than in any other major cereals). Carbohydrates are present in the wheat grain as starches and provide approximately 327 kilocalories of energy for every 100 grams. Additional important nutrients such as minerals and B vitamins are present in the husk of the whole grain, which also provides an important source of non-digestible fibre (or 'roughage').

What we recognise as modern wheat started out around 12,000 years ago as a type of wild grass—called Emmer[7]—in the Fertile Crescent and up into the southern part of Turkey. The three cereals recognised as 'founder crops' from this period also included another early variety of wheat, called Einkhorn, and barley. One of the important stages in domesticating these wild grasses as crops was to inbreed a mutation that prevented the grains from scattering and instead retained the seed head intact on the stalk. This mutation would not permit spontaneous propagation as the seeds would not germinate easily without their native dispersal mechanism. It allowed for easier harvesting, but meant that the seeds then had to been sown by hand for subsequent crops. Over the following millenia, selective breeding has resulted in over 25,000 varieties of modern wheat, with recent significant mutations such as dwarfing further improving yields and ease of harvesting. The degree of genetic tinkering that has taken place in wheat is staggering, and now in the era of genomic science we are trying to sift back through it to try to understand how it all works—several monthly scientific journals are dedicated solely to improving our understanding of cereal chemistry!

However, contrary to current thought (another coeliac myth), modern wheat varieties have *not* been bred to contain more gluten than the ancient strains, and in fact contain less protein overall and less gluten. Older varieties such as Emmer, Einkhorn and Spelt are still available and used in niche

[7] Emmer wheat was so named after the German—'Amelkorn'—where the word 'Amel' derives from the Latin 'Amylum' meaning starch. Einkorn (also German), on the other hand, derives from the fact that there is only one grain in each spikelet of the ear

culinary settings,[8] but it is a mistake to consider them lower in gluten content and therefore safe for people avoiding gluten.

Introducing Gluten

What then is this substance called gluten? Gluten is something of a generic name given to the protein fractions present in the storage part of the wheat seed. The wheat seed itself comprises three parts—the germ which grows into the wheat plant, the endosperm which provides the nutrients for the germ to grow into a seedling (before it can produce leaves and harness the energy of sunlight above the ground), and the fibrous husk surrounding it all. Flour can be produced using the whole grain, using just the wheat germ and endosperm, or just the endosperm (white flour). Mixing flour with water dissolves away the carbohydrates in the form of water-soluble starches, plus some of the proteins (globulins and albumins), and leaves behind the water-insoluble fractions called 'gliadins' and 'glutenins'. These are both complex mixtures of over 50 proteins. Together they are described as 'gluten' and make up around 75% of the total protein in the wheat endosperm.

The gluten proteins have unique properties. Whereas most proteins have specific functions for which they require a three-dimensional shape, the gliadins and glutenins appear only to provide a nutrient storage use. They are quite special in that they maximise the amount of nitrogen that is packed into a small space. This is achieved by being enriched in two particular amino acids which gives these proteins their alternative name of 'prolamins'—from **pro**line and glut**amin**e. It is precisely the unusual chemical composition of these gluten prolamins that endows them with the viscoelastic properties that bring about the baking properties of dough. This is its ability to stretch and deform whilst remaining 'sticky', hence enclosing air bubbles to provide lightness in the risen dough. Indeed, the very word 'gluten' is derived from the Latin word meaning 'glue'. It is also, as we shall see later, the peculiarities of the chemical structure of gluten proteins that lead to their relative indigestibility and results in relatively large fragments passing into the gut that can then be recognised by the immune system.

[8] Bread made out of Emmer wheat is called 'Pane di Farro' in Italy, where it is also used as a whole grain in soup. Einkorn is used as a form of 'Bulgur'—when milled and parboiled—and finds use in Middle Eastern dishes including Kibbeh (Levantine meatballs) and Tabbouleh (Lebanese salad). Bread made from spelt wheat is found as 'Dinkelbrot' in Germany, and the dried, un-ripened grains are sometimes eaten as 'Grunkern'. Beer is brewed from spelt in Holland and distilled from spelt for vodka in Poland.

Diverticulum #8.1 Foods that contain gluten

'Gluten' is present in wheat, barley and rye. These are the foods that need to be avoided by people with coeliac disease. This might seem straightforward, and whilst foods containing rye are usually obvious (such as 'ryebread'), barley is more difficult to exclude and wheat flour has myriad uses in catering.

The gluten-containing foods that initially spring to mind are of course bread, biscuits, pasta, pizza, cake and beer. However, flour is often used for thickening in sauces and gravy. It is present in many microwaveable 'ready' meals. It is also present in surprising places, including some chocolates and ice creams.

Gluten cannot be destroyed by heating and therefore chips that have been fried in the same oil as batter (fish and chips) are a source of gluten. Many fish and chip shops in the UK are now doing 'gluten free' nights where the oil is not contaminated.

The process of distilling removes gluten, so spirits (such as whisky) are gluten free, as is malt vinegar. Barley malt extract is used as a flavour enhancer in many foods and in large amounts it can be toxic to people with coeliac disease—therefore the safety of such foods depends on the amount of extract present in individual foods.

Maintaining a gluten free diet is not always intuitively easy. For instance, until recently eating a baked potato with grated cheese used to risk gluten ingestion. Cheese and potatoes are of course naturally gluten free. However, catering packs of ready-grated cheese used to have wheat flour added to stop the cheese from clumping. Thankfully, potato starch is more frequently used for this purpose nowadays.

Gluten as a Toxin

The discovery of the limited repertoire of class II MHC molecules associated with coeliac disease initiated a search for the toxic component of wheat. Clearly it had to be a protein, but which, and what part of the protein? Again, this was relatively straightforward to work out, as damage to the intestinal lining in coeliac disease is entirely reversible when gluten is taken out of the diet, and gets worse again if gluten is reintroduced. It was therefore possible to challenge individuals with coeliac disease on a gluten free diet (whose intestinal lining had recovered) with different protein fractions from wheat, and other cereals such as barley and rye, to see if any of them caused damage to occur over time. Later on, it became possible to keep small biopsies of the intestine alive long enough in organ culture experiments in the laboratory to challenge them in the same way and measure the strength of the immune response. In this way it was clearly identified that gluten was the culprit in all of these cereals. The study could be further refined by breaking up the gluten proteins into small fragments to see which parts were the most likely to be recognised by the immune system. The most reactive parts of the protein turned out to be exactly those that made gluten special—those which have a high proportion of the amino acid glutamine (although glutamine itself is not in any way toxic).

The next step was then to take individual lymphocytes out of the biopsies and stimulate them in the laboratory with the protein fragments. (In fact, doing the same thing with lymphocytes in the blood has very recently been used as a highly specific test for coeliac disease). However, as so often happens in science when the results *should* be entirely predictable, something surprising showed up. The peptide (protein) fragments that brought about the greatest responses in the intestinal biopsies did *not* for some reason manage to stimulate the lymphocytes on their own in a test tube.

When researchers were able to recreate the three-dimensional structures of the DQ2 and DQ8 molecules that bind to the toxic gluten protein fragments identified in the above experiments, they were similarly puzzled. Try as they might, they could not find a way to make the peptides fit into the 'presenting grooves' of these MHC molecules. Specifically, it was always the glutamine parts of the protein chain that got in the way—they were in fact actively repelled from it!

Why this turned out to be the case proved to be very instructive as we shall shortly see.

Diverticulum #8.2 What about oats?

It has been conclusively shown that some people with coeliac disease can react to a protein in oats that is similar to gluten—called 'avenin'. The proportion of people that do so is unknown but is likely to be less than one in 20 people with coeliac disease.

The issue is complicated by the potential for contamination of oats growing in the fields with other nearby cereal crops, and also by manufacturers handling different cereal grains in factories that produce predominantly oat-based foods. As oats are an important source of fibre and energy in the diet it is recommended that people with coeliac disease should continue to eat them in the first instance, as long as they are certified to be free of contamination with other cereals.

The Antibody Response in Coeliac Disease

The first tests for coeliac disease relied on the presence of antibodies in the blood against one of the gluten proteins, called 'gliadin'. However, it was quickly realised that such blood tests were not very accurate—as we have already experienced in the previous chapter, antibodies against food proteins are commonly found in people without any disease due to the constant activity of the gut immune system.

To see whether circulating antibodies could account for some of the features of coeliac disease, the blood of affected individuals was tested against a wide variety of different tissues. Some conditions that are described as 'auto-immune' conditions are caused by a misrecognition of 'self' and a resulting auto-destructive immune response. In auto-immune thyroid disease, for instance, an antibody can be identified that sticks to the cells of the thyroid gland. In coeliac disease, no such antibody was found that could conceivably be considered to be attacking the cells of the intestine. This confirmed suspicions at the time that the damage to the intestinal lining was not caused by antibodies.

Intriguingly however, an antibody was found in most cases that appeared to bind to the connective tissue 'glue' that holds muscles together—the endomysium ('*endo*' = inside, '*mysium*' = muscle). Finding such antibodies became the basis for a highly specific clinical blood test for coeliac disease, which has remained in routine use to the present day. The actual molecule in the muscles that bound antibodies from the blood of people with coeliac disease was identified in the late 1990s, and significantly improved our understanding of the mechanism of the condition.

The Missing Jigsaw Piece

The protein in the connective tissue against which the anti-endomysial antibodies in coeliac disease were directed turned out to be a common and very important enzyme called tissue transglutaminase (or TTG). This enzyme is present in all supporting tissues in every organ of the body and crucially acts to cross-link the scaffolding proteins of the structure. It works by chopping off a nitrogen atom and two hydrogen atoms from glutamine molecules in the protein chain—making it reactive and able to form strong chemical bonds with amino acids in adjacent protein strands. However, in the absence of a nearby protein, water molecules can step in and react instead, thereby converting the glutamine to glutamic acid.

When gluten proteins undergo chemical modification by TTG they lose the glutamine residues that were repelled by the DQ2 and DQ8 molecules, and they can now bind to them. This allows them to be 'presented' to the T 'helper' cells and initiate the cascade of immune responses leading to coeliac disease.

The antibodies that are formed against TTG in coeliac disease occur simply because the TTG binds the gluten protein fragments strongly, and nearby B cells in the lamina propria are able to act as their own 'presenting cells' to the immune system. Given the importance of TTG in making the scaffolding that holds tissues together, it might be thought that if the enzyme was blocked by antibodies against it, then tissue damage might occur. In fact, injecting human anti-TTG antibodies into mice does generate a minor degree of intestinal inflammation, but it is not currently thought that the antibody response in coeliac disease is much more than just a side issue, albeit one that gives us a useful blood test for the condition.

Back to the Drawing Board!

At this stage, one could be forgiven for thinking that we can now close the gastro-archeological trench on coeliac disease without any more digging. We have found that the T lymphocyte response is fundamental to the condition, we have identified the specific protein fragments that are recognised by the immune system, and we have seen that these need to be modified by TTG before the antibody response is generated. However, there are many outstanding questions that this mechanism does not answer. For instance, we have yet to explain how the T lymphocytes cause the tissue damage that we see with the condition. Indeed, the lymphocytes and dendritic cells necessary for the T lymphocyte response that we have described reside within the lamina propria beneath the surface, whereas the damage in coeliac disease is predominantly within the epithelial layer itself.

Usually in such circumstances we look for evidence of any similar conditions in animals which might shed light on the human disease. In this case, a very similar condition is found in some horses, rhesus macaque monkeys and also a breed of dog called the Irish setter. In affected dogs, there is damage to the intestinal lining—with villous atrophy—just as in coeliac disease in humans, and it causes episodic diarrhoea. Furthermore, it responds to removal of gluten from the diet. However, in these dogs there is no association with class II MHC, and the condition appears to be inherited by a single gene. We cannot therefore draw too many parallels with the human condition.

Modern experimental techniques mean that we can now insert all of the human genetic machinery that we think is necessary to generate coeliac disease into mice in the laboratory, to see if they develop the condition. Hence, we can breed mice that have the specific human class II MHC molecules, we can inject them with human T 'helper' cells that recognise gluten peptides, we can make sure that they have the correct TTG to modify the gluten, and we can then feed them gluten. And they remain healthy! There is not even a whiff of intestinal inflammation! Moreover, their cells express a chemical signature that suggests that the gut immune system is actually tolerant to the gluten. It is time to get the trowel out again!

Down to the Next Level

Digging deeper into the evolutionary history of the immune system takes us to the strata where we find the earlier types of lymphocyte, before the immunological 'Big Bang' and the expression of highly variable molecules associated with T and B cells. These cells are concentrated among the immune cells patrolling the epithelium—the intraepithelial lymphocytes or IELs.

You will remember that the intestinal IELs can be broadly divided into two types: those that appear to have a 'housekeeping' function—the 'natural' IELs—and those that are responding to immune stimuli—the 'induced' IELs. The induced IELs are in the most superficial layers of the gastro-archeological trench, and are involved in precise interactions through the specificity of their T cell receptors. Even though we know that there are very particular proteins involved in the immune response in coeliac disease, it is not in fact the induced IELs that appear to be relevant. Instead it is the natural IELs that seem to be most important—as if the gluten has hijacked the natural maintenance mechanisms of the epithelium to cause disease. Whilst these cells may bear a T cell receptor, this would not seem to be one that recognises a gluten protein or is necessarily involved in the reaction. It is the natural killer functions of these cells that are invoked in coeliac disease.

The natural IELs in coeliac disease are activated and primed to kill the enterocytes—the lining cells of the gut. This is part of their usual housekeeping role in removing damaged or otherwise defective cells in order to maintain the barrier—the front line—of the gut defences. However, in coeliac disease it is the wholesale destruction of these cells that leads to the characteristic tissue damage of villous atrophy seen with this condition. Interestingly, the basement membrane on which they sit is kept largely intact.

Why the IELs respond in this apparently exaggerated fashion requires us to delve yet deeper into the layers of the gut immune system.

The Lowest Layers

We have seen how the enterocytes of the epithelium themselves—the most archaic of the cells that could be considered part of the gut immune system—are able to direct immune responses by signalling distress to nearby immune cells. We have also come across the interleukins, molecules which were initially thought to signal only between the white (defensive) cells in the blood. However, we now know that interleukins are used much more widely, and one—interleukin-15—is used by the enterocytes to talk to IELs. Interleukin 15 is a very interesting signalling molecule in that it is not fully released by the enterocyte but 'presented' on the cell surface, on one half of its receptor. Contact with a cell bearing the other half of the receptor is required for the message to be transmitted to it. The production of interleukin-15 by the enterocytes is dependent on very ancient mechanisms such as those used in the response to bacteria. These include the Toll-like receptors and their internal cell-signalling proteins that we first came across in the sponge, which are similar to the 'TirA' molecule found in the social amoeba!

A certain low level of interleukin-15 expression on enterocytes is required as a maintenance signal for the IELs—particularly those with the γδ type of receptor. Mice that grow up in a sterile environment, lacking bacteria, produce less interleukin-15 and thereby have fewer IELs. Interleukin-15 not only increases the survival of these IELs but also 'primes' them for cell killing.

Although interleukin-15 is expressed at a low level to maintain IEL populations, its secretion is increased in times of threat. It is therefore acting very much like an alarmin, the group of 'danger' signals released from gut epithelial cells which we looked at in the previous chapter. Furthermore, the surface molecules on the IELs that are upregulated by interleukin-15 in order to activate them for cell killing are those that recognise proteins expressed by damaged or threatened enterocytes. A potential threat to the intestinal lining therefore evokes a response that leads to an increase in interleukin-15 secretion to increase IEL numbers and activate them to kill injured enterocytes. The response is neatly tidied up by the IELs also secreting growth factors to encourage enterocyte cells to divide and repair the resulting defect in the barrier.

This explains some simple observations of the intestinal lining in coeliac disease. There is an increase in IELs, particularly those with the γδ type of receptor. There is also a feature called 'crypt hyperplasia' whereby cell division is increased at the base of the villi in order to try to restore the epithelium.

Interleukin-15 is over-expressed by the enterocytes of patients with coeliac disease. However, the expression of interleukin-15 remains high in people with coeliac disease who are following a good gluten free diet and have no other signs of intestinal damage. Furthermore, close relatives without any signs of coeliac disease may also show increased levels of interleukin-15, as well as other markers of enterocyte stress. It is possible then that genetic factors could lead to higher interleukin-15 expression in the epithelium which then predisposes to the development of coeliac disease, but requires additional triggers before it occurs.

Interestingly, when we return to the genetically engineered mice that have the correct human class II MHC to recognise gluten, and add in a further gene to make them overproduce interleukin-15 in the epithelium, they still do not develop coeliac disease when we feed them gluten!

'Stressing' the Epithelium

At this point in time we still do not know for certain the identity of the additional triggers that cause the stress in the intestinal epithelium. There is increasing evidence that a common viral infection caused by rotavirus in children is associated with the later development of coeliac disease in susceptible individuals. It is also apparent that both the number of infections in childhood and the amount of gluten ingested work together to increase the risk. This would fit with what we understand of the need to induce a degree of damage in the intestinal lining in order to initiate coeliac disease. Why, though, would this lead to a specific immune response directed against gluten itself? And why is there no other disease process that is in any way similar to coeliac disease, caused by other food proteins?

One of the answers to this conundrum may be that the gluten proteins themselves contain fragments that may directly lead to damage and epithelial stress. One particular such part of the protein appears to lead to damage to the intestine, when tested in laboratory conditions, without initiating the T lymphocyte response. This piece of the protein is quite distinctive from that which is recognised by the T cells. The exact mechanisms of the damage caused by gluten are currently unclear (in fact a whole range of different possibilities have been suggested), but they may interact with the responses seen in a viral infection. In effect this would mean that the gluten is working as its own 'adjuvant'—a means of increasing an immune response that we saw associated with mucosal vaccines in the previous chapter.

However, there are two particular facets to the peculiar actions of gluten on the gut that are of special interest. Firstly, it appears to make the epithelium 'leaky'. It does this by stimulating the production of a protein by the enterocytes that weakens the join between them. This permits quite large fragments of the gluten protein to seep into the underlying lamina propria where they can be recognised by the immune cells present there. Secondly, it appears to increase the expression of interleukin-15—the stress response messenger—in the lamina propria itself.

This would appear to be fundamental to the development of coeliac disease. As we have seen, we can genetically engineer mice to have the correct human class II MHC molecules which recognise the protein fragments of gluten. We can also make these mice express excessive amounts of interleukin-15 in the epithelium. However, they do *not* develop the characteristic features of coeliac disease when fed gluten. Only if they are made to *also* over-express interleukin-15 in the lamina propria as well as the epithelium do they develop a condition resembling coeliac disease in humans.

License to Kill

We are now in a position to begin to understand how coeliac disease occurs. Gluten—being relatively indigestible due to its unique amino acid composition—is present in the intestine in large fragments that are capable of being recognised by the immune system. Some parts of the gluten molecule have a direct effect on the epithelium to make it leaky and allow access to the underlying lamina propria. Some fragments also appear to 'distress' the epithelial cells which leads to them expressing cell surface markers of damage, and to produce increased amounts of interleukin-15. This stimulates and primes the IELs in the epithelium, but also increases the production of interleukin-15 in the lamina propria.

In those individuals who have the correct class II MHC molecules, gluten protein fragments (that have been appropriately modified by tissue transglutaminase) are recognised and presented to T 'helper' cells in the lamina propria. As a result, they stimulate the production of antibodies that are recognised in the blood stream as a test for coeliac disease (and may also have some role in causing tissue damage). The T 'helper' cells also produce chemicals that 'licence' the primed IELs to kill the stressed enterocytes. This results in the damage to the intestinal lining characteristic of coeliac disease, and the majority of the clinical consequences and symptoms experienced (Fig. 8.2).

This extraordinary sequence of events only occurs with gluten and no other protein found in food. Why it should be that gluten has so many immunostimulatory properties is unclear—certainly it is not just the fact that it is poorly digested and present in large fragments in the gut. It appears to have

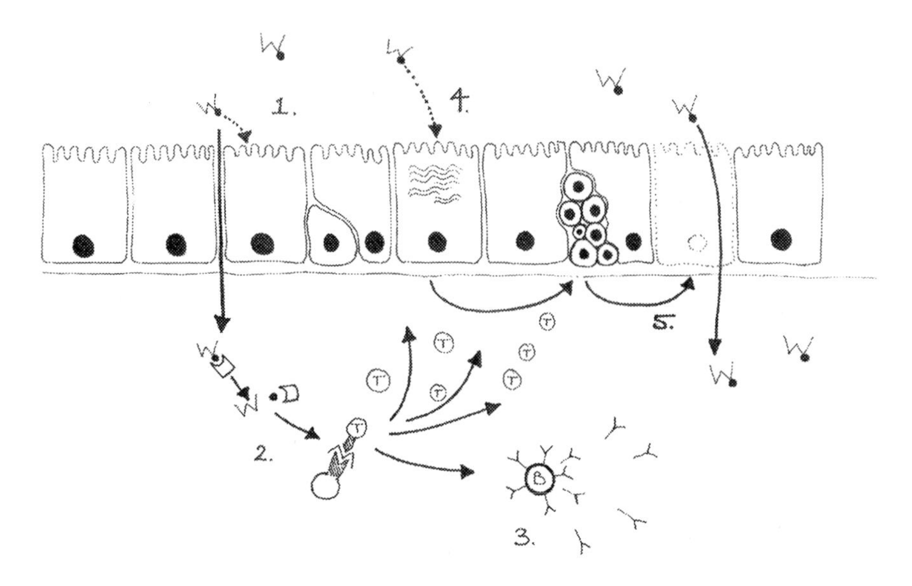

Fig. 8.2 Our current understanding of the mechanisms of coeliac disease. (1) Large peptide fragments of gluten are present in the gut due to its relative indigestibility as a result of the unusual amino acid composition of the protein. Gluten fragments act on enterocytes to open up the tight junctions between the cells and allow the large protein fragments through into the lamina propria. (2) An enzyme present in the lamina propria binds to the gluten fragments and modifies them to a form where they can bind to specific MHC class II molecules on dendritic cells and thereby activate T cells via the 'protein' sandwich. (3) Activated T helper cells assist B cells in making antibody against the gluten fragments. However, because tissue transglutaminase binds strongly to gluten and is taken up by phagocytosis inside B cells, T helper cells also stimulate the production of antibodies against tissue transglutaminase, which can be detected in the blood stream as a useful marker of the disease. (4) Gluten fragments (or other wheat components) cause enterocytes to become stressed and secrete interleukin-15. This leads to proliferation and activation of cell-killing intraepithelial lymphocytes (IELs) (in concert with cytokines produced by activated T cells in the lamina propria). The IELs kill stressed enterocytes by virtue of their 'housekeeping' function as they recognise molecules produced on the cell surface of damaged cells. (5) Impaired epithelial barrier function permits the passage of more gluten fragments across the epithelium

'self-adjuvanting' properties and the stimulation of the interleukin-15 production in the lamina propria also seems to overcome the usual tolerance to ingested proteins. These special features of gluten may mimic the molecular effects of a viral infection in order to produce this effect.

One other special feature of the immune response to gluten is that the usual immune pathway is bypassed. Proteins to which the gut is tolerant are recognised by presenting cells in the lymph nodes of the mesentery where the appropriate tolerant T cell response is engendered. However, in coeliac disease the B cells in the lamina propria of the gut appear to be capable of presenting gluten to the T cells themselves, without trafficking to the lymph nodes. It is probably because of the interaction of gluten with tissue transglutaminase that this occurs in this site and causes a 'short circuit', which results in licensing the primed IELs to kill the stressed enterocytes.

'Refractory' Coeliac Disease

Usually, when gluten is removed from the diet, the intestine fully recovers. As we have seen it does not go entirely back to normal—there remains an increase in the level of interleukin-15 produced by the enterocytes and an ongoing change in the composition of the lymphocytes present in the epithelium. However, the inflammation and damage (villous atrophy) recover over a period of several months. In some cases however, the intestine fails to recover when gluten is taken out of the diet. This is called 'refractory' coeliac disease—in other words it is 'refractory' to gluten withdrawal. Given how difficult it is to remove all of the gluten from the diet, it is not uncommon for this condition to be over-diagnosed. Refractory coeliac disease is separated into two distinct types. The first type is indistinguishable from active coeliac disease and may therefore reflect an exquisite sensitivity to gluten such that even trace amounts of gluten present (when every effort has been made to take it out) is enough to trigger a response. On the other hand, type 2 refractory coeliac disease is characterised by the presence of a very special population of IELs within the epithelium.

Diverticulum #8.3 How much gluten is too much?

In most countries where a substantial proportion of the population is at risk of coeliac disease, there is stringent legislation in place regarding the labelling of gluten-free foods. In most instances—Europe, Canada and the USA—foods can be labelled as 'gluten free' if they contain less than 20 parts per million of gluten. In Australia and New Zealand however, the law is much tougher and foods are only allowed to be described as 'gluten free' if they have no detectable gluten present in them and do not contain oats.

To put this into context, 20 parts per million is the weight equivalent of 1 teaspoonful of flour out of 4 metric tons. In practical terms, the detection limit of most assays for gluten in food is around 10 parts per million.

Challenge studies—giving people set amounts of gluten in capsules over a period of time—have been used to evaluate the safety of different amounts of gluten in the diet. These have shown that a total level of up to 50 mg a day is safely tolerated by the majority of individuals with coeliac disease. However, one patient was noted to respond to as little as 10 mg a day. There is a therefore quite probably a spectrum of individual sensitivities to gluten with different people tolerating more or less around this limit.

To put this into context, 50 mg is the amount of gluten present in 1/100th of a slice of normal bread.

A Window on the Past?

One of the most dreaded complications of coeliac disease is a type of cancer of the immune system called a lymphoma, occurring in the small intestine. Usually, lymphocytes proliferate when they are stimulated by a trigger that they respond to, but at the same time they initiate a 'self-destruct' pathway to limit their growth when the threat has passed. However, when the growth controls fail to work the cells continue to divide and grow into a mass called a tumour, which can then spread to distant sites as a cancer. The lymphoma associated with coeliac disease is—perhaps unsurprisingly given what we know of the disease—a tumour of the T lymphocytes. It is extremely difficult to treat when established, and as with other cancers early detection may be the key to cure.

In the late 1990's, a team in Paris discovered that the lymphoma in coeliac disease is caused by proliferation of an unusual type of lymphocyte, which can often be identified diffusely within the intestinal epithelium for some time before the cancer develops. Patients with these cells have ongoing severe changes of coeliac disease in the intestine despite the withdrawal of gluten and are therefore refractory to treatment. The presence of these unusual IELs defines a new form of refractory coeliac disease, known simply as 'type 2'. These patients need to be kept under close surveillance for the development of lymphoma, which occurs in about half of such cases. Thankfully however, type 2 refractory coeliac disease is exceptionally rare, and occurs in as few as 1 in every 1000 patients with coeliac disease. Unfortunately, many do not know that they are likely to have had coeliac disease for many years beforehand without showing or recognising the symptoms, as the refractory condition tends to occur later in life.

The particular cells that expand in type 2 refractory coeliac disease show features of both T cells and natural killer cells, and are capable of direct killing of enterocytes to cause the damage seen in this condition. However, unlike T lymphocytes they do not express T cell receptors on their cell surface but only within the cell itself. In this regard they appear similar to immature T cells which are still undergoing development. Analysis of the pattern-recognising portion of the T cell receptor genes show that they have undergone RAG-mediated rearrangement of the relevant DNA. Such cells are not unique to type 2 refractory coeliac disease and are found in normal active coeliac disease and also in non-coeliac individuals where they comprise up to 10% of the IELs.

However, the unique feature of these unusual cells in type 2 refractory coeliac disease rather than healthy individuals is that, rather than demonstrating a wide range of pattern recognition against different proteins, the cells all appear to share the same individual T-cell receptor. This suggests that they have all developed as 'clones' of one cell in which this DNA change took place, and where growth restraints became inactivated to allow it to proliferate wildly. These 'precancerous' cells wreak havoc in the epithelium by killing enterocytes, in much the same way as activated IELs do in active coeliac disease. In type 2 refractory coeliac disease the cells appear to have developed a mutation in their internal machinery that leads them to make an exaggerated response to the normal background levels of interleukin-15 which activates them. However, they then set up a vicious cycle by damaging the epithelium, which leads to greater interleukin-15 production, more proliferation and cell death.

In the healthy individual without coeliac disease, the function of these cells is unknown. Recent studies suggest that they arrive in the intestinal epithelium directly from the bone marrow, without passing through the thymus. They appear to have the potential to mature into $\gamma\delta$ T lymphocytes, although this would appear to be a minority route of IEL production in humans. They receive the necessary signals and growth factors to do so from the intestinal epithelial cells, just as the majority of maturing T cells are instructed by the epithelium of the thymus. This would appear to echo the dim and distant past where the gut epithelium may have been the primordial site of immune cell generation, as we surmised in Chap. 5. In refractory coeliac disease these maturing IELs appear to be subverted into becoming killer cells by a now familiar arch-culprit in the epithelium—interleukin 15 itself.

Diverticulum #8.4 Possible treatments for coeliac disease

Currently, the only treatment for coeliac disease is to remove gluten from the diet. However, given the number of critical steps in the development of coeliac disease, there are a number of possible targets for drug developers that might prevent the immune response to gluten.

Firstly, it might be possible to genetically engineer wheat to remove the gluten protein from it. This is actually much more challenging than it might sound due to the genetic complexity of wheat, which has resulted in many copies of protein genes being present. Gene 'silencing' techniques whereby genes are stopped from translating into protein have proven capable of reducing the gluten content of wheat by between 75 and 90% without adversely affecting the baking quality. Unfortunately, this is not enough of a reduction for the wheat products to be safe for people with coeliac disease to eat.

Next, one could ensure that the gluten is fully digested, as its relative indigestibility leads to large fragments being recognised by the gut immune system. Enzymes from some moulds and bacteria are capable of completely digesting the protein fragments into small enough chunks to evade detection and work is ongoing to either pre-treat food with them before eating, or to introduce these enzymes in capsule form along with the food. However, at the present time these approaches still do not work well enough.

If the intestinal epithelium were more resistant to the protein fragments passing across it then this would prevent the immune reaction occurring. There is one drug already in trials that blocks the increased leakiness of the gut lining with coeliac disease. The effects on patients whilst eating gluten are clearly discernible but are still inadequate to allow people with coeliac disease to eat gluten.

At the next step in the pathway towards coeliac disease, blockade of the tissue transglutaminase enzyme which prepares the gluten fragment for recognition by the immune system might work. TTG has a number of essential roles in the body and producing a medicine that safely blocks the enzyme in humans is potentially challenging, but nevertheless there are candidate molecules that are ready to enter trials.

Finally, interleukin-15 can be blocked using an antibody made against it. Using artificial antibodies to block chemical messengers is a tried and tested technique in medicine and many such 'biological' treatments are available. Unfortunately, blocking interleukin-15 by itself does not appear to have a dramatic effect in coeliac disease: it appears that other interleukins may help play a part in the inflammation.

Finally, there was much hope until recently that 'immunising' by regular injections of gluten peptides into the skin would tolerise the immune response and provide an immunological cure—much like desensitisation from allergies. Unfortunately, this strategy has proved unsuccessful in clinical trials.

Unfortunately, then, we are still quite a long way away from a cheap, safe and effective non-dietary treatment for coeliac disease.

The Bottom Line: Intestinal Bacteria

Once again, when we follow the trench to its deepest extent we come to the relationship of the gut immune system with the bacteria that live within it. We are now a long way from the most modern, upper layers of the coeliac dig where so much of the mechanism of the disease appeared to be present at first sight. Is it possible that bacteria could have a role in the development of coeliac disease as well?

We are familiar with the presence of large numbers of bacteria in the colon (the large intestine) but it is clear that the small intestine also has its own 'microbiome' of bacteria that varies between individuals. Bacteria play a role in the maturation and complexity of the gut immune system, and factors that they produce nurture the immune cells and alter their function. We have already seen with food allergies how the presence of bacteria might have a tolerising effect on the gut immune system, which seems to have been lost in coeliac disease.

Simple experiments in rodents would suggest that the bacteria of the intestine can reduce inflammatory responses. For instance, rats and specific strains of mice that have been brought up in a sterile environment develop intestinal damage when fed gluten over a prolonged period of time, but not when colonised with normal bacteria. However, whereas some bacteria may lead to a reduction in inflammation, others may exacerbate it. Interestingly, gluten-induced damage can be re-initiated in the mice by introducing a common strain of bacterium (*Escherischia coli*, commonly known as *E. coli*) from the gut of a patient with coeliac disease. Human studies have also implicated the early use of antibiotics in children as leading to a higher risk of later coeliac disease.

The nature of the specific interactions of bacteria and the immune system that modulate the response to gluten are unclear. Certain direct effects of bacteria could be postulated. For instance, some may produce enzymes that can digest gluten without it becoming visible to the immune system, whilst others may increase the leakiness of the gut epithelium.

At the present time we have not been able to identify particular species of 'good' or 'bad' bacteria in coeliac disease and it is too early to dash out and buy a probiotic remedy for the condition. Unfortunately, our understanding of the interplay of the bacteria and the gut immune system is still in its infancy—particularly in the small intestine where sampling can be challenging. Most of the studies of the gut bacteria in coeliac disease are of the colonic flora rather than that of the small intestine and we do not yet know how the two relate. The picture is further complicated by the effects of diet on the bacterial composition of the gut (including a gluten-free diet) and a bi-directional effect of the host itself on the bacteria.

However, there are clear differences seen in the types of bacteria present in patients with coeliac disease compared to those without. In particular, there appears to be a higher proportion of types of bacteria such as our old friends the bifidobacteria that produce short chain fatty acids (butyric acid for instance) and appear to downregulate inflammatory responses in the gut by a direct action on regulating T lymphocyte function. Butyric acid also has a direct effect on preventing stress within the enterocytes themselves. Further understanding of the role of the gut bacteria in the development of coeliac disease is a very active area of current research and will hopefully produce some breakthroughs in treatment. At the beginning of this chap. I noted that the most obvious mechanisms of coeliac disease reside in the most superficial layers of our gastro-archeological dig—the B and T lymphocytes, and for many years we looked no deeper. It is ironic then, and testimony to the concept, that possible explanations and cures should come from improving our knowledge at the very bottom of trench!

Dissecting the Layers

There are still many unknowns about coeliac disease. As we have seen, the response to gluten involves all of the layers from the most ancient levels of the gut immune system—the enterocytes signalling danger responses, through the levels of ILCs and natural killer cells to the most evolutionarily recent immune pathways of T and B cells in the lamina propria and the immune recognition of specific protein fragments.

Just as we did with food allergies, we are now in a position to look at each of these levels and ask what happens if not all of the necessary components are present to mount the full response.

Many genes have been found to be associated with coeliac disease in some way and most of them relate to the initiation or regulation of the inflammatory response more generally. It is therefore highly likely that whilst the clinical picture of 'coeliac disease' remains characteristic, different individuals may differ in their sensitivity to gluten, depending on the particular genetic mutations they carry.

Diverticulum #8.5 The genes involved in coeliac disease

There is clearly a strong genetic component to coeliac disease. Identical twins share all of their DNA genetic code, and if one has coeliac disease then it is most likely (at least 80%) that the other has it or will develop it. In contrast, first degree relatives of a patient with coeliac disease such as siblings, parents or children, who share only 50% of their DNA, have a 1 in 10 chance of developing the condition.

The correct class II MHC molecule (DQ2 or DQ8) is clearly a prerequisite and this is genetically determined, but it only contributes about a third of the heritability of the condition. Nearly 100 additional genes have been implicated in coeliac disease in some form or other but all of those identified (including the MHC association) still add up to less than 50% of the genetic heritability. In such 'polygenic' diseases, common mutations in lots of genes (or the 'on' and 'off' switches that control their translation) add up to cause the condition. This means that there may be significant variability in the condition. However, no particular patterns have been identified in coeliac disease to suggest that there are subtypes.

When a disease is caused by a mutation in a single gene then the mechanism of the disease is usually easy to understand. Where many genes are involved one can instead build up a picture of pathways in which their proteins are involved to try to understand the disease. In the case of coeliac disease, however, there are few surprises—the genes involved largely implicate T and B cell functions, early immune pathways and also genes associated with autoimmune diseases such as type 1 diabetes or rheumatoid disease. Surprisingly however, none as yet identified are specifically involved with the gut itself.

Sometimes patients have the coeliac disease-associated antibody present in the blood (the 'anti TTG' antibody), yet the intestine shows no damage. Clinicians call this situation 'potential' coeliac disease. They have struggled to understand whether this represents a very mild form of coeliac disease, or whether the presence of the positive blood test is an early sign of disease which will occur later. It could be either. Interestingly, the genetically engineered mouse model of coeliac disease (in which interleukin-15 over-expression is required in both the epithelium and the lamina propria) may provide a clue. When interleukin-15 is only expressed in the lamina propria but not the epithelium, the characteristic anti-TTG antibodies are produced but the epithelium remains intact—a mimic of 'potential' coeliac disease. Furthermore, a study of unaffected relatives of coeliac disease patients showed that some showed subtle features of 'stress' in their gut epithelium without having any features of the disease itself. It is possible therefore that coeliac disease involves both a genetically sensitive epithelium that is easily stressed, along with over-production or over-responsiveness to interleukin-15 in the lamina propria. If only the latter is present without the epithelial weakness, then antibodies may be produced without the intestinal damage—'potential' coeliac disease. Clearly immune activation is required in both sites—the lamina propria and the epithelium—for the full-blown condition of coeliac disease to coour.

Taking Away the Surface Layers: Gluten 'Intolerance'

We should now perhaps envisage what might happen if we lacked the top layer of the immune response to gluten. This could happen quite easily—for instance if people do not have the correct class II MHC molecules to recognise and 'present' the gluten peptides to the immune system. This removes the B and T cell responses that require MHC to work, similarly to knocking out RAG in the mice experiments on food allergy in the last chapter.

We are just beginning to understand that the deeper layers of the immune response to gluten may be capable of producing disease by themselves, without the inflammatory augmentation from the upper levels. For many years, patients have complained of symptoms relating to gluten but have shown no signs of the intestinal damage or antibody response associated with coeliac disease, or lack the correct MHC molecules. Whereas such patients were initially dismissed, recent studies are beginning to shed light on this newly described condition of 'non-coeliac gluten sensitivity' or 'NCGS'.

We are still in the early stages of understanding this condition. Indeed, many people who consider themselves to have significant symptoms after eating wheat do not actually experience any when challenged with gluten in a trial setting and may therefore not be reacting to gluten at all, but perhaps something else in the wheat. For this reason, some are describing this syndrome as 'non-coeliac gluten/wheat sensitivity'—'NCGWS'.

The symptoms of NCGWS predominantly arise from the abdomen—bloating, diarrhoea, abdominal pain or constipation—but 'extra-intestinal' symptoms are common. These may be difficult to characterise or measure but include muscle aches and pains, fatigue and headaches. It appears that self-reported NCGWS is substantially more common than coeliac disease—for instance, as much as one quarter of the Australian population restrict gluten for assumed health benefits. The popularity of a gluten-free diet owes much to its promotion by celebrities and sports personalities, and undoubtedly overstates the number of true NCGWS sufferers—by up to five-fold when compared to experimental trials where patients have received wheat without their knowledge.

The Other Culprits in Wheat

Complex carbohydrates in wheat starch may lead to gastrointestinal symptoms that are not related to the immune system. Being poorly digested, such 'fructans' and 'fructose oligosaccharides' present in wheat can pass through the upper intestine without being broken down and absorbed. When they reach the colon, the plentiful bacteria there have the enzymes capable of digesting them and produce a variety of by-products including gas. Therefore, such components in wheat can lead chemically—non-immunologically—to bloating and abdominal pain characteristic of the irritable bowel syndrome (IBS). Not surprisingly then, treatments aimed at reducing such IBS symptoms often rely on reduction or withdrawal of wheat intake.[9]

[9] Dietary components that lead to increased fermentation (and gas production) in the colon have been lumped together in the expression 'FODMAP' coined by the Australian gastroenterologist Peter Gibson in 2005. FODMAP stands for 'Fermentable Oligo-Di- and Monosaccharides And Polyols' and includes sugars and starches that are poorly digested or absorbed in the upper gastrointestinal tract. They lead to irritable bowel symptoms of bloating, abdominal pains and change in bowel habit through increasing the fluid content of the small intestine, increasing the amount of gas produced in the colon and stimulating colonic movement to cause diarrhoea. There are concerns that the use of a low FODMAP diet in the long term could lead to nutritional inadequacy and weight loss through restriction of carbohydrates, and detrimental alteration of the colonic bacterial composition. It is also complex and difficult to follow unless through the tutelage of experienced dietitians. Other forms of low-fibre diet may be easier to follow and equally efficacious if used in the short term.

However, there is another component of wheat that appears to cause problems. This is called ATI, which stands for 'Amylase/Trypsin Inhibitor'. ATI is a natural inhibitor of enzymes that digest starch (amylase) and proteins (trypsin). It is present in quite large amounts in the storage compartment of the grain, where it functions to counteract digestive enzymes produced by potential invasive organisms such as bacteria and fungi and thereby protects the seed from rotting in the ground. Wheat ATI has been shown to stimulate some of the oldest components of the gut immune system and acts on the same Toll-like receptor (TLR4) that recognises the lipid component of bacterial membranes. Activation of this Toll-like receptor can cause a stress response in enterocytes and increase interleukin-15 production. In the laboratory, inflammation can be induced in samples of intestinal lining taken from patients with coeliac disease on a gluten free diet by the addition of ATI.

Interestingly, the activation of TLR4 leads to the release of molecules into the blood-stream that serve as an 'immune signature' when detected on blood tests. This immune signature is present in both coeliac disease and NCGWS and suggests that TLR4 activation plays a role in both conditions, potentially by ATI. Furthermore, the same immune signature has been identified in some patients with chronic fatigue but no gastrointestinal symptoms.

Stepping Back: a View of the Whole

It is time now to take a last step back and look at the strata that we have uncovered in our dig. Interestingly, and in complete contrast to what we always considered to be the mechanism of this condition, the sequence of events appears to start from the bottom up.

The very lowest level is that of the intestinal bacteria and their interaction with the epithelial lining and the immune system of the gut. We are still unclear how they may modulate the immune response to gluten, but there are increasing signs that the gut bacteria are capable of tolerising or activating the immune system of the gut depending on their composition and activity.

A pathway that enterocytes use to respond to harmful bacteria appears to be subverted and switched on either by gluten itself or by ATI, a protein commonly found along with gluten in cereals. This leads to enterocyte damage and the expression on the cell surface of 'stress' molecules. It also leads to the secretion of interleukin-15, which activates the lymphocytes of the next level up in our trench. Interleukin-15 also acts in the lamina propria to silence the 'regulatory' T cells that might provide tolerance.

There are several substrata of IELs present in the trench. The lowermost have been identified through their abnormal proliferation in type 2 refractory coeliac disease and appear to develop in the intestinal epithelium itself. This is perhaps a throwback to ancient times—as far back as the jawless fishes (agnathans)—when lymphocytes are postulated to have developed entirely in the intestinal lining. The different strata of IELs involved in the response include (in upwards order) the 'innate lymphoid cells', natural killer cells, $\gamma\delta$ T cells and 'natural' $\alpha\beta$ T cells. They are primed by interleukin-15 and have cell surface markers that recognise the stress markers on the enterocytes, stimulating the IELs to kill them. This leads to increasing leakiness of the surface epithelium that allows large fragments of protein into the lamina propria.

The gluten fragments are modified by an enzyme called tissue transglutaminase (TTG) that allows them to be recognised by the immune system. Usually 'professional' presenting cells (such as dendritic cells) carry proteins to the lymph node to generate a tolerant response, but in coeliac disease there appears to be a 'short circuit' whereby B lymphocytes in the lamina propria present the gluten fragments directly to T 'helper' cells. This brings about the production of anti-TTG antibodies which can be detected in the bloodstream and used as a diagnostic test. They may also have an additional role in causing disease. The T 'helper' cells also secrete chemical messengers that further stimulate the IELS to kill the surface enterocytes.

All of these layers work together to generate the full-blown response. However, only a small proportion of people with the necessary class II MHC molecules who eat gluten suffer from the condition. Other genes are clearly critically important—as shown by the twin studies—and many are yet to be uncovered. Those that we do know about modulate the inflammatory response of the uppermost layers (the T and B cell responses) and molecules involved in the antiviral response. It would therefore appear that there may be a 'threshold' effect dictated by the background genetic makeup of the individual determining the level of their immune response to environmental triggers such as bacterial or viral infection, occurring also during the ingestion of gluten. The genetic effects probably permeate all the way through the layers of the trench—involving genes that modify the bacterial species present in the gut, the TLR based response to them and the amount of enterocyte stress shown.

There are tantalising hints of the ways in which genes may separate out the strata. For instance, unaffected family members of patients with coeliac disease show signs of increased epithelial stress, suggesting that this may be inherited separately from other components required for the condition to occur. Possibly the opposite of this is those patients that have antibodies to TTG but do no damage in the epithelium ('potential' coeliac disease)—maybe as a result of not inheriting the epithelial stress response. Evidence to support this separation of the layers comes from the genetically modified mice that only show a coeliac disease-like condition when they have mutations affecting both the epithelium and the lamina propria.

Finally, we are just beginning to consider the possibility that there are patients whose disease affects only the lowest layers—what we now consider as 'non-coeliac gluten or wheat sensitivity'. Furthermore, these patients may also include some with no gastrointestinal symptoms at all but features of aches and pains and chronic fatigue—a condition that has long been poorly recognised and treated by the medical profession.

All in all, the gastro-archeological yield from the coeliac trench, whilst leaving quite a few unknowns has clearly brought up some useful gems!

9

'Back to the Very Beginning'

The Inflammatory Bowel Diseases

Summary In which we move from the small intestine and the immune mechanisms involved with its relationship to food and encounter the inflammatory bowel diseases affecting the other end of the gastrointestinal tract and its association with commensal bacteria. Once again, we come across the 'great schism' in the sidewall of the gastro archeological trench that divides the antibody-type from the cell-killing type of immune response with the conditions known as ulcerative colitis and Crohn's disease lying on respective sides. We identify common pathways of these two conditions in the upper strata of the dig relating to the 'on' and 'off' switches of the adaptive immune system and we see how these are in turn controlled by deeper and older mechanisms. At the very bottom of our trench we reveal how ulcerative colitis and Crohn's disease differ, by respectively dealing with invasion of microbes across the epithelial containment barrier, and the cellular containment barrier. Ultimately we dig down to the molecular switch that separates the two sides of the great schism and encounter some ancient secrets of life through association with a surprising location.

J. Woodward, *The Gastro-Archeologist*, https://doi.org/10.1007/978-3-030-62621-1_9

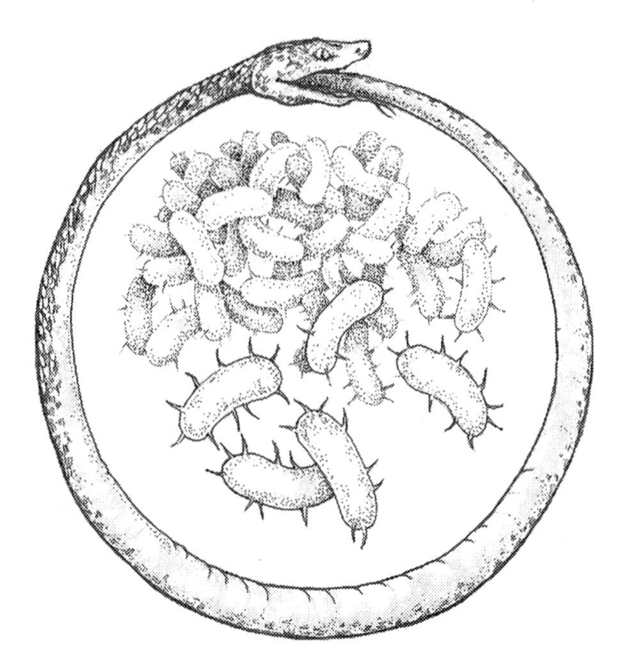

Ouroboros—the serpent eating its own tail as a symbol of the cycle of life, death and rebirth

'In my beginning is my end....in my end is my beginning'

'What we call the beginning is often the end. And to make an end is to make a beginning. The end is where we start from'.

From 'Four Quartets' by T.S. Eliot[1]

Life, Death and Rebirth

Demeter was the Greek goddess of agriculture and harvest. Given the central role of this figure in mythology, it is not surprising to find that references to her date back to at least the Mycenean period, over 1000 years BCE. The principal role of Demeter was as a goddess of fertility and life. However, her

[1] Thomas Stearns Eliot (188-1965), better known as 'T.S.Eliot' was an American-English poet, born in Boston but who moved to England in 1914 and revoked his US citizenship. He was awarded the Nobel prize for literature in 1948. These quotes are from his acclaimed 'Four Quartets'. The first are the opening and ending words of 'East Coker' the second of the poems, a village in Somerset where his ashes were taken after his death, being the place where his family originated before emigration to America. The second is from the fourth poem 'Little Gidding'. This is a village in Cambridgeshire noted for its seventeenth century religious community with whose approach to faith Eliot sympathised. There are themes in this poem that relate to the subject matter of this chapter, not least the 'fiery death' necessary for purgation, the importance of sacrifice for living, and the unity of past, present and future!

importance grew over time and in many regions she became seen as a Gaia-like 'Earth-mother'. Other deific functions become conflated with her primary purpose and she came to represent death as well as life—perhaps relating to the sprouting of wheat from buried seed under the ground. The Athenians called their dead '*Demetrioi*' in reference to her. This cycle of life and death is a central theme of many theologies. The associated symbol of the Ouroboros—the snake eating its own tail—has mythological links in diverse and separate ancient cultures. Indeed, the earliest dated finding of such an image was in the tomb of Tutankhamen in Luxor, from the fourteenth century BCE.

One story in particular relates to Demeter's oversight of life and death in their many forms. A king of Thessaly called Erysichthon is said to have ordered his men to cut down all of the trees in Demeter's sacred grove. They refused to fell a mighty oak, garlanded with the wreaths representing all the wishes granted by Demeter to men. Erysichthon grabbed an axe and felled it himself, killing a dryad nymph in the process. Demeter punished Erysichthon by inflicting Limos, the spirit of insatiable hunger, on him. Thenceforth, Erysichthon could never satisfy his appetite for food regardless of how much he ate. In order to afford to feed himself he sold his daughter, Mestra, into slavery. In fact, he did so repeatedly as, being released initially by the god Poseidon (a former lover), Mestra was granted the gift of shape shifting which Erysichthon was able to use to his advantage! Nevertheless, this was not sufficient and he still found himself starving and unable to mollify his cravings. He met his doom 1 day by being so hungry that he ate himself, such that the following morning, nothing remained of him!

The relevance of 'self-eating' to our story will become apparent when we revisit Erysichthon's tale later.

Mike's Story

Mike and Karen met at college and remained together afterwards, eventually deciding to get married in their late twenties. Karen persuaded Mike to give up smoking and over the 2 years that they were saving up for their wedding, they put the money that he would have spent on cigarettes into their honeymoon fund. As a result, they were able to afford a fabulous trip to India—Karen had always wanted to see the Taj Mahal since seeing the photographs taken of Princess Diana there!

Three days before the end of their trip they were both struck down with traveller's diarrhoea. They had a miserable journey home and spent the rest of the week recovering rather than visiting their friends and family as newly-weds. Whilst Karen recovered completely, Mike's condition just got worse and worse. At the end of the first week he started passing blood with his bowel motions and experienced worsening of his abdominal pains—he was completely off his food and losing weight.

After 10 days he went to his general practitioner, who took blood tests and a stool sample to look for bacterial infections such as *Salmonella* or *Shigella*. The results revealed increased markers of inflammation but there were no causative organisms found in the stools. Mike's symptoms continued to deteriorate and after 3 weeks he had lost nearly a stone in weight and was feeling extremely unwell. When he collapsed on his way to the toilet one night, Karen called an ambulance and he was admitted into hospital.

The doctors in the hospital put Mike on an intravenous drip for rehydration and carried out X-rays and further blood tests. They passed a flexible endoscope into the lower bowel. The lining appeared inflamed and they took some samples of tissue to examine under the microscope. As a precaution they started him on high-dose steroids to try to suppress the inflammation. The biopsies were reported by the pathologist as showing signs of acute inflammation which could be due to infection, but suggested that inflammatory bowel disease could not be ruled out.

After 3 days of high-dose intravenous steroids, Mike's symptoms began to settle—he was not having his bowels open so often and there was no blood in the stools. He was able to go home after a week and gradually recovered his weight and appetite. His treatment was changed to a tablet that he took once a day that did not cause side effects like the steroids.

However, Mike's symptoms never completely settled. On a repeat camera examination of the entire colon that was undertaken as an outpatient 3 months later, the doctors found that there was ongoing inflammation and ulceration in the colon. On this occasion the biopsies showed chronic changes that were now suggestive of a condition called ulcerative colitis (one of the inflammatory bowel diseases). Over the following year he experienced two further episodes of quite severe diarrhoea with bleeding that required treatment again with steroids, and on the second occasion the hospital doctors started him on an immunosuppressive drug that would prevent further flares of the condition. This kept the condition under control and Mike had no further problems over the next few years, although he had to have regular blood tests and was told that he was at increased risk of skin cancer: he had to wear sunscreen if he went out. Mike and Karen took all their subsequent holidays in the UK.

Interestingly, when he eventually managed to catch up with family and friends to show them the wedding and honeymoon photographs, Mike's aunt Cath told him that she had had to undergo an operation to remove part of her small intestine 20 years earlier due to a condition called Crohn's Disease—another type of inflammatory bowel disease.

Cath's Story

Mike's aunt Cath left university with a history degree and embarked on a promising career as a journalist that saw her travelling around the world covering individual human stories of tragedy in the aftermath of conflict. She developed severe abdominal pain whilst on assignment in the Gulf and was diagnosed as having appendicitis by a British military surgeon in Saudi Arabia. After her appendectomy in the field hospital (set up inside a tyre factory!), she wrote about the experiences of some of the injured soldiers she had met and decided she would follow up with their stories and how they adjusted to life back home after the war.

After returning to the UK, Cath had ongoing pain in the lower right side of her abdomen and started to experience frequent diarrhoea. Her doctor treated her for possible infections that she could have picked up whilst on assignment, but her symptoms worsened. When she developed a high fever and extreme tenderness in the abdomen, her puzzled GP (who knew that this could not be appendicitis given her previous surgery but thought that it bore all the hallmarks) referred her to the local hospital emergency department. A CT scan of the abdomen showed that the end of the small intestine was thickened and involved in an inflammatory mass with a probable abscess within it. A separate area of thickening of the small intestine was noted further up. The radiologist reporting the scan suggested that this most likely represented a condition called Crohn's disease. Cath was admitted into hospital and fed a liquid diet through a tube placed into her stomach, whilst receiving antibiotics intravenously.

Cath's symptoms settled over the next week in hospital and her diarrhoea and fevers abated. She was started on a course of treatment with steroid tablets, allowed to eat a soft diet and discharged home after 10 days. However, when Cath reduced the steroid tablets as instructed, she found that her abdominal pains worsened again and she was experiencing abdominal bloating. She was started on an immunosuppressant drug (the same one as Mike was prescribed) and she was told to have regular blood tests through her GP to watch for potential side effects. Unfortunately, whilst the pains resolved her bloating sensation became worse, and a few months later she found that she was able to eat less and less and was losing weight. Her hospital gastroenterologist arranged an MRI scan, which showed that the inflammation had resolved but that the thickened small intestine was scarred and narrowed. The intestine above it was distended, showing that the contents were being held up at the strictured area. The gastroenterology team wanted to carry out a camera-tube test to confirm the diagnosis of Crohn's disease but felt that they could not safely give Cath the necessary purgative to clear the colon because of the partial blockage.

Instead, Cath's case was discussed with the surgical team who decided that given her initial presentation with an abscess and the relatively small amount of small intestine involved, an operation would be appropriate. They removed the diseased part of the bowel as well as a short stricturing segment upstream in the intestine which they found during the operation. Although they were able to join up the ends of small intestine and colon, the surgeons took the precaution of forming a 'defunctioning stoma' where the bowel contents were brought out above the joins they had made so as not to result in a leakage and peritonitis should the surgical connections fail. Cath had to wear a bag attached to her abdomen to collect her bowel contents for 6 months until the surgeons carried out a further, smaller, operation to close this opening and restore her intestinal continuity.

Following the operation, Cath continued to take the immunosuppressant medication for over 10 years and she quit smoking. She decided to give up her foreign assignments and took a desk job with a publishing company. Cath also wrote her book about the experiences of the soldiers she met—and ended up marrying a major from a tank regiment, who had been in hospital at the same time as her.

The Inflammatory Bowel Diseases

At first sight, Mike's story looks quite different from that of his aunt. They have different diseases—Mike suffered from ulcerative colitis whilst Cath's condition was Crohn's disease. Their diseases affected different organs—Mike's large intestine (colon) and Cath's small intestine. Why then should we combine their stories into the same chapter and their diseases together as 'Inflammatory Bowel Diseases'?

We actually consider ulcerative colitis ('colitis' is short for 'colonitis'—meaning simply 'inflammation of the colon', in this case with ulcers) and Crohn's disease to be different ends of a spectrum of conditions causing inflammation in the intestine or colon. There are many differences between them as we shall discover, but the principal distinction is that whereas Crohn's disease can affect any part of the gastrointestinal tract (from the mouth to the anus), ulcerative colitis only affects the colon, in whole or part. However, when Crohn's disease does just affect the colon it can be difficult to distinguish it from ulcerative colitis. Recent genetic studies suggest that this probably represents a third type of inflammatory bowel disease that is separate from either Crohn's disease that just affects the small intestine, or ulcerative colitis. Around one in ten cases of inflammatory bowel disease are considered to be 'indeterminate' colitis—meaning that they might be 'ulcerative' or 'Crohn's' in nature. It is likely that they are only distinguishable on the basis of specific genetic differences between them.

There is in fact a very strong inherited component to both conditions—more so for Crohn's disease than ulcerative colitis. Over 200 separate disease associated genes have been identified to date, and more than half of these are common to both conditions. You have a high risk of developing the disease if you share much of your genome with someone else who has it. That risk can be as high as 30% for identical twins where one has Crohn's disease. Similarly, in families that have more than one affected individual, around a third are 'mixed' ulcerative colitis and Crohn's families as in the case of Mike and Cath.

A common link between the two conditions is that they both have a predilection for the lower gastrointestinal tract (the end of the small intestine in Crohn's disease and the colon in both —Fig. 9.1). This is in contrast to the conditions described in the previous two chapters that are linked to the relationship of food with our gut immune system and affect the upper part of the small intestine where food is encountered. The inflammatory bowel diseases, on the other hand, are associated with our gut immune system's relationship with its bacterial passengers—the 'microbiome'. In this chapter we will revisit some of the ideas and principles that we came across right at the beginning of this book, including familiar concepts such as 'containment' and the feeding

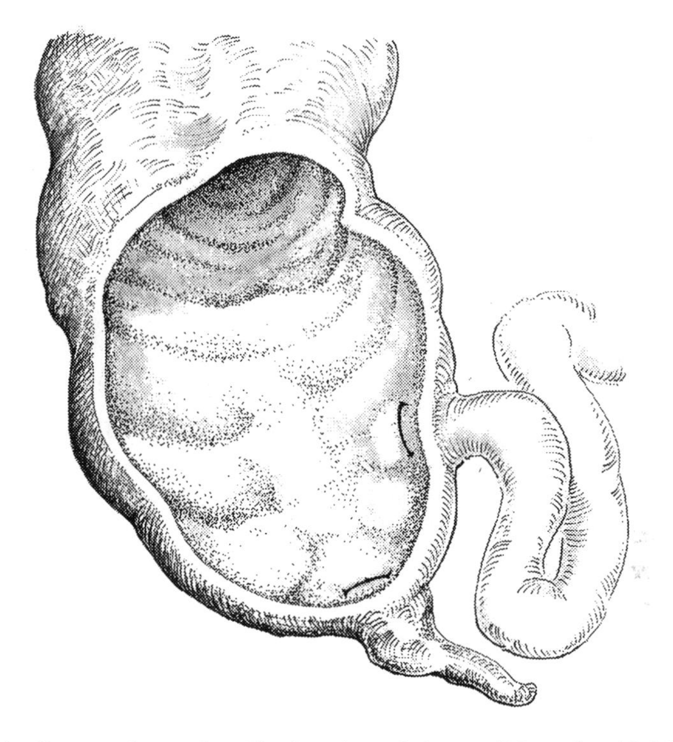

Fig. 9.1 The ileocaecal complex. The junction of the small intestine (right) and the large intestine (left) is a challenging immunological environment. The small intestinal immune mechanisms are not really equipped to deal with the huge bacterial load of the colon and a valve mechanism between them may help reduce contamination. This region of the gastrointestinal tract is peculiarly prone to inflammation and a common site for the development of Crohn's disease. The appendix (at bottom) is packed with immune cells and may play a role in tolerance towards our passenger bacteria or maintain them away from the main flow of gut contents in order to restore normal symbiotic populations after an infection

mechanisms of single cells to overcome it. We will also come to understand finally how eating, housekeeping, immunity and the holobiont (us and our bacteria) are intimately related throughout evolution, by discovering the 'master switch' within us that links them all.

Ulcerative Colitis

We shall start our investigation of the inflammatory bowel diseases by focussing on ulcerative colitis. Inflammation of the colon—whether due to infection or ulcerative colitis—results in frequent bloody diarrhoea and symptoms of illness. Throughout our past history and until recent times, bacterial infection (dysentery) has been by far the more common cause of these symptoms. Hence it is difficult to know when this condition was first separately identified. Some have even attributed the first ever description to Aretaeus of Cappadocia, whom we met in the previous chapter due to his association with coeliac disease. We do know, however, that ulcerative colitis was given its name by Sir Samuel Wilks in 1859,[2] who identified the hallmark features during an autopsy on a 42-year old woman that succumbed after many months of symptoms.

Ulcerative colitis typically presents in young adulthood, and as many as 1 in 300 Europeans are affected, of both sexes equally. Intriguingly, whilst it can affect all or part of the colon, it does so in a way that it always affects the lowest part—the rectum—and then a variable distance upstream. In over half of all patients with ulcerative colitis the condition is in fact just limited to the rectum, and in about 10% it affects the whole of the colon. Very often we see a sharp demarcation between the affected part and the normal colon. However, even in individuals with disease limited just to the lower or 'left' colon, we can see a patch of inflammation around the opening of the appendix in the 'right' colon. This is of particular interest as there is good evidence that having had appendicitis (and then having the appendix removed, as opposed to having a normal appendix removed) actually protects from the development of ulcerative colitis in the future.

Usually, ulcerative colitis causes only superficial damage to the lining of the bowel, but when severe and affecting the whole colon it can lead to distension and weakening of the colon wall. This can cause perforation and peritonitis—infection of the abdominal cavity—which carries a high risk of mortality.

We have still not discovered the 'cause' of ulcerative colitis. This is undoubtedly because we have been asking the wrong questions. There are certainly some diseases that can be attributed to a single very specific cause—for instance infective diarrhoea caused by the *Salmonella enterica* bacterium, or cystic fibrosis caused by a mutation in a single protein. But we have failed to identify a single particular bacterium or gene defect that causes 'ulcerative colitis'. Instead, as we have already seen, there are probably several hundred different genes associated with multiple mutations in individuals that each contribute small amounts to the development of the condition or protection from it.

[2] Sir Samuel Wilks (1824-1911) was a physician who trained and worked at Guy's Hospital in London and became 'Extraordinary physician' to Queen Victoria in 1897. It is ironic that ulcerative colitis is not known as 'Wilks' disease' given that he wrote biographies of his colleagues—Thomas Addison, Thomas Hodgkin and Richard Bright—and promoted the use of their names as eponyms associated with diseases they first described. The patient in whose autopsy ulcerative colitis was first described was immortalised by being named in the publication—*'Morbid Appearances in the Intestine of Miss Bankes'*—a practice that would be forbidden by modern medical ethics. In retrospect, many suspect that Miss Bankes actually had Crohn's Disease rather than ulcerative colitis!

Instead, we need to consider the inflammatory bowel diseases in a slightly different way—a bit like asking what causes a heart attack (technically known as a myocardial infarction). Most people know that this occurs when an artery supplying part of the heart muscle with oxygenated blood becomes blocked. The affected region of muscle dies as a result of lack of oxygen, the heart is permanently damaged and its pump function impaired. The arterial occlusion results from calcium and fat deposits called 'plaques' building up in its lining, which lead to narrowing and hardening of the vessel. A blood clot forming on the plaque then blocks the artery there or breaks off and obstructs a smaller branch further down. This is the mechanism, but if we were to ask what 'causes' the plaques that lead to a heart attack we would struggle to limit ourselves to just one thing. We are all familiar with the many environmental risk factors for heart disease such as smoking, lack of exercise and poor diet. There are also strong genetic factors as well: around 200 genes have been implicated in the development of this condition, a similar number as for the inflammatory bowel diseases. Analysis of the effects of these genes points towards distinct pathways, some of which relate to known risk factors such as the ways in which we metabolise fats, how our blood clots and how we regulate our blood pressure. The interplay of these genetic and environmental risk factors contributes to the progression of the disease process.

In the same way, looking at the large numbers of genes involved in inflammatory bowel diseases and their roles in different pathways can give us clues to the mechanisms involved. In common with heart disease, and in addition to the genetic component, there is also a strong environmental component to ulcerative colitis. For instance, the prevalence of ulcerative colitis has increased dramatically throughout the last half century and appears to be associated with 'western' lifestyles. People moving from areas of low prevalence to Europe or the United States of America seem to share the same risk as the local population, rather than that from which they came. This could relate to factors such as diet, pollution, stress or improved hygiene. However, even within Europe there is a north-south divide with ulcerative colitis being commoner in the north. This might hint at an association with lower levels of vitamin D, which is made in the skin by exposure to sunlight. Finally, there is a big clue in that cigarette smoking makes Crohn's disease worse but seems to have a protective effect in patients with ulcerative colitis, who may develop the condition after quitting (as did Mike). Again, it is undoubtedly the complex relationship between genetic and environmental factors that 'causes' ulcerative colitis. Note that in our example of heart disease, the genetic and environmental factors collectively contribute to the progression of arterial plaques. The ensuing blockage leads to the heart attack as a single disease 'entity'. However, when it comes to ulcerative colitis it is quite possible that there are actually several forms of the disease that share common pathways and risk factors but are similar enough to be given the same name.

Diverticulum#9.1 the Appendix

'With respect to the alimentary canal, I have met with an account of only a single rudiment, namely the vermiform appendage of the caecum...not only is it useless but it is sometimes the cause of death'.

So wrote Charles Darwin in the 'Descent of Man' in 1871. If only he knew what we know now—much of which is based on his understanding of natural selection!

The 'vermiform' (= *'wormlike'*) appendix has long thought to be a vestigial leftover of evolution—a bit like the skeletal limb remnants of the boa family of snakes. There are more than a few hints, however that this is not the case. Humans are rarely born without an appendix. The appendix is of variable length in us, but always of similar girth—7-8mms, with a central channel of 1-2 mms across. Looking across different species, it may vary in shape and size, but is always placed in the right side of the colon and as a diversion from the main faecal stream. Finally, rather than appearing once in evolution and then gradually petering out, it appears that the appendix has evolved independently more than 30 times in animals! All in all, it looks as though it really is doing something useful—although we all know people where (as Darwin suggested) it causes more harm than good.

The appendix is placed between the small intestine and the colon and out of the flow of the bowel contents. It is in an ideal situation to act as a 'repository' for the bacterial flora (microbiome) that inhabits the colon. When we suffer from a gastrointestinal infection, the 'weep and sweep' and inflammatory response may clear all of the culprit bugs out of the bowel, but along with them the normal resident bacteria—which have grown up with us and are unique to us. The store of our own bacteria kept in the appendix—a biological 'vial'—can then divide and repopulate the colon when the infection has passed.

The best evidence for this being fact rather than conjecture is that after a course of antibiotics that affects the equilibrium of our microbiome, we are peculiarly prone to a toxin-producing bacterium called *Clostridium difficile*. This can cause severe diarrhoea and even inflammatory colitis and can be notoriously difficult to get rid of thereafter, leading to recurrent infections. It appears that our normal microbiome keeps these rogue bacteria in check. Repeated 'C diff' infections are nearly 4 times as common in those who have had their appendix removed as in those who have kept it—presumably due to early reconstitution of our familiar bacterial populations.

The appendix may also be in a useful location for educating the immune system to tolerate its bacterial passengers. For instance, there is now good evidence to suggest that appendicectomy earlier in life prevents the later development of ulcerative colitis. One possible explanation could be that this is a 'priming' location for a (defective) gut immune system in this condition and leads to anti-microbiome immune reactions, whereas a normal gut immune system might lead to bacterial tolerance.

Scraping the Surface

We can start as before by looking at the most advanced, recent and therefore 'superficial' levels of the gastroarcheological dig. This layer includes the specific variable pattern receptors embodied in antibodies and T cell receptors.

We could also consider on which side of the 'great schism' ulcerative colitis lies—whether or not it is related to the antibody pathway leading from phagocytosis and cell-eating, or whether it is related to the cytotoxic or pore-forming toxin pathway.

A superficial analysis of the mechanisms of ulcerative colitis suggests that the condition lies on the antibody side of the great schism. For instance, measurement of the immune cell signalling molecules reveals a very similar profile to that of food allergic responses, with increased levels of interleukin-4 and other closely-related chemicals. Examination of the inflamed colon under the microscope shows that there are 'germinal centres' found in the lamina propria, where lymphocytes are being instructed to make high-affinity antibodies. Moreover, whereas the gut B cells usually produce the form of immunoglobulin known as IgA, it is a different type that predominates in ulcerative colitis, called IgG. We know that under normal circumstances much of the IgA that is produced in the colon recognises commensal bacteria and may protect them from immune attack. This is because the 'adaptor' end of the IgA molecule (the other end from the 'pattern recognising' part of the antibody) does not bring about inflammatory responses. In ulcerative colitis, IgG antibodies are also raised against bacteria that are normally quietly resident in the bowel. However, the adaptor end of the IgG molecule binds to phagocytic cells through a receptor and leads them to secrete powerful inflammatory signalling molecules. These increase the inflammation in the lining of the bowel and attract other immune cells to the site. As a final clue in this regard, one of the most important genes (specifically identified in Japanese people with ulcerative colitis) appears to increase the effect of the binding of IgG to the receptor on phagocytic cells, which produce greater amounts of inflammatory messenger molecules. Thus, it would appear that ulcerative colitis involves the gut B cells mounting an abnormal—IgG based—antibody response against the normal bacterial inhabitants of the colon, causing damage to the gut lining through an inflammatory response.

In addition, there has long been a question as to whether the damage caused by ulcerative colitis may be due to antibodies that are 'auto-reactive' and recognise components of the patient's own body. A few such antibodies are variably found in sufferers of ulcerative colitis including some that appear to recognise goblet cells in the lining of the colon, and others that bind to internal components of 'neutrophils'—white blood cells that actively phagocytose bacteria. There is also a self-directed antibody that recognises a molecule shared between the colon and other sites that can be affected in ulcerative colitis such as the joints and the eyes. Indeed, it would be entirely feasible for ulcerative colitis to represent an auto-immune condition as many of the genes associated with ulcerative colitis are shared with other such conditions.

However, the finding of 'autoantibodies' is by no means universal in patients with ulcerative colitis and may just be a secondary effect of the tissue damage—a form of 'housekeeping' response rather than an underlying cause of the condition.

Gene 'Knockout' Models and Ulcerative Colitis

As we have already seen, genetic engineering allows us to insert or remove specific segments of DNA from the genome of an entire animal, or just from specific cell types and under certain conditions. The earliest and easiest technology in this regard is the 'knockout' experiment where a gene is entirely deleted. For coeliac disease (Chap. 8), we found that it was actually extremely difficult to generate an experimental model in the mouse despite using genetic engineering techniques. For ulcerative colitis, it is quite the opposite! So far around 75 different individual gene knockout models have been developed, all of which spontaneously develop the condition. However, the universal requirement for inflammation to occur is the presence of bacteria in the gut, without which nothing happens. The implication is that such gene mutations lead to disruption of a fragile and complex immune network that regulates the relationship between the host and its bacteria.

The first such gene knockout models modified the T-cell receptor—the highly variable pattern detecting molecule on T cells that is the equivalent of the antibody for B cells. The T cell receptor comes in two separate halves that are assembled on the cell surface, and each has a separate coding gene. Knocking out one or the other half of the T-cell receptor leads to the spontaneous development of colitis in the majority of mice after about 6 months of age. These animals still produce T cells, but with T-cell receptors where both halves are identical, mirroring each other. These abnormal T cells do not undergo the usual education process in the thymus that removes cells carrying dangerous, self-recognising T cell receptors. In addition, the thymus is usually responsible for the generation of 'regulatory' T cells that suppress immune responses—and their production is also impaired by the presence of the defective abnormal T cell receptor. These findings would appear to fortify the suggestion that ulcerative colitis was indeed caused by an auto-immmune B cell response that was normally held in check by regulatory T cells. Such a premise was only strengthened by the finding that knocking out RAG—and therefore all B and T cells—did *not* lead to spontaneous colitis in mice.

In fact, all of the early genetic experiments in colitis were interpreted in the same way—that potentially even 'normal' B cells would cause colitis when not under T cell control. For instance, very severe early onset spontaneous colitis could be caused experimentally in mice by knocking out a specific immune signalling chemical called interleukin-10. This molecule is secreted by regulatory T cells and underpins the mechanism by which they control immune responses. A rare genetic mutation in humans affecting the interleukin-10 receptor (and therefore the signalling function of the molecule) also leads to severe early onset colitis in children.

At this stage we might be forgiven for thinking that we can explain ulcerative colitis simply by examining the most superficial and recent layers of 'gastro-archeology'—which show us that it is an uncontrolled B cell and antibody-mediated condition. However, this would be extremely naïve as we know that the relationship between host and bacterial symbionts is one of the core foundations of immunity and has been evolving since the time of the earliest multicellular organisms. Deeper, and older, strata will undoubtedly need to be examined.

Magic Bullets...and Back to the Drawing Board

The production of antibodies by B cells can be protective to an organism or cause disease processes by being misdirected, as postulated for ulcerative colitis above. Reminiscent of Paul Ehrlich's 'magic bullets', antibodies can also be engineered in the laboratory and on a large scale by pharmaceutical companies to target specific cells or molecules for treatment purposes. Because they have the ability to recognise very specific molecular shapes, antibodies can be made that stick to molecules—such as interleukins, or cell surface receptors—and block their effects. They can also be manufactured to attach to cells recognised by a specific cell surface marker and trigger the immune system to destroy them. Such 'monoclonal antibody'[3] treatments have spawned massive growth in a new area of drug development called biologics.[4] The use of this

[3] Each B cell produces only one specific type of antibody. By fusing an antibody-producing B cell with cell from a patient with myeloma (a cancer of B cells), Cesar Milstein (1927-2002) and Georges Kohler (1946-1995) were able to generate immortalised cell cultures producing large amounts of the same antibody that could then be purified for use. These are therefore clones of the initial antibody-producing cell, hence the expression 'monoclonal antibody'. For this work carried out in the Laboratory of Molecular Biology in Cambridge, they were awarded the Nobel Prize in Physiology or Medicine along with Nils Jerne in 1984.

[4] The first commercially available biologic was a humanised form of insulin called 'Humulin' produced in 1982, developed by Genentech and marketed by EliLilly. In this case, the gene for human insulin was inserted into a bacterium—*E. coli*—for large scale production. Many hundreds of such biologically engineered proteins are now available including hormones, growth factors, blood clotting factors and monoclonal antibodies. The global biopharmaceutical market is projected to reach a value of $526 trillion by 2025.

term covers the therapeutic use of all manufactured proteins and includes specific whole antibodies, as well as antibody fragments and genetically modified antibodies to alter their biological effects. Such agents are usually given by direct infusion into the veins or by injection at intervals of 4 to 8 weeks, in order to bind to and neutralise individual targets.

Returning to ulcerative colitis, there are a number of monoclonal antibody biologics that could be used to target and treat the B cell response, if this were indeed the root cause of the problem. Therapeutic monoclonal antibodies have been generated to recognise and block antibody-response signalling molecules such as interleukins 4 and 13 (which are closely related) and are helpful in treating allergic conditions such as asthma and eczema. However, they appear to have absolutely no effect in ulcerative colitis. Needless to say, this finding came as something of a surprise!

Furthermore, it appears that in the original mutated T cell receptor mice that developed colitis, additionally removing all B cells actually made the condition worse! Similarly, a therapeutic monoclonal antibody used in humans to destroy B cells appears to have no benefit in ulcerative colitis and may have the potential to make it worse. Ulcerative colitis has even appeared as a side effect of such B cell depleting therapy when used for other conditions. By way of corroboration from animal experiments, it is possible to induce colitis in an immunodeficient mouse lacking all B and T cells by simply transferring just 'naïve' T helper cells (that have not already been activated) into it from a normal mouse. The process can then be blocked by also transferring 'educated' T cells that have previously encountered their specific peptide and therefore include 'regulatory' T cells. This would suggest that B cells are *not* necessary to cause colitis after all and that the condition can be caused by T cells and *not* B cells. How confusing!

It transpires that there are 'regulatory' B cells as well as 'regulatory' T cells, and they also secrete interleukin-10 to suppress immune responses. Therefore, removing *all* B cells takes away another dampening component to the inflammatory condition just as taking away all the T cells removes their regulatory immunosuppressive cohort. Hence, whilst B-cells and antibody production clearly play a part in the inflammation of ulcerative colitis they are unlikely to be causative and may even be a part of the remedial response to some other process occurring.

The clear message that arises is not that the condition is caused by B cells and antibodies, but that there is actually a defect in the immune system's 'off switch' when it comes to colitis. As a result, inflammation inappropriately continues despite the apparent removal of the initial trigger. A defective 'off switch' explains why the normal response to an infection (bacterial dysentery) and ulcerative colitis appear very similar—the only difference being that the infection resolves spontaneously whereas ulcerative colitis leads to ongoing chronic inflammation. Indeed, as we have already seen, lymphocytes have an inbuilt self-destruct apoptosis (cell death) signal that is primed when they are activated, leading to a self-terminating and short-lived response. In order to lead to chronic inflammation, these populations must either be resistant to the self-destruct sequence or be replenished by an influx or replication of cells caused by an ongoing trigger. In keeping with this, studies have shown unusually long-lived lymphocytes that prolong the inflammatory response in both ulcerative colitis and Crohn's disease. Furthermore, of the 200 or so genes that have been identified to have effects that contribute towards inflammatory bowel diseases, a large number code for proteins that are involved with suppressing or regulating the immune response.

In order to find the regulator switch that balances the activating and suppressing immune functions of T cells, we need now to get our trowels out and dig a little deeper, back in time, to the layer of early lymphocyte-like cells. These are the cells that lack the highly variable 'adaptive' receptors of T and B cells and include the natural killer cells, innate lymphoid cells and T cells with invariant receptors such as the γδ T cells. Recent evidence suggests that these cells play a crucial function in modulating the normal immune response in the gut, as well as the dysfunctional response in inflammatory bowel disease. However, before we start digging we need a slight preparatory pause.

Cytokine Swarm

The 2019–2021 COVID-19 pandemic led to the appearance of the expression 'cytokine storm' in the lay press.[5] This refers to an overwhelming secretion of immune-related chemicals (called 'cytokines') that leads to catastrophic

[5] The term 'cytokine storm' was first coined in 1993 to explain some of the features of the immune response to bone marrow transplantation. It is thought to be responsible for the paradoxically high mortality seen in otherwise fit adults due to infections with coronavirus (whereupon they develop COVID-19, severe acute respiratory syndrome) and possibly also the 1918/19 influenza pandemic. The name also became widely known due to the 2006 Northwick Park Hospital clinical trial of Theralizumab, a monoclonal antibody against a T-cell surface marker called CD28 in which 6 young males who received the drug almost died. The aim of the trial was to generate regulatory T cells. In order to be activated, T cells require signals generated internally through both the T cell receptor and binding of a second surface molecule, CD28. The thought was that by only partially activating the T cells through CD28 binding rather than receiving both signals, the T cell would change into a regulatory cell and suppress immune responses. Unfortunately, this was not the case and massive stimulation of T cells, producing maximal amounts of cytokines, occurred instead. In retrospect, the preliminary animal safety experiments in macaque monkeys were flawed: T cells in this species do not actually produce CD28!

organ failure and death. The name 'cytokines' (from '*cyto*' = cell, '*kinesis*' = movement) is now used to refer to all molecules that signal immune messages between cells. This term therefore includes the interleukins that we have already come across, comprising a diverse array of hundreds of different messengers secreted by cells to interact in the network of immunity. Cytokines act to alter cell functions in different ways –activating and suppressing, stimulating or inhibiting replication or cell death, or attracting cells to sites of damage. Not only are there very many different such chemicals involved in immunity, both in 'maintenance' and 'threat' situations, but individual cytokines may be released by different cell types under different circumstances and have diverse effects on a wide range of other cell types.

The way in which cells involved in the immune response interact through cytokines is best appreciated as a 'network' rather than individual pathways. The consequence of this is that some cytokines can have both 'good' and 'bad' effects depending on the context, and that small changes to one part of the interacting network can lead to major responses elsewhere. Once again, this is evidence that looking for a single 'cause' of the inflammatory bowel diseases is doomed to failure. Changes that could perturb the fragile network that keeps the immune system and symbiotic bacteria in check may include any or all of the genetic or environmental factors that we have identified as contributing to the inflammatory bowel diseases. Furthermore, blocking individual cytokines (for instance with therapeutic monoclonal antibodies used as drugs) may have unexpected results, or little effect at all in such a complex interplay of factors.

We have already come across interleukin-10 as a key regulatory cytokine in the gut, and seen how impairment of its function can lead to an uncontrolled immune response that leads to chronic inflammation. This is just one of at least 60 different cytokines that have been demonstrated to have altered levels in the immune response in inflammatory bowel diseases. However, this undoubtedly represents just a relatively small proportion of the hundreds of cytokines involved in maintenance of the normal immune network. Things could get rather complicated if we were to try to describe how each of these chemicals plays its part in the normal peaceful response to bacteria and in the dysregulated chaos of inflammatory bowel disease (there are plenty of long and difficult-to-read reviews that attempt to do just that!) Instead, we will limit ourselves to just a small number of (what we now think are) the key cytokines as we continue our dig.

The 'Peacekeepers'

Just as we seen before—for instance with food allergies—we can dissect the immune response and take out the top layer of our trench by knocking out a single gene—RAG. This codes for the gene responsible for the highly variable molecular pattern recognising receptors of B and T cells, without which these cells are entirely absent.

Once again, it should come as no surprise to hear that RAG 'knockout' mice are capable of mounting a characteristic defensive inflammatory immune response to bacteria that cause infections. The crucial cells once again appear to be types of innate lymphoid cell (ILC)—lymphocyte-like cells that lack the variable surface receptors of T and B cells but mimic them in many other ways. Just as with T and B cells, a 'regulatory' subset of ILCs has recently been identified that suppresses inflammatory responses. However, whereas we have seen how the type 2 innate lymphoid cell (ILC2) was critical in the development of allergic responses (and defence from parasites), and the natural killer cell subset of type 1 ILCs in coeliac disease, it is the type 3 (ILC3) that appears to regulate our relationship with bacteria in the colon.

ILC3s play a significant role in the maintenance of the normal, healthy intestinal lining. Among the cytokines that they secrete under normal conditions are two (interleukins 17 and 22) that have significant effects on the epithelium. Interleukin-22 signals the goblet cells to produce mucus to provide a protective layer, and stimulates the production of invariant pattern-recognising defensive molecules ('defensins') by the ancient Paneth cells and enterocytes. Interleukin-22 also stimulates the replication of enterocytes to replace defects in the continuous cell lining. These functions appear to be essential to protect against microbial invasion in the intestinal lumen, as interleukin-22 knockout mice succumb easily to certain bacterial infections.[6] Interleukin-17 (which is also produced in normal conditions by $\gamma\delta$ intraepithelial T cells) maintains the integrity of the epithelium by strengthening the junctions between the cells.

In normal times, ILC3s also secrete a chemical that prompts the presenting cells such as phagocytic macrophages and dendritic cells to produce retinoic

[6] Unlike for interleukin 10 there are no known cases of deficiency of interleukin-22 or its receptor arising from genetic mutations in humans. However, in chap. 5 we saw that mutation of the gene called AIRE that is expressed in the thymus leads to a wide array of auto-immune conditions. These patients experience chronic fungal infections of the finger and toe nails, and mucus membranes such as the mouth. The reason for this is that they produce their own blocking antibodies against both interleukin-17 and interleukin-22 (such as drug companies develop for therapeutic use against different cytokines), which impairs their resistance to fungal infection.

acid and interleukin-10. These molecules help generate regulatory T cells and suppress the active T helper cells. Interestingly, ILC3s are also capable of acting as presenting cells by themselves as they carry Class II MHC molecules on their surface. They have recently been shown to use this in a unique way by *killing* any T cells that are capable of binding the peptide presented by the MHC molecule on their surface—instead of activating them as normal presenting cells would do. In this way, T cells that might have mounted a response to the normal passenger bacterial flora are destroyed to prevent them from doing so. ILC3s are so important in the maintenance of a healthy gut lining that animals experimentally depleted of them are highly susceptible to infections or chemical damage to the epithelium.

ILC3s therefore seem to represent key 'peacemakers' of the gut, moderating the immune response to the normal symbiotic bacteria that live there and keeping the aggressive potential of the T and B cells in check.

A Little Is Good: A Lot Is Not

However, there is an altogether different side to ILC3s. As well as maintaining the status quo between the host and its symbiotic passenger bacteria, they are also fundamental to the immune response that rids the colon of troublesome disease-causing bacteria. In so doing, they lead to the development of the self-limiting colitis which causes dysentery symptoms. And they can do so without the back-up of T and B cells. Hence, RAG knockout mice develop a protective inflammatory response when infected with certain pathogenic bacteria, but fail to do so when an additional mutation is added that removes the ILC3s.

As we have already seen in our investigation of other diseases in this book, the usually peaceful immune cells of the gut can be activated by signals of danger. In the case of coeliac disease an important trigger was the cytokine, interleukin-15. With regard to colitis, the signal that changes the mood of ILC3s appears to be a different cytokine, this time interleukin-23. Interleukin-23 is usually produced in small amounts in the gut by phagocytic macrophages which clean up debris, as well as dendritic cells which sample the contents of the gut and present the protein fragments on their cell surface to the immune system. This low-level production is enough to stimulate the ILC3s to produce small amounts of two further cytokines—interleukins-22 and 17—for their beneficial effects on the gut epithelium.

However, in the setting of infection or inflammation, dendritic cells and macrophages significantly upregulate the amount of interleukin-23 produced.

This changes the behaviour of the ILC3s. They begin to secrete much larger quantities of both interleukin-22 and 17. Whereas low level production is beneficial, when present in greater amounts interleukin-22 has different effects. One of these is to stimulate the enterocytes to secrete a range of other cytokines, which attract immune cells out of the blood stream and into the epithelium where they can cause further damage.

Furthermore, when stimulated by large amounts of interleukin-23, ILC3s can change their nature completely and turn into a different type of ILC—the type 1 ILC (ILC1). This is another 'Jekyll and Hyde' moment in the gut immune system such as we came across in chap. 7. When they change into type 1 ILCs, these cells no longer secrete the cytokines that lead to the development of regulatory ILCs and regulatory T cells. Instead, they now secrete activating or pro-inflammatory cytokines. Without the effects of the suppressive cytokines produced by ILC3s, the regulatory T cells also change their nature to stop suppressing immune responses and instead produce large quantities of interleukin-17 along with a range of other activating cytokines. Once again, whereas small amounts of interleukin-17 help to regulate the epithelium, when present in such high quantities, interleukin-17 has an entirely different effect—to activate the entire range of immune cells present in the gut and recruit macrophages and neutrophils from the blood stream.

It appears therefore, that a small amount of interleukin-23 leads to the production of only small amounts of interleukins 17 and 22 and helps to maintain a healthy epithelium and a peaceful environment. However, excess production of interleukin-23 by dendritic cells and macrophages acts as the switch that works through ILC3s to toggle between a regulatory, suppressive response and active inflammation mediated by high levels of interleukin-17 production. This occurs through the plasticity of the ILCs that can change their nature from being the (benign) type 3 to the (aggressive) type 1, and the regulatory T cells that can change their nature from suppressing responses to activating them. Interestingly, our diet may also play a role at this stage as chemical products of foods that we identify as 'healthy'—particularly the brassica family of vegetables such as broccoli, cabbage and sprouts—act through a receptor to keep the switch on the peaceful 'ILC3' rather than the inflammatory 'ILC1' setting.[7]

[7]The receptor is called the aryl hydrocarbon receptor', and is present on enterocytes, IELs and ILCs. It recognises a variety of chemicals produced by bacterial metabolism of dietary components, but also toxins such as dioxin from cigarette smoke. It leads to the production of interleukins-10 and 22 and downregulates the production of interleukins 8 and 17, thereby altering the regulatory/activated 'helper' cell ratio in the intestine.

This switching mechanism between 'peace' and 'aggression' allows the gut to protect against invasion by bacteria—as long as it is capable of being turned off. We have already seen how the lack of an effective 'off switch' can lead to the prolongation of this response and the development of inflammatory bowel diseases. There are also clues to suggest that dysregulation of the 'on switch' that we have just described is important in the development of inflammatory bowel diseases. For instance, carrying a gene mutation of the interleukin-23 receptor that downregulates its effect appears to reduce the likelihood of a person developing inflammatory bowel disease. Vitamin D also appears to reduce the inflammatory effect of interleukin-23—and this may be relevant to the geographical distribution of the diseases, as they are more common in northern latitudes where less vitamin D is made in the skin from the effect of sunlight.

Interleukin-23 has also been the focus of much research by the drug companies that have produced therapeutic monoclonal antibodies (such as 'Ustekinumab', or 'Risankizumab') to block its receptor and are used to treat both ulcerative colitis and Crohn's disease. However, as an example of the complexities of interfering with the immune cytokine network, monoclonal antibodies that block interleukin-17 appear to have no effect whatsoever—and may actually cause colitis as a side effect when used to treat other diseases such as the skin condition, psoriasis. This undoubtedly reflects the dual role of such cytokines—being beneficial in small amounts yet harmful when present in excess. For drug developers, this poses special problems: a molecule cannot be seen simply as either good or bad, but depends on the amount present (and also perhaps where it is present). Therefore, blocking its actions could have either—or both—beneficial and detrimental effects. As we shall now see, the maintenance of integrity of the epithelial layer is fundamentally important to prevent inflammation, and the loss of this function of interleukin-17 is probably why completely blocking its effects creates more problems than it solves.

Containing the Threat

Analysis of the genetic markers associated with inflammatory bowel diseases show that most are shared, with few that are specific to either Crohn's disease or ulcerative colitis. So far, the pathways that we have explored—namely the 'on switch' regulated by interleukin-23, and the apparent lack of an effective 'off switch'—are shared by both conditions. However, when we get down to the oldest layers of our trench, we find that the two ends of the inflammatory bowel disease spectrum differ with regard to potential mechanisms and the associated genetic pathways.

Gene targets that appear to be associated more specifically with ulcerative colitis than Crohn's disease include several that encode proteins modulating the function of the epithelial barrier itself. The notion that a primary defect in the epithelial containment occurs in ulcerative colitis is supported by a number of observations. For instance, in close relatives of patients with the condition, there is a measurable increase in the 'leakiness' of the intestine, and in the patients themselves, increased permeability appears to precede the onset of inflammation. Examination of the colon (even in uninflamed areas) of patients with ulcerative colitis reveals a defective mucus barrier, both in terms of thickness and chemical composition, as well as a reduction in the number of goblet cells that produce it. In mice, the use of a chemical called dextran sodium sulphate (DSS) in drinking water leads to thinning of the mucus layer and to a condition mimicking ulcerative colitis. Furthermore, mice lacking the necessary genes for mucus production also develop a similar colitis. This may provide a clue as to the potential effect of a 'westernised' diet on the development of ulcerative colitis: many processed foods contain emulsifying agents such as polysorbate or carrageenan[8] that can chemically interact with mucus. In mice carrying mutations that predispose to colitis, the use of standard emulsifying agents from human food can make it appear earlier or to a worse degree. In all of these models, the bacteria that live in the colon are found in close proximity to the surface of the epithelial lining cells rather than being held at 'arm's length' on the other side of the mucus layer.

This gives us two further avenues to explore. Firstly, the possibility that bacteria could infiltrate through the epithelial layer (breaking 'containment' by the epithelium) to prime a response that leads to ulcerative colitis, or (secondly) that there may be some interaction between bacteria and the enterocytes themselves that leads to the subsequent responses. We are now back at the deepest levels of the trench and will revisit molecules, pathways and mechanisms that we first encountered in our discussion of the immune responses of the sponge!

---

[8] These molecules can work in completely different ways. Carrageenan is a polysaccharide developed from the red edible seaweeds found in Ireland (the name is derived from the word for 'little rock' in Gaelic), which acts as a thickener and binds strongly to food proteins. It emulsifies by stabilising the lipid suspension in a gel. Polysorbates, on the other hand, are artificially produced long chains of hydrocarbons linked to a sugar. One end is therefore water soluble and the other fat soluble.

The Other 'Shape of Life'

We came across two molecule superfamilies in chap. 4 that dictate the shapes of life—namely the immunoglobulin superfamily that begets antibodies, MHC molecules and T cell receptors, and the leucine rich repeat (LRR) family which spawned the similar pattern recognition receptors of lymphocyte-like cells in the lamprey. The Toll-like receptors are representative of the latter superfamily.

Toll-like receptors are 'fixed pattern receptors' that identify basic components of bacteria such as the flagella protein, or unique lipids of the bacterial membrane (such as lipopolysaccharide or LPS characteristic of particular bacteria). Others recognise the specific differences between the DNA of bacteria and ourselves, viral RNA or other specific components of bacterial coatings. How the gut immune system interacts with its bacteria through the Toll-like-receptors (TLRs) is only now being worked out. For instance, mice that are given antibiotics are more prone to colitis after DSS has been given to weaken the epithelial defences, but can be protected by activating TLRs on the epithelial surface. One Toll-like receptor in particular—TLR 9—recognises bacterial DNA, so in effect notices the presence of dead bacteria from which the DNA has escaped. TLR 9 is expressed on both sides of gut epithelial cells. When activated on the surface in contact with luminal contents, it suppresses the epithelial cell's inflammatory response. When expressed on the border of the cell facing into the body it can instead activate the cell to stimulate inflammation. In this way, it can act as a signal for bacteria that are in the wrong place and have penetrated through the epithelium, as a first line of defence.

There is one TLR in particular that may be of interest in ulcerative colitis—TLR 4. This is the Toll-like receptor that identifies the LPS present in some bacterial cell membranes. Just as in the sponge, detection of bacteria on the epithelial cell surface through TLR 4 can signal through an intermediate (in fact exactly the same molecule as found in the sponge, called 'MyD88') to the intracellular effector molecule. There are several intracellular effects of the signalling pathway which lead from engagement of the surface TLR 4 with LPS. However, one in particular results in binding to specific regions of DNA in the nucleus, to stimulate the production of a range of inflammatory cytokines. In addition, in just the same way as we have already seen in early animals, the TLR 4 signal can activate a member of the NOD-like receptor (NLR) family proteins (which we encountered in chap. 4) inside the cell, leading to cell death and the release of interleukin-1. This mode of cell death is called pyroptosis or 'fiery death' and incites an inflammatory response, attracting cells out of the blood stream and into the gut lining. Once again, in moderation such a mechanism may be beneficial to prevent the ingress of bacteria. However, in excess it might overcome the 'housekeeping' ability of the normal gut to tidy up the debris and patch up the holes in the epithelium caused by dying cells, leading to the potential for bacteria to cross over. Indeed, one of the hallmark features of colitis is increased evidence of cell death and cellular debris in the colonic epithelium.

Evidence to suggest a role for TLR 4 in inflammatory bowel diseases comes from genetic association studies that implicate the gene that encodes it. Analysis of the colonic lining in patients demonstrates higher than normal levels of TLR 4 and also lower levels of the chemicals that regulate its actions inside the cell. Finally, there is a gradient of expression in the normal colon such that TLR 4 is found in increasing quantities along the length of the colon and is maximal in the lowest part—the rectum. This *might* provide a clue as to why ulcerative colitis tends to affect a varying length of colon but always from the bottom up.

TLRs are, of course, also found on the presenting cells such as dendritic cells and macrophages, and increased levels of TLR expression or activity could additionally modulate their effects in inciting an inflammatory response.

Diverticulum #9.2—Necrotising Enterocolitis, another 'TLR 4' Disease?

A devastating condition of premature infants, necrotising enterocolitis, results in death of the tissues of the bowel. It can require emergency surgery within days of birth, leading to removal of large amounts of intestine and often leaving the baby critically dependent on intravenous nutrients for nourishment. It occurs predominantly with formula-fed infants, and can be prevented by the use of breast milk. Antibiotics and acid suppression can predispose to the condition as well as episodes of low oxygen levels in the blood.

Mutations of the TLR 4 encoding gene have been described that increase its activation level and such mutations are more commonly found in infants with necrotising enterocolitis. Furthermore, mice can be genetically altered to be protected from the condition by knocking out TLR 4. It also appears that there are specific natural inhibitors of TLR 4 present in breast milk which are being developed as potential drugs for treating or preventing the condition.

The Second Wave

Cytokines released from damaged enterocytes send a signal to recruit blood cells from the circulation in order to stem the breach in the intestinal lining and control the inward flood of bacteria. In particular, cells called neutrophils are attracted to the site of the leak. These bear similarities to some of the oldest immune cells that we have come across in evolution—the sentinel cells of the social amoeba. In the same way, they actively phagocytose any bacteria that cross the epithelial membrane layer and they also produce sticky DNA 'nets' laced with bactericidal chemicals. In peacetime, small numbers of neutrophils actively patrol the gut epithelium in order to deal with any minor border transgressions. Their importance in this role is demonstrated by the development of inflammation—colitis—in cases of neutrophil deficiency such as can occur in cancer patients receiving chemotherapy. However, once again we see that a normal regulatory response required to keep the gut healthy can become pathological in excess. Thus, a major influx of neutrophils into the gut lining can overwhelm the housekeeping systems and induce inflammation and cell damage. In particular, neutrophils 'chase' the infiltrating bacteria back out into the intestinal lumen, passing through the colonic epithelium. In doing so they damage the tight junctions between the cells that maintain the epithelial integrity and potentially lead to yet more ingress of bacteria.

The other cell type that is attracted from the blood stream is the 'activated' macrophage. These lead to an inflammatory response, quite unlike the effects of the normal quiescent, tolerising macrophages and dendritic cells that are present under more peaceful conditions. Once again, these cells serve to magnify the inflammatory response.

Gut Bacteria: Friends or Foes?

Given the similarity between the early stages of ulcerative colitis and self-limiting bacterial dysentery, it is of course appropriate to look at the bacteria to see whether there are potential culprits present that could cause the conditions. Our colons harbour over 100 trillion bacteria, weighing around 1.5kgs and comprising hundreds of different species, such that it is estimated that there are at least 100 times more genes in the DNA inside our bowels than in the nuclei of our own cells! However, around 30 species make up 99% of the bacteria in our colons. A key feature of all the inflammatory bowel diseases is that there is a 'dysbiosis'—meaning that the normal composition of the bacterial flora is altered. It has proven difficult to identify specific bacteria involved in causing the disease process, although certain types of bacteria have been associated with a 'healthy' constitution and others with disease. There is usually less diversity, and smaller numbers of bacteria are present in the colon of patients with ulcerative colitis than normal, but the patterns of different bacteria are very individual and vary geographically. Whether or not the dysbiosis is a primary feature or secondary to the inflammation is at present difficult to establish, as there is a two-way communication between bacteria in the gut and the host immune system.

There are certainly many predisposing features of our current lifestyles that can lead to an altered microbiome such as lack of breast feeding, a diet low in 'fibre' and medications including antibiotics. The genetic make-up of the host also affects the type of bacteria present in the colon. In turn, the composition of the bacterial flora can influence the balance between immune regulation and inflammation in the host. Chemical products of bacterial metabolism signal through specific receptors on immune cells such as the aryl hydrocarbon receptor (see footnote 7 above) and act particularly in the layer of innate lymphoid cells and intraepithelial lymphocytes to regulate inflammation and maintain the cell populations that are responsible for epithelial repair and immune tolerance. In turn, inflammation in the bowel can also influence the composition of the bacterial flora!

Given such complexities, we have only just begun to try to modulate the bacterial flora in a meaningful way. Many of the probiotic strains of bacteria commercially available in supermarkets in products such as yoghurt drinks have little scientific evidence to support their use, and in many cases contain few viable organisms able to gain a foothold in the already crowded colon. However, there is one notable exception with a long track record.

Alfred Nissle (1874-1965) was a German physician with an interest in infectious diseases of the bowel. A favourite practical experiment that he carried out with medical students was to grow out the bacteria from their own faeces on agar plates. Sometimes he would introduce disease-causing *Salmonella* bacteria onto the plates to see how well they grew. In most cases they were unhindered by the other bacteria present, but in some cases the growth of the *Salmonella* bacteria was completely inhibited. Nissle suggested that this was due to 'antagonism' by other 'stronger' bacteria. He was able to pursue his hunch in 1917 when he came across a corporal in the German army serving in a field hospital in Freiburg, which was near to his laboratory. This soldier had recently returned from Dobruja (on the Romanian Black Sea coast), where his unit had suffered greatly from dysentery due to the *Shigella* bacterium. The corporal, however, appeared to be immune to it. Nissle was able to grow a species of *Escherichia coli* (*E.coli*[9]) bacteria from his faeces that

[9] Interestingly the common bacterium of the colon—*Escherichia coli*—was given its name from its discovery Theodor Escherich only in 1919, 2 years after Nissle isolated his special brand of *E coli*.

suppressed the growth of *Salmonella* and *Shigella* in the laboratory. Rather than investigate why this was, Nissle filed for a patent and sold the bacterium (called '*E.coli* Nissle 1917') as a potential remedy called 'Mutaflor', and it has been continuously produced and marketed ever since.

E. coli Nissle 1917 has been tested in numerous conditions with variable and usually limited success, including irritable bowel syndrome and infectious diarrhoea. In ulcerative colitis it does indeed appear to have some benefit in maintaining remission comparable to other medications. Its effects are mediated by substances it produces to kill or suppress other bacteria, but also through beneficial modulation of the gut immune system leading to a more tolerant rather than inflammatory state. However, the effect is not dramatic enough for it to be considered by many as a routine treatment for the condition.

The other way in which we are now trying to modulate the gut microbiome is by introducing a whole new microbiome directly into the gut from someone else. Faecal microbial transplantation[10] (FMT) can be undertaken using liquidised donor faeces introduced through a tube into the intestine or into the colon. This treatment has seen most success in its use for recurrent infection with *Clostridium difficile* (see Diverticulum #9.1 above). Trials in ulcerative colitis have shown some benefit, with up to a third of patients with mild to moderate disease entering remission. However, this is less effective than standard treatments and whilst perhaps demonstrating that altering the bacteria can modulate the disease, it does not yet represent a practical treatment. The clinical trials have been limited though, and we may yet chance upon a 'super-donor' such as the German corporal whose stools provide greater benefit!

The Evolving Picture of Ulcerative Colitis

The precise way in which ulcerative colitis develops is challenging to establish, as a cascade of immunological events occurs in quick succession. However, we can surmise that the condition arises as a combination of environmental and genetic insults to a very finely-balanced network of immune regulation that maintains our relationship with our colonic bacteria. This is

[10] The ancient Chinese used a concoction of human faeces that they called 'yellow soup' to treat a variety of conditions, and faecal ingestion has been used by a variety of cultures for treatment purposes ever since. Its first use in diarrhoea caused by *Clostridium difficile* was in 1958, but the first clinical trial only as recently as 2013.

what we discovered in part 2: the umbrella of immunity extends to our symbiotic passengers and the immune system regulates the rules of social engagement, rather than just policing the transgressions. In keeping with this concept of immunity, the inflammatory bowel diseases appear to arise from vicious cycles of unregulated immune responses, which under normal circumstances work in balance to maintain the *status quo*. This leads to some confusion in working out the mechanisms, as we can see both 'good' and 'bad' effects of different immune mechanisms at play: 'good' in that they are required to maintain the relationship such that defects can *lead* to colitis, and 'bad' where excess, unregulated activity leads to colitis and blocking this activity may actually *improve* the condition. Furthermore, the same immune defects that may lead to unregulated inflammation in colitis can also predispose to some forms of cancer, the risk of which is increased in this condition. In this instance the immune system is regulating the relationship between cell types of the host, rather than between the host and its passengers.

By going back to the earliest stages of immune evolution (those that developed in concert with symbiotic bacteria), we have seen that a breakdown in the epithelial barrier defences can lead to disruption of the normal interaction with our bacterial flora in the colon. Increased leakiness of the epithelium leads to loss of 'containment' and results in normally harmless bacteria passing through the lining, whereupon a significant immune response occurs. The disease process at this stage may be no different to the type of response that is seen when the body is under attack from a disease-causing bacterium. However, it becomes prolonged and self-perpetuating, either by an inability to repair the mucosal breach and prevent further bacterial influx or by a defective 'off switch' at the level of the innate lymphoid cells. Many of these secondary defects in regulation are shared between ulcerative colitis and Crohn's disease and so occur across the spectrum of the inflammatory bowel diseases. To a large extent therefore, the sequence of events in ulcerative colitis follows the evolutionary pathway that we have followed throughout this book, from mechanisms seen at the lowest levels of the gastro-archeological trench through to the surface. This pathway has deviated to the 'antibody' side of the great schism as a result of the lower-most layers dealing with extracellular threats—outside the cell. It will come as no surprise then that our next focus on Crohn's disease leads us inside the cell which takes us along the 'cytotoxic' side of the schism.

Crohn's Disease

Probably first described in 1759 by the Italian pathologist Giovanni Morgagni, and then in a series of nine cases by the Scottish surgeon, Thomas Kennedy Dalziell, this condition did not receive its eponymous epithet until 1932.[11]

Despite the fact that ulcerative colitis and Crohn's disease share over 100 associated genes in common, as well as the inflammatory pathways from the upper gastro-archeological layers, they can appear surprisingly different. Crohn's disease can affect any part of the gut from mouth to anus. Unlike ulcerative colitis, which is continuous in distribution, Crohn's disease can affect short stretches of intestine or colon with normal bowel in between—so-called 'skip lesions'. In particular, it has a peculiar predilection for the end of the small intestine just before it joins the colon (Fig. 9.1). This may not be too surprising as this is where two different immune systems merge—that of the small intestine that requires a tolerant approach to food proteins, and that of the colon that requires both protection from its passenger bacteria and tolerance of them. At this point there is a one-way valve—the ileo-caecal valve, named from the end of the small intestine (the terminal ileum) and the first part of the colon (the caecum). However, this valve is not always very effective and it is common to find colonic bacteria at the end of the small intestine. Hence patients with ulcerative colitis can experience ulceration and damage in the end of the small intestine (called 'backwash ileitis'), without us necessarily regarding this as Crohn's disease.

The terminal ileum is immunologically stressed under *normal* circumstances as it is really not designed to provide the same degree of protection from bacteria as the colon does. In the normal gut, this is the place where we see the highest background level of some inflammatory cytokines, such as interleukin-23. With the help of modern technology, gastroenterologists have been examining the terminal ileum more frequently during colonoscopy (camera inspection of the colon) or wireless capsule enteroscopy (a 'camera pill' that films the intestinal lining whilst passing through). In doing so we have come across large numbers of people who have mild inflammation and

[11] Leon Ginsburg and Gordon Oppenheimer put together a collection of 12 patients from Mount Sinai Hospital in New York with a condition they described as 'regional ileitis'. They were about to publish but were introduced to the gastroenterologist, Burrill Crohn (1884-1983), who had two additional cases to add. Having written the majority of the manuscript they were then surprised to find it published with Crohn's name listed first in the October 1932 *Journal of the American Medical Association*. However, rather than stealing the limelight, it appears that Crohn's name took first place (and thus entered immortality) as a result of the editorial policy of listing authors alphabetically. A famous sufferer of Crohn's Disease was Dwight D. Eisenhower, who was treated by Crohn himself and underwent emergency surgery whilst US President at the age of 65 in 1956.

small ulcers in this part of the small intestine, which do not cause symptoms and do not progress into Crohn's disease. Whether this is due to an inefficient ileo-caecal valve allowing some colonic bacteria into the small intestine, resulting in a degree of inflammation, remains to be seen.

Unlike ulcerative colitis, which only affects the superficial lining of the gut, Crohn's disease can affect the whole thickness of the intestinal wall and lead to the formation of holes right through to the outside surface. The leakage of contaminated fluid out of the gut and into the abdominal cavity can then cause life-threatening peritonitis. However, more usually the process of erosion is slow and leads to the formation of an abscess on the outside of the gut wall, or gradually erodes through into another organ or through the skin. This new connection between bowel and other organs is called a 'fistula' and can result in gastrointestinal short circuits (where fistulation occurs between bowel loops), chronic urinary infections (when into the bladder), or leakage of bowel contents through the vagina or onto the skin.

Another unpleasant feature of Crohn's disease is that chronic inflammation can lead to scarring, including the gradual production of fibrous tissue that shrinks over time (like any scar). When circumferential around the intestine, this fibrotic tissue can lead to narrowing and ultimately blockage of the bowel. Under these circumstances, medications can do little and a surgical operation is usually required to remove the affected blocking segment. Over time and with recurrent operations to remove portions of small intestine, the patient may be left with insufficient intestine to meet their nutrient and fluid absorption requirements and require lifelong intravenous supplementation.

One characteristic feature of Crohn's disease is seen by examining involved areas under the microscope. Structures called granulomas[12] are tiny aggregations of chronic inflammatory cells including macrophages. These do not always appear in Crohn's disease but when present they give a strong clue to the diagnosis.

Granulomas are also found in other conditions including infections and around indigestible foreign objects such as sutures or splinters. They form as an attempt by the immune system to get rid of something that is resistant to

[12] Strictly grammatically speaking, this should be 'granulomata'. They were first described by the German pathologist, Rudolf Virchow (1821-1902). Virchow is known as the 'father of pathology' for all of his discoveries and theories, and was a maverick genius. He was an anti-evolutionist (all the more incredible given that he held no religious belief), and did not believe in anti-sepsis. He held strong political views and even founded his own political party, which brought him into conflict with the great Otto von Bismarck. My favourite story about him was that the pugilistic Bismarck challenged Virchow to a duel over a falling-out. Virchow, being the person challenged, was allowed to choose the weapons and apocryphally chose two pork sausages, of which one was laced with pork tapeworm eggs! Supposedly Bismarck declined and withdrew the duel...

all the usual mechanisms and thereby excites a chronic inflammatory response. This ultimately results in the source being surrounded and 'locked away' in a fibrous cocoon. Of most significance to our story is the relationship of granulomas to tuberculosis and leprosy—both infections caused by unusual bacteria called mycobacteria, which can evade being engulfed and digested by the cells of the immune system. On which note, let us get our trowels out!

Into the Land of NOD[13]

Let us start at the bottom of the trench that we have already dug to look at the gastro-archeological strata of ulcerative colitis, as we know that many of the upper layers are shared between ulcerative colitis and Crohn's disease. At this stage we should revisit some of the earliest immune mechanisms that are present even in sponges and possibly date back to pre-Cambrian explosion times. These are the surface Toll-like receptors, and the NOD-like receptors within the cell, both of which are based on the leucine-rich repeat motif rather than the immunoglobulin superfamily. Our interest in the NOD-like receptors (NLRs) came about as recently as 2001 as a result of genetic association studies that showed a very strong link between Crohn's disease and a gene encoding an NLR protein called NOD-2. Having mutations in both copies of your NOD2 genes carries almost a 20-fold increased risk of developing the condition compared to having two 'normal' NOD-2 genes. Intriguingly however, none of the associated NOD-2 mutations have been identified in any Japanese patients with Crohn's disease—once again highlighting the diverse nature of inflammatory bowel diseases.

In the sea urchin, NLRs are only expressed in the lining cells of the gut, which suggests an evolutionary role in the host-microbe interface. There are 23 NLR proteins in humans, of which NOD-1 and NOD-2 are both expressed in intestinal tissues. Interestingly, NOD-2 is normally found only in Paneth cells—these are the cells in the small intestine that are specialised to secrete antibacterial proteins. However, under conditions of inflammation, NOD-2 is also made in the common absorptive enterocyte cells that line the gut. Thus, when we knock out the NOD-2 gene in mice, we see no particular changes under usual conditions, but they do become more susceptible to certain infections.

As we would expect as a result of their leucine-rich repeats, NODs 1 and 2 are each capable of recognising specific molecular patterns inside the cell. In

[13] The 'Land of Nod' was where God banished Cain in the Old Testament book of Genesis, chap. 4 verse 16. It was 'east of Eden'. The word derives from the Aramaic root related to 'wandering'. It has been used to describe desolate areas. Its alternative popular use derives from the pun with 'nodding off' meaning to go to sleep, in which setting it was first used by Jonathan Swift in *Gulliver's Travels*.

the case of NOD-2, the molecular motif identified is called 'muramyl dipeptide', a component specific to bacterial cell walls. Over 90% of the mutations of NOD-2 found in Crohn's disease affect the region of the gene encoding the leucine-rich repeat end of the molecule. This suggests that defective recognition of the bacterial wall product, and in turn loss of normal NOD-2 functions, can underly the inflammation of Crohn's disease.

Inside the cell, the activation of NOD-2 resulting from binding its uniquely recognised bacterial molecule leads to a number of knock-on events which are of potential interest to us. One of these is the stimulation of the secretion of specific anti-microbial substances by Paneth cells. Paneth cells isolated from patients with Crohn's disease are indeed deficient in producing these molecules, as are mice that lack NOD-2, and this predisposes to 'dysbiosis'—a change in the normal bacterial flora of the gut. NOD-2 also stimulates the production of a range of proinflammatory cytokines. In addition, NOD-2 influences the survival of intestinal stem cells, necessary for maintaining cell division in the intestinal lining.

NOD-2 is also found expressed in presenting cells such as the dendritic cells and macrophages, where it plays a key role in the production of inflammatory cytokines which drive the immune response. Activation of NOD-2 in these cells also alters the nature of the immune response to veer towards the 'antibody' type response—therefore in the absence of effective NOD-2 (for instance by a genetic mutation), the switch tends to the 'cytotoxic' side of the 'great schism' in the wall of our trench. This is what happens in Crohn's disease, as opposed to ulcerative colitis.

Another recently-discovered role of NOD-2 in the enterocytes lining the gut has transformed our understanding of how Crohn's disease comes about, but in order to understand this we must return to the tale of Erysichthon's unfortunate demise, as I promised earlier.

'Self-Eating': The Internal Housekeeping System

A further astonishing result to come out of genetic studies of Crohn's disease was the link with another protein called ATG16L. In this case the letters ATG are an abbreviation of 'autophagy'[14]—literally translating as 'self-eating'.

[14] The concept of autophagy and the name were first coined by Christian de Duve in 1963 on the discovery of lysosomes and their function in digestion within the cell. However, it was Yoshinori Ohsumi from Tokyo who earned the 2016 Nobel Prize in Physiology or Medicine for elucidating the underlying mechanisms.

We have already come across the fundamental role of the immune system in the multicellular organism as a 'housekeeper', tidying up cell debris and microbes. The discovery of autophagy led to our realisation that there is a similar need for a housekeeping system within the cell itself. This first arose in evolution in eukaryotic cells—those with internal membranes and organelles such as the mitochondrion (the intracellular energy generator). Through increasing our understanding of autophagy, we have come to learn that this process is fundamental to many processes including aging and senescence, and invoked in disease processes as diverse as Crohn's disease, Parkinson's disease and Alzheimer's dementia.

In Chap. 1 we saw the importance of the cell membrane in defining the cell itself, and how eukaryotes have used the unique properties of the membrane in overcoming the 'containment paradox' and establishing digestive machinery inside the cell. We saw how the surface cell membrane can invaginate and then pinch off to form bubbles or vesicles within the cell—a process called endocytosis. The internal membrane structures of the cell also act as a platform for proteins to interact together. In autophagy, the cell goes one step further and builds its own membranes internally using a protein scaffold made up of ATG proteins. In this way, structures within the cell such as mitochondria can be encased by membrane, without them needing to have entered from the outside through endocytosis. It effectively wraps internal constituents to keep them separate—a bit like cellophane on separate items kept in the refrigerator. However, the benefit of autophagy is that these membrane-bound components can then merge with the lysosomes, other membrane-bound bodies within the cell which contain digestive enzymes. On fusing with a lysosome, the contents of the resulting larger vesicle will be broken down into separate molecules that can then be reused. Autophagy is therefore the equivalent of a cellular recycling mechanism. Given that old or damaged parts of the cell can be selectively disposed of and reprocessed internally, the cell is capable of rejuvenating itself from the inside—hence autophagy is closely related to mechanisms of aging and senescence.

Starvation is also a potent trigger of autophagy as it allows internal structures to be recycled and digested in order to provide nutrients to sustain the cell. It is effectively eating itself, but in a way that allows it to survive and regenerate (unlike poor Erysichthon).

Autophagy also provides the energy for cell division and replication, and is therefore of importance in rebuilding and wound healing—and therefore in maintaining the intact epithelial cell lining of the gut. Indeed, this role in replication may have been its primary purpose as proteins that are similar to those involved in autophagy are found in early single-celled prokaryotes, where they act to recycle damaged proteins and are implicated in cell division.

Diverticulum #9.3 Autophagy and Aging

Autophagy, by removing effete structures within the cell and effectively rejuvenating it, also preserves the cell's ability to replicate itself. Thus, autophagy is essential for the maintenance of stem cells, which are cells that continue to divide. The products of this division are generally another stem cell (to keep the process going) and a daughter cell which will differentiate into a specialised cell type, which might go on to replace an older version of that cell. This is one example of how autophagy reduces 'aging'. Suppressing autophagy leads cells to differentiate away from the basic stem cell type.

Simple experiments performed in fruit flies (*Drosophila*) and the *Caenorhabditis* worm have shown that activating autophagy is not just involved in cell maturation but actually affects the aging of the entire animal. The mechanism is closely related to how the hormone insulin works, and one mutation of the insulin signalling pathway can almost double the life expectancy of the fruit fly through enhancing autophagy! A drug that is used in type 2 diabetes in humans called metformin can also increase life expectancy when given to fruit flies, as it alters insulin signalling pathways. Its side effects however prevent us from using it to prolong human life expectancy except in diabetics where its benefits outweigh its risks! One way of switching on autophagy is through diet and calorie restriction, and indeed simply starving fruit flies or mice makes them live longer. The restriction should not lead to nutritional deficiency, hence short-term fasting is thought to have similar benefits for all animals—including ourselves.

Eating Strangers

Given that autophagy can digest and recycle internal organelles such as mitochondria (which originally evolved from microbes), it is no surprise that autophagy is also capable of disposing of bacteria which evade the extracellular defences and manage to breach the cell membrane. This goes by the name of xenophagy, literally 'eating strangers'. The missing link between NOD-2 and autophagy-affecting mutations in Crohn's disease became apparent only recently when it was discovered that NOD-2 attaches to the inside of the cell membrane at the location where a bacteria is trying to enter and it attracts the membrane-building machinery of autophagy to that spot. Effectively, through its pattern recognition function, it identifies the bacterium entering the cell and connects it to the membrane formation system that can then isolate the bacterium from the cell contents within a special vesicle, and ultimately digest it by fusing that vesicle with enzyme-containing lysosomes. Thus NOD-2 and autophagy act in concert in this regard.

Only certain bacteria have the ability to enter cells in this way and survive within them. The quest to identify a bacterium that causes Crohn's disease has been long and fruitless, just as with ulcerative colitis. As in ulcerative colitis, an altered bacterial flora (dysbiosis) is found in Crohn's disease and the types of bacteria that predominate tend to be those that incite inflammatory responses through their interaction with the gut immune system, rather than those that suppress it. However, there are many mechanisms present in the host that could lead to the dysbiosis (such as altered Paneth cell function), and it is difficult to know whether the change in the gut bacteria is causative or an effect of the condition.

There is one particular type of bacterium that may well play a role in some cases. We have already come across *Escherischia coli* (*E. coli*) thanks to Alfred Nissle. *E. coli* bacteria are very commonly found in the colon and also in the small intestine, even though they make up only about 1 in 1000 of our gut bacteria. They are extraordinarily diverse bacteria, sharing as little as 20% of genetic information between different strains. Whilst most are harmless (or even beneficial, such as the Nissle 1917 strain), they can cause problems when in the wrong place, such as when they enter the blood-stream to cause septicaemia, or the bladder as a common cause of urinary tract infections. *E. coli* are closely related to the familiar disease-causing *Salmonella* found in birds and reptiles. In fact, their evolutionary divergence occurred about 100 million years ago, at about the same time as their hosts evolved along separate lines. Some varieties of *E. coli* cause disease in the gut by producing toxins, or by finding ways of evading the gut immune defences and invading the epithelial lining cells. These latter types of *E. coli* are called adhesive-invasive *E. coli*, or AIEC for short.

'Stealth' Bacteria

The adhesive-invasive *E. coli* bacteria that have been associated with Crohn's disease possess a number of special mutations that allow them to evade the gut's defences. They produce enzymes that digest the mucus so that they can penetrate down to the level of the epithelium. They also control the appearance of a propulsive flagellum so as to do so only when necessary—this makes it more difficult for the host to detect the flagellar proteins using its Toll-like receptors. It also appears that the AIEC include a supercharged model that is able to work its flagellum quicker and propel it faster to its target. They also carry gene mutations that make them resistant to defensins, the fixed pattern-recognition molecules that are secreted by Paneth cells and lace the mucus. They have surface molecules that identify and bind to specific proteins on the membrane of the epithelial cells, to allow them to enter. Once inside the cell they thrive—particularly in conditions where the cell is stressed. They can also survive within macrophages that are attracted by the inflammatory process from the blood-stream into the intestinal wall, where they lead to the over-secretion of a particularly important cytokine called tumour necrosis factor (TNF). This actually *encourages* their replication within the cell!

The process of autophagy normally helps to clear bacteria from animal cells. Thus, although AIEC can be found in the gut of normal individuals without Crohn's disease, such individuals are able to kill the bacteria within their cells through autophagy. However, as we know, there are specific mutations in patients with Crohn's disease that diminish this response and thereby allow the bacteria to replicate within the cell. Macrophages derived from blood cells of patients with the condition are also unable to clear the bacteria, unlike macrophages from patients with ulcerative colitis or people without disease.

There is plenty of evidence to support the concept that intracellular persistence of bacteria such as AIEC could be a factor leading to the development of Crohn's disease. High powered microscopy can identify AIEC within cells in the gut of many patients with Crohn's disease. AIEC are found to inhabit the end of the small intestine out of preference—the terminal ileum—which is a common location of Crohn's disease lesions and is a site where *E. coli* and similar bacteria tend to concentrate. In experimental situations, AIEC can cause inflammation and colitis as part of the body's response to clear them, but when persistent this leads to fibrosis and scarring, just as in Crohn's disease itself. It should be noted, however, that AIEC can be detected in the gut of up to 20% of normal individuals and their presence is by no means universal in Crohn's disease, having been detected in only around 50-60% of cases. This may mean that they are associated with a specific form of Crohn's-like inflammatory bowel disease (given that this is likely to represent a wide spectrum of similar conditions) or that there are other bacterial culprits that we still have to identify.

Of course, if this is the case then antibiotics would be of benefit in treating Crohn's disease—and indeed they are! One particular antibiotic called metronidazole is of such value that it is used to prevent recurrence in patients who have undergone surgery to remove affected segments of intestine. However, antibiotics themselves carry risks of side-effects including dysbiosis and are not ideal in the long term. Nor are they particularly effective, as they may not reach the bacteria or the bugs may develop resistance to them. Furthermore, simply diverting the faecal stream (by surgically creating an opening from the gut on the skin surface and draining the intestinal contents into a bag, so that they do not reach the diseased area) has a beneficial effect in Crohn's disease but not ulcerative colitis. The implication is that something—most likely bacteria—plays an important ongoing role in the disease process.

Diverticulum #9.4 Crohn's—Or 'Johne's'—Disease?

The hunt for a causative bacterium behind Crohn's disease has been a long one. As previously mentioned, special bacteria called mycobacteria are capable of living within cells and lead to the long-lasting infections associated with tuberculosis and leprosy. When grown in the laboratory they appear similar to mold, hence the 'myco' prefix often used to refer to fungi. They have a very thick, waxy cell wall that makes them highly resistant to the usual immune defences and to antibiotics, and they divide very slowly—taking up to 20 days rather than the usual 20 minutes for bacteria to replicate. Their ability to live within cells leads to the formation of granulomas where the immune system tries but fails to get rid of them. This led early investigators in the field of Crohn's disease to look for a mycobacterial cause of the condition.

There is a very similar condition that affects the small intestine in farm animals—goats and cattle—called Johne's disease.[15] This affects the animal in much the same way as is seen in humans with Crohn's disease, and the appearance of the intestine when examined under the microscope is remarkably similar. Johne's disease is known to be caused by a mycobacterium called *Mycobacterium avium paratuberculosis* or 'MAP' for short. This infection is highly prevalent amongst cattle and is found in the majority of herds in the USA, although only 5-10% of affected animals develop the disease. It is found in meat and milk, it is resistant to pasteurisation and can survive up to a year in soil. Humans are therefore exposed to MAP in their food and in the environment.

The similarity of Johne's disease to Crohn's disease was noted some 20 years before Crohn's paper. MAP can be detected by molecular techniques in tissues and blood cells in patients with Crohn's disease—more commonly than in healthy individuals or patients with ulcerative colitis. Similarly, antibodies against MAP are more commonly found in patients with Crohn's disease. Clusters of outbreaks or coincidental development of Crohn's disease in married couples may hint at an infectious cause, and there is some epidemiological evidence to suggest a higher prevalence in livestock agricultural areas (such as Winnipeg in Canada) than in neighbouring locations. Furthermore, the increase in Johne's disease in cattle and introduction of the condition into areas appears to precede a rise in the incidence of Crohn's disease. However, individuals in potentially high risk occupations such as farmers do not appear to be at increased risk of Crohn's disease. Prolonged antibiotics treatment for up to 2 years with multiple drugs is required to treat MAP in humans, but trials do not appear to show any benefit for Crohn's Disease.

The jury is still out therefore on whether Crohn's disease is or can be caused by infection with MAP, and we may not know for certain until efforts to eradicate MAP in cattle are successfully undertaken!

[15] Named after Heinrich Johne (1839—1910), a German pathologist instrumental in introducing regular inspection of meat used for human consumption. He described this condition in 1895. Being German, his name is pronounced 'Yoh-na' rather than rhyming (coincidentally) with Crohn…!

The Evolution of Crohn's Disease

We have now arrived at the bottom of our dig and can trace the story of Crohn's disease back through the layers of time. A defect in the fundamental process of autophagy results in impaired clearance of specialised bacteria and effete mitochondria that could damage the cell. The persistence of bacteria inside the cells leads to a chronic inflammatory state that can result in fibrosis and scarring, and can cause the narrowing and blockages seen in Crohn's disease. Alterations in NOD-2 and autophagy-related proteins also affect the responses of immune cells that enter the gut wall from the blood-stream, leading to an exaggeration of the resultant inflammation. Knock-on effects include increased leakiness of the epithelium allowing bacteria to enter the body through the gut, and an alteration of the bacterial flora which provides an increased inflammatory signal to the gut immune system. This may be further enhanced by additional defects in the immune 'on' and 'off' switches, which we have seen are controlled by the products of genes associated with both ulcerative colitis and Crohn's disease. However, as we have seen, not all individuals with this condition show genetic mutations in NOD-2 or have identifiable intracellular bacteria. At the current time we do not know whether further mutations involving the same pathways will be identified in such cases, or whether they are suffering from different forms of Crohn's disease that are caused by separate mechanisms.

This pathway stretches from the earliest molecules and processes involved in the immune system (including the intracellular housekeeping mechanisms) to the surface layers of recently-evolved immune mechanisms along the 'cytotoxic' path, rather than the 'antibody' path of ulcerative colitis. This is only to be expected as a result of Crohn's disease deriving from an inability to clear an invader *inside* the cell, whereas ulcerative colitis results from an immune response against aliens crossing the epithelium that remain *outside* the cell. From the perspective of the earlier chapters in this book, Crohn's Disease results from microbial invasion across the containment barrier of the cell membrane, whereas ulcerative colitis results from breach of the epithelial containment barrier of multicellular animals.

Drug Treatments for Inflammatory Bowel Diseases—The First 50 Years

There is one more facet of the story that we must visit before we end our epic journey. However, in order to get there we should take a diversion to consider some of the available treatments for the inflammatory bowel diseases, and how they work. In fact, for more than half a century, our therapeutic armoury comprised only three drugs, all of which were developed in the 1950's.

In September 1948, a 29 year old woman in Rochester, Minnesota, who was crippled with rheumatoid arthritis received three injections of a trial drug called 'Compound E'. The effect was nothing short of miraculous and 2 days later she went for a three-hour shopping spree! The drug was developed by a rheumatologist, Philip Hench, and a biochemist, Edward Kendall, after nearly 20 years of investigation,[16] working on the hormone products of the adrenal glands. During the Second World War, high priority was given to research into the adrenal hormones (greater even than investigation into antibiotics such as penicillin), following false information that Luftwaffe pilots were able to fly at high altitude without oxygen due to injections of adrenal extract! Compound E was renamed 'cortisone', and in various modifications has become the mainstay of treatment for inflammatory conditions ever since.

The adrenal glands are found sitting on the top of each kidney and are small (about the size of a walnut in its shell) but highly active hormone-producing glands. They produce hormones that regulate the blood pressure, the 'fight or flight' hormone adrenaline, and sex hormones. They also produce the hormone cortisol, which is the naturally-occurring version of the synthetic drug cortisone. Cortisone and cortisol are together known as 'glucocortico-steroids' (or glucocorticoids for short) as they are steroid hormones from the adrenal cortex which affect the metabolism and release of glucose within the body. Every cell in the body has receptors for glucocorticoids which bind to DNA in the cell nucleus to affect the transcription of many genes into proteins. Interestingly, glucocorticoids have effects on genes that are involved in immunity as well as metabolism. The immune effects of the glucocorticoids are universally suppressive as they act to reduce the production of all of the inflammatory cytokines that we have previously discussed—and more. They are therefore the panacea for all inflammatory conditions! However, as is well known, their use comes with a price. Because of their metabolic functions, glucocorticoids increase the blood sugar glucose leading to diabetes, increase appetite and lead to weight gain and fat deposition. They cause thinning of the skin, acne and reduced bone density. Therefore, whilst extraordinarily useful to suppress immune reactions, their side effects preclude their long-term use.

---

[16] For which they were awarded the 1950 Nobel Prize in physiology or medicine, along with the Swiss chemist Tadeusz Reichstein. Possibly Kendall's greater achievement was the isolation of thyroxine from thyroid glands.

Only 6 years after they were first used to treat rheumatoid arthritis, Sidney Truelove and Leslie Witts[17] published their results of using cortisone to treat ulcerative colitis. During a first attack of ulcerative colitis, treatment with cortisone resulted in remission or improvement in over three quarters of patients, compared to less than a quarter of those who received a placebo. 66 years later, we still treat patients suffering from severe ulcerative colitis with glucocorticoids as our first line treatment. Side effects of glucocorticoids limit their benefit in long-term use. However, newly-developed preparations are formulated to be released in the small intestine or colon but are then deactivated as they pass through the intestinal lining and the liver, thereby significantly reducing their detrimental actions.

In the 1950's it was thought that rheumatoid arthritis resulted from a bacterial infection and attempts were made to treat it with antibiotics—of which few were available at the time. The first effective antibacterial drugs were based on sulphur and called sulphonamides.[18] Nanna Svartz, a rheumatologist working in the Karolinska Institute in Stockholm, worked with a Swedish company to produce a drug called sulphasalazine, which was found to be effective in her arthritic patients. This substance was a sulphonamide antibiotic chemically attached to a small molecule quite similar to aspirin called 5-amino-salicylic acid, or 5-ASA for short. Fortuitously and coincidentally it was found to be effective in ulcerative colitis and not only helped to resolve an acute flare-up but also worked to keep the condition under control when used on a daily basis.

Subsequent studies showed that—to many people's surprise—it was not the antibiotic but the 5-ASA linked to it that was effective in treating colitis. This was again lucky as the antibiotic part of the combined drug was found to be responsible for the frequently-occurring allergic reaction to its use. The 5-ASA portion was thereafter manufactured separately and Sidney Truelove in Oxford was once again instrumental in designing trials to demonstrate its value. However, the delivery of 5-ASA (or mesalazine as it is now known) was

[17] Leslie Witts (1898-1982) first made his name as a haematologist rather than a gastroenterologist, and became the first Nuffield Professor of Medicine in Oxford. He co-founded the British Society of Gastroenterology in 1937.

[18] Sulphonamide-based antibiotics were the only antibiotics available before penicillin and were used widely in the late 1930's and early years of World War II. I remember one dive on the wreck of the *President Coolidge*, an American troopship that sank in Vanuatu (then the New Hebrides) in 1942. On entering the medical storeroom inside the wreck, I was met with the surreal site of hundreds of glass vials floating at different levels in the water and making magical tinkling chimes as they collided when pushed around by my exhaled bubbles. They were all filled with a yellow powder that I presumed to be a sulphonamide antibiotic.

problematic. Taken by mouth it was ineffective, as it was absorbed or metabolised before reaching the colon. It could therefore initially only be given within enemas into the rectum. The reason that sulphasalazine is effective as a tablet whereas 5-ASA alone is not is that the larger, combined molecule is not absorbed in the intestine and enters the colon, where bacteria break the join between the two molecules and release them separately. This finding spawned a variety of ingenious ways of delivering the 5-ASA into the colon when taken by mouth. These included linking 5-ASA to an inert carrier molecule, joining two molecules of 5-ASA together, or coating the drug-containing capsule in resins that dissolve only at the correct acidity to liberate the active drug in the colon. Again, over half a century later, mesalazine is still used as standard treatment in oral or rectal formulations for mild or moderate ulcerative colitis, and it is extremely safe with few recorded side effects.

It is only recently, though, that we have begun to understand how mesalazine works. It appears to act predominantly on the surface layers of the intestine and prevents the increase in leakiness of the epithelium that occurs with colitis. It does this by mopping up some of the damaging reactive chemicals (previously known as free radicals but now called reactive oxygen species or ROS) produced during the inflammatory response. Levels of inflammatory cytokines are also reduced by mesalazine, and it probably works in this way through suppressing the pathway from TLR 4 that leads to their production. Unlike steroids, mesalazine is only effective for ulcerative colitis and not for Crohn's disease.

The third of our trio of drugs that have been used since the 1950's is called azathioprine, and was the immunosuppressive drug prescribed to both Mike and Cath in the vignettes at the beginning of this chapter. However, unlike the fortuitous discoveries that led to the use of glucocorticoids and mesalazine, azathioprine was developed through a completely different route.

'Designer' Drugs: And Unexpected Results

Azathioprine was one of the first drugs to be invented and chemically synthesised with a particular molecular target in mind. This revolution came about through simultaneous advances in many scientific fields and heralded the modern era of medicine discovery and development. Gertrude Elion and George Hitchings appropriately, but rather belatedly, shared the 1988 Nobel Prize for their work together in rational drug design. Their work started in the late 1940's and focussed on imitating natural molecules to see if they could alter biological functions. In particular, they were interested in making mimics of the components of nucleic acids such as DNA to see if they could inhibit the replication of rapidly-dividing cells in cancers. Ironically, azathioprine is not an effective cancer treatment and its use can actually result in cancers, of the immune system or skin. However, it is not just malignant cells that reproduce quickly but also immune cells such as T and B lymphocytes. As a result, azathioprine (and its precursors) found their greatest utility as immunosuppressant agents—initially enabling organ transplantation through the suppression of immune rejection, and subsequently in the treatment of inflammatory and autoimmune diseases such as rheumatoid arthritis and inflammatory bowel diseases.

Whilst it is tempting to assume that azathioprine works through suppressing the massive proliferative expansion of lymphocytes in an immune response, recent evidence suggests that it has a completely different effect in treating inflammatory bowel diseases. After modification inside the cell, the azathioprine molecule binds to and inhibits a protein that prevents the process of apoptosis and hence is instrumental in cell survival. As you will remember, following activation, lymphocytes trigger a self-destruct sequence. This 'off-switch' appears defective in the inflammatory bowel diseases, but azathioprine seems to correct it. This may explain why the effect of azathioprine is somewhat delayed and its use in treating ulcerative colitis and Crohn's disease is to maintain remission bought through the use of steroids.

The 'Biologic' Revolution

Over the last two decades, rational drug design has entered an entirely new phase thanks to the development of monoclonal antibodies as 'magic bullets' that can be directed against almost any protein we choose. As we have seen, when it comes to the inflammatory bowel diseases we are spoilt for choice of molecules to target. The first such monoclonal antibody to be used was raised against the inflammatory cytokine tumour necrosis factor (or TNF). This protein is produced in large quantities in Crohn's disease by lymphocytes and macrophages, and acts to enhance the inflammation by increasing the secretion of a range of other cytokines. It also increases apoptosis (cell death) of epithelial cells, thereby increasing the permeability of the gut lining, and reduces the number of defensive Paneth cells. The introduction of the monoclonal antibody against TNF as a drug called infliximab in the closing years of the twentieth century was an extraordinary advance in the management of Crohn's disease. In some patients, infliximab makes ulcerations and even fistulae heal up completely. There are some downsides: being a protein, it has to be given by injection rather than by mouth. Also, as infliximab was developed from mouse antibodies, it could generate antibodies against itself that would block its effects: recent developments include 'humanising' the molecule to improve its efficacy. Patients are more prone to infection when receiving infliximab and in particular, tuberculosis can be reactivated even in people who were not aware that they had it previously.

Through only examining the uppermost strata of gastro-archeology, it was thought that infliximab would benefit Crohn's disease (being a 'cytotoxic' type response) and not ulcerative colitis (being an 'antibody' type response). It took several years before it was also trialled in ulcerative colitis but was surprisingly found to be effective. It has now become a second-line agent when glucocorticoids fail to help this condition.

Given that infliximab was specifically designed for its purpose—to 'mop up' TNF and stop it from exerting its effects on immune cells—it came as quite a surprise to find that other drugs designed to stick to TNF do not seem to work for Crohn's disease, even though they are effective in other inflammatory conditions such as rheumatoid arthritis. It turns out that it is not just the TNF secreted outside the cells that matters, but also TNF molecules that are embedded in the cell membrane. The TNF binding agents that improve inflammatory bowel diseases are capable of attaching to the TNF on the cell membrane, whereas the agents that are ineffective cannot. Once the antibody is bound to the TNF, the cell membrane TNF triggers an apoptosis signal within the lymphocyte—effectively restoring once again the immune response 'off switch', which appears lacking in the inflammatory bowel diseases.

There are now a large number of monoclonal antibodies in development for the treatment of inflammatory bowel diseases, including some that inhibit cytokines such as interleukin-23 and others that block the 'homing' molecules that direct lymphocyte traffic to the gut. However, given that the web of immune interactions is so finely balanced in the gut, it is perhaps not surprising that some 'designer' treatments completely fail to deliver their promised results and can even be detrimental, as we have seen in the case of antibodies against interleukin-17.

We now have one last treatment to explore on our journey. This agent is not one of our main tools in the treatment of inflammatory bowel diseases, although it is effective and we occasionally resort to using it. The story of this drug is a classic of 'reverse' discovery that uncovered a crucial mechanism within our cells—a master switch that regulates metabolism and immunity and even life and death decisions. The protein that acts at the core of this fundamental role was named after the drug that blocks it. And the drug was named after the place it was discovered—Easter Island in the South East

Pacific, also known as '*Rapa nui*'.[19] The drug we are interested in, called rapamycin, was isolated by investigators from a bacterium found on the island in 1964 as part of the global hunt for new antibiotics. The protein it attaches to is called 'mTOR'—short for the 'mechanistic target of rapamycin'.

Enter mTOR: The End of the Road

The metabolism of nutrients to generate energy evolved initially in the absence of oxygen. However, the 'invention of eating' through phagocytosis resulted in the symbiotic existence of living bacteria within the cell that became the mitochondria and harnessed the power of oxygen. These two mechanisms of energy production—with and without oxygen—are retained within our cells and serve separate purposes. The metabolism of glucose without oxygen (called glycolysis) provides as a by-product many of the essential basic building blocks required to make a cell, but generates relatively little energy. On the other hand, the use of oxygen for metabolism by mitochondria furnishes a large amount of energy but produces reactive oxygen species (ROS) as a dangerous by-product. These can damage proteins and lipids within the cell.

mTOR is like a microchip within the cell that integrates signals received from cell surface receptors, sensors of nutrient supply and oxygen levels. It is the switch between catabolism (breaking down cell structures) and anabolism (cell building and growth). Insulin, for instance, is a major anabolic hormone and acts through mTOR in increasing the uptake of glucose, storing it in cells and increasing lipid and protein (and muscle) synthesis. Body builders are well aware that mTOR can be activated and thereby increase their muscle bulk, through loading their diet with amino acid or protein supplements, independently of insulin. mTOR, stimulated by persistent nutrient excess, is the fundamental switch responsible for the diseases of the rich—obesity, heart disease, hypertension, fatty liver and type 2 diabetes.

[19] Easter Island was so-called by its first European visitor, Jacob Roggeveen, who landed there on Easter Day 1722. It is thought to have been continuously inhabited by its indigenous Polynesian people from around 1200 AD. It is truly the remotest inhabited place on earth, being over 1500 miles from the nearest settlement with more than 500 inhabitants and 2000 miles from the Chilean coast—it was at one time called 'The Navel of the World' in view of its remoteness. Its local name Rapa Nui ('Big Rapa') separates it from 'Rapa iti' (small Rapa), which was thought to have been colonised by Easter islanders subsequently. Over-population is thought to have led to deforestation and catastrophic societal failure, as outlined by Jared Diamond in his 2005 book 'Collapse: How Societies Choose to Fail or Succeed'. The population of around 4000 was to decline to just 111 indigenous inhabitants through war, disease (smallpox) and abduction by Peruvian slave traders.

In times of deficiency—as we have seen—autophagy is activated to increase the intracellular recycling and supply of nutrients. The mTOR 'switch' usually suppresses autophagy, but in starvation its activity is reduced and the control is released. Autophagy promotes cell survival and negates the need for a cell to programme its own death through apoptosis. Reduction of mTOR action through starvation is therefore the mechanism by which calorie restriction increases life expectancy—the use of rapamycin to block mTOR can double the life expectancy of fruit flies! Autophagy also prevents the form of cell death called pyroptosis whereby the cell releases large amounts of pro-inflammatory chemicals (such as interleukin-1) when dying. Inhibiting mTOR reduces its suppression of autophagy, thereby making such inflammatory responses less likely.

mTOR is also part of the mechanism through which stem cells differentiate to turn into specialised cells, such as the absorptive enterocyte, goblet cell or Paneth cell in the gut epithelium. mTOR blockade therefore preserves the 'stemness' of cells and regenerative capability, and may therefore seem to be the key to youthfulness as well as longevity.

At this stage one might well ask why we do not all take rapamycin pills as a daily supplement! The Canadian team that found this chemical at the 'navel of the world' may just have chanced upon the elixir of eternal youth. However, biology as we know is rarely so straightforward and rapamycin is sadly not the answer to all our problems. Firstly, at a cellular level, prolonging life and youthfulness and preventing apoptosis means preventing the specialisation of cells through differentiation and all of the accrued benefits of physiological apoptosis. Therefore, suppression of mTOR may not be altogether beneficial for complex organisms. Secondly, it turns out that there are two forms of protein complex containing mTOR which have different and sometimes opposing actions, and rapamycin blocks them both. As a result, side-effects of the drug even *include* diabetes. Furthermore, mTOR appears to have a key role in regulating immune processes, such that rapamycin is potently immunosuppressive and indeed our main use for it currently is in organ transplantation.

mTOR exerts its activity on the immune system by regulation of metabolism. In order to proliferate, lymphocytes require rapid metabolism of glucose through the 'non-oxygen' using pathway (glycolysis) which is stimulated by mTOR, whilst the mitochondrial 'oxygen-using' pathway is suppressed. Blocking mTOR therefore impairs the massive expansion of lymphocytes seen in the immune response. However, the effects of mTOR on immunity are even more impressive on closer inspection. mTOR blocks the production of regulatory lymphocytes, such that mice lacking mTOR altogether produce only regulatory T lymphocytes even under conditions that usually lead to inflammatory responses. mTOR also acts on the presenting cells to inhibit the uptake of proteins through phagocytosis and their subsequent expression on the cell surface in order to show them to the immune system. In essence, mTOR is the underlying immune switch, right at the bottom of our gastro-archeological trench. One of the two separate complexes made up of mTOR stimulates immune responses along the 'antibody' side of the great schism, whilst the other activates the 'cytotoxic' pore-forming toxin pathway.

mTOR therefore lies at the fork in the path between the ulcerative colitis and the Crohn's disease ends of the spectrum of inflammatory bowel diseases. Indeed, it may even be the key to understanding why smoking has protective effects in one whilst exacerbating the other as cigarette smoke (and nicotine in particular) suppresses mTOR activity and stimulates autophagy. Autophagy is undoubtedly a good thing for ulcerative colitis as it prevents the loss of enterocytes, maintains stem cells and the intestinal epithelium and reduces inflammatory cytokine responses. However, activation of 'defective' autophagy as seen in Crohn's disease is potentially a bad thing as it simply exacerbates the underlying disease process.

Beginning...and End

We are now at the end of our trail, and can stand on our prominent position, surveying the landscape that we have uncovered.

We started our trail at the beginning of this book in the 'Lost City' of alkaline hydrothermal vents under the middle of the Atlantic Ocean. Here, in equivalent ancient structures we recreated the birth of cells, defining life itself as what is bound within the confining cell membrane. We saw how this membrane concentrated the chemicals of life yet also generated a unified entity on which natural selection could act for it to evolve. However, we realised that the membrane also formed a barrier that prevented the influx of the necessary nutrient raw materials for energy, growth and division. In order to overcome this 'containment' paradox, the cell developed mechanisms to transport nutrients (food) into itself. One of these was by engulfing larger food particles, effectively eating by the process called 'phagocytosis'.

The evolution of phagocytosis created a selective pressure on the living cells that would been consumed, leading to the concept of 'predation'. In order to avoid becoming prey, cells needed the advantage of size. Thanks to Galileo's 'square-cube' law, there was a limit to the size that single cells could achieve.

This was overcome initially by cells aggregating in colonies, and ultimately by becoming multicellular.

Phagocytosis also enhanced the concept of infection, as a result of bacteria evolving to escape digestive mechanisms within the cell and ultimately killing it.

Multicellular animals still faced the same necessity of maintaining a constant internal environment as did single celled creatures and once again achieved this by containment, in this case through epithelia that lined their surfaces. They then shared the challenges of taking in nutrients across a barrier which resulted in the need for a specialised organ—the gut—for digestion and absorption. Competition for the availability of nutrients to fill the gut drove the evolution of a nervous system and sensory organs, and the ability to move by the development of appendages such as fins or limbs. Further differentiation of the gut into devolved organs led to the ability to store nutrients in the liver, digest foods by pancreatic secretions and even absorb oxygen through the lungs.

Multicellularity also required a 'rule book' for housekeeping and social cohesion between the different cell types of the animal. The evolution of the gut therefore coincided with the advent of multicellularity and immunity. Given that bacteria and other small creatures were always present in the surrounding fluid at the time of the formation of the gut in multicellular animals, the 'umbrella' of immunity covered not only the cells of the organism itself but also those enclosed within its gut cavity—its symbiotic bacteria. The immune system had to control the tendency to selfishness, of body cell types wanting to outgrow their limits to become cancers or misbehaving microbes that wished to invade and feed off the host. Through the need to retain a degree of permeability to nutrients through its intestinal epithelial lining, this thin layer of cells 1/50th of a millimetre thick (but covering a huge surface area) became the key interface between the creature and the external world—and therefore the site where immunity principally evolved.

From its earliest stages, the immune system developed two principal housekeeping tools—phagocytosis and 'pore forming toxins' that punctured the membrane of other cells. The former could be used to dispose of unwanted debris outside the cell—including other living cells. The latter evolved as a means of digesting prey that had been engulfed. The gut and immune system co-evolved, adding layer upon layer of complexity over time—the very strata of 'gastro-archeology'. However, immunity is constrained by the ability of small and single-celled creatures such as bacteria rapidly to evolve new ways to evade defensive mechanisms through their extremely fast reproduction rate. Fixed 'pattern recognition' defences that identified the signatures of organisms not obeying the housekeeping rule book could only change gradually in animals, generation by generation. It was only by allowing (at considerable risk) a rapid rate of mutation in immune cells through the 'immunological Big Bang' that multicellular immunity caught up with the ability of bacteria to adapt. The result was the generation of T and B lymphocytes with their infinitely adaptable receptors. Notably, the thymus gland (for the immunological education of T cells) and the avian bursa of Fabricius (for the generation of B cells) were both derived from the gut lining.

As a result of its history, the gut is enriched in all of the layers of immunological evolution. Our previous understanding of immunity came from analysis of the blood, in which the most recently evolved components are mainly represented. Therefore, by looking at the gut and its diseases and taking a necessary sideways step to dissect the different layers over time we learn much about those diseases, but also about the immune system and life itself. This is the concept of 'gastro-archeology'.

By adopting this approach, we have seen that the division between B and T cells is far from a recent invention but dates right back to the beginning. The 'great schism' that separates them predates the separate T and B cell receptors and stretches right back to the fundamental principles of controlling threats outside the cell (the phagocytosis-opsonin-ILC2-B cell-antibody pathway) and those within (the pore-forming toxin-ILC1-NK cell-T cell-cytotoxicity pathway). We have seen how this schism divides inappropriate immunological responses of the gut towards food that affect the small intestine (allergy vs coeliac disease) and also inappropriate responses towards bacteria affecting the lower gut (ulcerative colitis vs Crohn's disease). Revisiting our diagram of the wall of our trench in chap. 5, processes involved with food allergy and ulcerative colitis lie to the left-hand side of the schism, those involved with coeliac disease and Crohn's disease lie to the right.

Ultimately, through our discoveries, we have come to the bottom of the great schism and found mTOR, the master switch between the two pathways. In doing so we have uncovered a fundamental process that writes the immunological rule book of social cohesion between the constituent cells of a multicellular organism and its symbiotic passengers. Even more astonishingly though, this same mechanism constitutes the basis of all cell fates, governing decisions of life and death and even influencing how long—and how well—we live. We have no idea just where this journey will take us next.

It is fitting, however, that we should end our voyage (for the time being) at this point, on the remotest island in the world. There are two aspects of the history of Easter Island for which it is renowned. Much has been written about the strife of its warring tribes in the 17th and 18th centuries, and it has

Fig. 9.2 Moai of Easter Island

been used as an example of the collapse into chaos and anarchy that can occur through a breakdown of normal social interactions. Possibly as a result of depletion of natural resources, conflict rather than co-operation led to a descent into a fragmented society, living in caves and even resorting to cannibalism. However, Easter Island is best known for the monolithic stone effigies called Moai that stand up to 10 metres tall and were created between the 13th and 15th centuries by the indigenous Rapa Nui people. They represent deified deceased ancestors that watch over the living and have their backs turned towards the 'Spirit World' of the ocean. The Rapa Nui people believed in a synergy between the dead and the living. The dead provided for and watched over the living, who supported the ancestral spirits through sacrifice and offerings. A continuum of life and death.

These two themes—the ongoing interacting cycle of life and death and the rules of social cohesion that govern it—are those that we have seen played out in the pages of this book through the evolution of life, the gut and the immune system. They are written large in the gastro-archeological strata of time (Fig. 9.2).

Recommendations for Further Reading[1]

Chapter 1

1. Charles Darwin (1859) 'On the origin of species by natural selection, or the preservation of favoured races in the struggle for life'. Cambridge Library Collection, Cambridge University Press. *Not just for the scientific foresight but the beauty of its observations and clarity of description.*
2. Richard Dawkins (1976) 'The selfish gene'. Oxford Oxford University Press. 1987. *A classic of biological philosophy presented in plain language.*
3. David Hull (1981) 'Units of Evolution: A Metaphysical Essay'. In Jensen EJ, Harre R (eds). The Philosophy of Evolution. Harvest Press 1981. *A further exploration of Dawkins' views and an initial exposition of an ongoing debate.*
4. Martin WF, Garg S, Zimorski V. (2015) Endosymbiotic theories for eukaryote origin. Phil. Trans. R. Soc. B 370 *A review of current theories and ongoing debates of the origin of intracellular organelles*
5. Martin W, Russell MJ. (2007) On the origin of biochemistry at an alkaline hydrothermal vent. Phil. Trans. R. Soc. B 362 1887-1925 *A detailed but clear exposition of biochemical processes that could have led to the creation of life.*
6. Mariscal C and Doolittle WF (2015) Eukaryotes first: how could that be? Phil. Trans. R. Soc. B 370 *A radical argument proposing that LUCA was eukaryotic and archaeans and bacteria have subsequently shed intracellular complexity. Thought provoking ideas on the origin of cellular life.*

[1] In writing this story I have consulted reams of sources from ancient mythologies, through works of philosophy to scientific review articles, original scientific papers and current news media. Given that this is not a textbook (or a scientific paper) I have not listed them all, as this would probably double the length of the book. I have therefore limited myself to my favourite sources (usually books or review articles but with occasional original research publications) for each chapter.

J. Woodward, *The Gastro-Archeologist*, https://doi.org/10.1007/978-3-030-62621-1

7.Yutin N, Yolf MY, Wolf YI, Koonin EV (2009). The origins of phagocytosis and eukaryogenesis. Biology Direct 4: 9 *A good argument for phagocytosis being the origin of endosymbiosis. I have watered this down in my text in light of the following.*

8.Martin WF, Tielens AGM, Mentel M, Garg SG, Gould SB. (2017) The physiology of phagocytosis in the context of mitochondrial origin Microbiol Mol Biol Rev. 81(3) e00009-17 *A strong counterargument to phagocytosis leading to mitochondrial endosymbiosis on bioenergetic grounds. How I wish we knew how the mitochondria got there …!*

9.Gray MW (1998) Rickettsia, typhus and the mitochondrial connection. Nature 396; 108-109 *Editorial following the publication of the Riskettsial genome by Anderson SGE et al Nature 396; 133-140 – highlighting similarities with the mitochondrial genome.*

10.Weiss MC, Sousa FL, Mrnjavac N, Neukirchen S, Roettger M, Nelson-Sathi S, Martin WF. (2016). The physiology and habitat of the last universal common ancestor. Nature Microbiology 1: 1-8 *A powerful review of genetic, chemical and biochemical data outlining the likely origin and location of LUCA*

Chapter 2

1.Jay Gould, Stephen. Ontogeny and Phylogeny (1977). The Belknap press of Harvard University. Cambridge, Massachusetts. *A comprehensive overview of theories relating the evolution of structure to embryological development including Haeckel.*

2.Cavalier-Smith, Thomas. (2017). Origin of animal multicellularity: precursors, causes, consequences – the choanoflagellate/sponge transition, neurogenesis and the Cambrian explosion. Phil. Trans. Soc. B 372 *An exposition of the development of multicellularity from choanoflagellates to sponges and cnidarians.*

3.Nielsen C. (2013) Life cycle evolution: was the eumetazoan ancestor a holopelagic, planktotrophic gastraea? BMC Evolutionary Biology 13:171 *An alternative viewpoint based on the evolution of cnidaria from neoteny of a larval form.*

4.Grosberg RK, Strathmann RR. (2007) The evolution of multicellularity: a minor major transition? Annu. Rev. Ecol. Evol. Syst. 38: 621-54 *Ideas that link true multicellularity and cellular differentiation to clonal replication rather than colonial aggregation – this underpins my thoughts on the origins of immunity.*

5.Parfrey LW and Lah DJG. (2013) Multicellularity arose several times in the evolution of eukaryotes. Bioessays 35: 339-347. *At least 25 times according to this lineage analysis*

6.Smith CL, Pivovarova N, Reese TS (2015). Co-ordinated feeding behaviour in Trichoplax: an animal without synapses. PLOS one 10: 1-15 *My favourite animal. Need I say more?*

7.Wijgerde T, Diantari R, Lewaru MW, Verreth JAJ, Osinga R (2011). Extracoelenteric zooplankton feeding is a key mechanism of nutrient acquisition for the scleractin-

ian coral Galaxea fascicularis. J Exp Biol 214: 3351-3357 *Evidence of ongoing exo-digestion in a coral under conditions of excess nutrient supply*
8.Lanna E. (2015) Evo-devo of non-bilaterian animals. Genetics and molecular biology 38; 3: 284-300 *Summary of ideas on the development of bilaterians including the evolution of hox genes and the development of a second axis.*
9.Dunn CW, Leys SP and Haddock SHD (2015) The hidden biology of sponges and ctenophores. Trends in Ecology and Evolution 30: 282-291 *Not as simple as we may think.*
10.Cereijido M, Contreras RG, Shoshani L. (2004). Cell adhesion, polarity and epithelia in the dawn of metazoans. Physiol Rev. 84: 1229-1262. *Theories on the evolution of epithelia in multicellular animals.*

Chapter 3

1.Darwin, Charles. (1881) The formation of vegetable mould through the action of worms, with some observations on their habits. London UK: John Murray
2.Holland PWH (2015) Did homeobox gene duplications contribute to the Cambrian explosion? Zoological letters 1:1 *Perhaps by fashioning the body plan that permitted the development of a through-gut?*
3.Hejnol A and Martin-Duran JM. (2015) Getting to the bottom of anal evolution. Zoologischer Anzeiger 256: 61-74 *Still a hot topic of debate*
4.Graham A and Richardson J (2012) Developmental and evolutionary origins of the pharyngeal apparatus. EvoDevo 3:24 *Ontogeny and phylogeny arguments once again played out thanks to the pharyngeal slits*
5.Gillis JA and Tidswell ORA (2017) The origin of vertebrate gills. Current Biology 27, 729-732. *A comprehensive overview with excellent diagrams*
6.Evans DH, Piermarini PM, Choe KP (2005). The multifunctional fish gill: dominant site of gas exchange, osmoregulation, acid-base regulation and excretion of nitrogenous waste. Physiol Rev. 85: 97-117 *A similarly complete overview of gill physiology*
7.Pieler T and Chen Y (2006). Forgotten and novel aspects in pancreas development. Biol Cell 96: 79-88 *Ontogeny and phylogeny of the pancreas - of particular interest for the disposition of hormone secreting cells*
8.Hsia CCW, Schmitz A, Lambertz M, Perry SF, Maina JN. (2013) Evolution of air breathing: oxygen homeostasis and transitions from water to land and sky. Comparative Physiology 3(2): 849-915 *A detailed review of air breathing in all animals including invertebrates*
9.Ashley-Ross MA, Hsieh ST, Gibb AC, Blob RW (2013) Vertebrate land invasions – Past,Present and Future. *Thought provoking ideas on the morphological adaptions required for life on land*

10.Laudet V. (2011) The origins and evolution of vertebrate metamorphosis. Current Biology R726-R737 *A fascinating review*
11.Schmidt-Rhaesa A. (2007) The evolution of Organ Systems. Oxford University Press, Great Clarendon St, Oxford UK

Chapter 4

1.Tauber, Alfred (2017). Immunity, the Evolution of an Idea. Oxford University Press, Great Clarendon St, Oxford, UK *A challenging perspective on the ways of perceiving immunity*
2.Rinkevich, Baruch (2004) Primitive immune systems: are your ways my ways? Immunological reviews 198: 25–35 *Thoughts on the original functions of immunity*
3.Chen G, Zhuchenko O, Kuspa A. Immune-like phagocyte activity in the social amoeba. Science 2007 2317: 678-681 *An introduction to sentinel cells, TirA and phagocytosis as an immune mechanism*
4.Garderes J, Bourguet-Konracki M-L, Hamer B, Batel R, Schroder HC, Muller WEG (2015) Porifera lectins: diversity, physiological roles and biotechnological potential. Marne Drugs 13: 5059-5101 *Insight into the extraordinary complexity of sponge lectins*
5.Ghebrehiwet, Berhane (2016) The complement system: an evolution in progress. F1000Research 2840 1–8
6.Wilmanski JM, Petnicki-Ocweja T, Kobayashi KS (2008). NLR proteins: integral members of innate immunity and mediators of inflammatory diseases. J Leukoc Biol 83: 13–30
7.McCormack R, Podack ER (2015) Perforin-2/Mpeg1 and other pore-forming proteins throughout evolution. J Leukoc Biol 98: 761–768
8.Iacovache I, van der Goot FG, Pernot L. (2008) Pore formation: an ancient yet complex form of attack. Biochimica et Biophysica Acta 1778: 1611–1623
9.Fujita T (2002) Evolution of the lectin-complement pathway and its role in innate immunity. Nature Rev. Immunol 2: 346–353
10.Oikonomopoulou K, Ricklin D, Ward PA, Lambris JD. (2012) Interactions between coagulation and complement – their role in inflammation. Semin Immunopathol 34: 151–165
11.Roach JM, Racioppi L, Jones CD, Masci AM. (2013) Phylogeny of Toll-like receptor signaling: adapting the innate response. PLOS One 8(1) e54156
12.Huang S, Yuan S, Guo L, Yu Y, Li J, Wu T, Liu T, yang M, Wu K, Liu H, Ge J, Yu Y, Huang H, Dong, M, Yu C, Chen S, Xu A (2008). Genomic analysis of the immune gene repertoire of amphioxus reveals extraordinary innate complexity and diversity. Genome Research 18: 1112–26 *The end of the road for innate immunity…*

13.Buchmann K (2014) Evolution of innate immunity: clues from invertebrates via fish to mammals. Frontiers in Biology 459 1-8 *A solid wide ranging review*

Chapter 5

1.Flajnik MF (2016) Evidence of G.O.D's miracle: unearthing a RAG transposon. Cell 166: 11-12 *Where G.O.D stands for Generation of Diversity!*
2.Litman GW, Rast JP, Fugmann SD. (2010) The origins of vertebrate adaptive immunity. Nat Rev. Immunol 10: 543-553 *A useful overview*
3.Zhu L-Y, Shao T, Bie L, Zhu L-y, Xiang L-x, Shao J-z. (2016) Evolutionary implication of B-1 lineage cells from innate to adaptive immunity. Molecul Immunol 69:123–130
4.Parra D, Takizawa F, Sunyer JO (2013) Evolution of B cell Immunity. Annu Rev. Anim Biosci 1:65–97
5.Castro CC, Luoma AM, Adams EJ. (2015) Coevolution of T-cell receptors with MHC and non-MHC ligands. Immunol Rev. 267: 30–55
6.Sunyer JO (2012) Evolutionary and functional relationships of B cells from fish and mammals: insights into their novel roles in phagocytosis and presentation of particulate antigen. Infect Disord Drug Targets 12: 200–212
7.Adams EJ, Gu S, Luoma AM (2015) Human gamma delta T cells: evolution and ligand recognition. Cell Immunol 296: 31–40
8.Geenen V (2012) The appearance of the thymus and the integrated evolution of adaptive immune and neuroendocrine systems. Acta Clinica Belgica 67: 209–213
9.Hsu E (2011) The invention of lymphocytes Curr Opin Immunol 23: 156–162
10.Boehm T (2012) Evolution of vertebrate immunity. Current Biology 22 R722–732
11.Boehm T, McCurley N, Sutoh Y, Schorpp M, Kasahara M, Cooper MD.(2012) VLR-based adaptive immunity. Annu Rev. Immunol 30:203–220

Chapter 6

1.Sankaran-Walters S, Hart R, Dills C. (2017) Guardians of the gut: enteric defensins. Frontiers in Microbiology 8:647 1–8
2.Chairatana P, Nolan EM. (2017) Defensins, lectins, mucins and secretory immunoglobulin A: microbe-binding biomolecules that contribute to mucosal immunity in the human gut. Crit Rev. Biochem Mol Biol 52: 45–56
3.Haber AL, Biton M, Rogel N et al (2017) A single-cell survey of the small intestinal epithelium. Nature 551: 333-339 *The application of exciting new techniques reveals unexpected findings*

4.Birchenough GMH, Johansson MEV, Gustafsson JK, Bergstrom JK, Hansson GC. (2015) New developments in goblet cell mucus secretion and function Mucosal Immunol 8: 712–719
5.Vivier E, van de Pavert SA, Cooper MD, Beiz GT (2016) The evolution of innate lymphoid cells. Nat Immunol 17: 790-794 *A useful review of the then state of the art in a fast moving field*
6.Hepworth MR, Mauter M, Hartmann S. (2012) Regulation of type 2 immunity to helminths by mast cells. Gut microbes 3:476–481
7.Gassler N. (2017) Paneth cells in intestinal physiology and pathophysiology. World J Gastrointest Pathophysiol 8: 150–160
8.Steele SP, Melchor SJ, Petri WA (2016) Tuft cells: new players in colitis. Trends Mol Med 22: 921–924
9.Cheroutre H, Lambolez F, Mucida D. (2011) The light and dark sides of intestinal intraepithelial lymphocytes. Nat Rev. Immunol 11:445-456 *An excellent overview of these fascinating but complex cells*
10.McDole JR, Wheeler LW, McDonald KG, Wang B, Konjuca V, Knoop KA, Newberry RD, Miller MJ. Goblet cells deliver luminal antigen to CD103+ DCs in the small intestine. Nature 483: 345–349 *The images and videos accompanying this paper are quite stunning - as is the novel concept!*

Chapter 7

1.Tordesillas L, Berin MC, Sampson HA. (2017) Immunology of food allergy. Immunity 47: 32–50
2.Rezende RM, Weiner HL. (2017) History and mechanisms of oral tolerance. Seminars in Immunology 30: 3–11
3.Bryce PJ. Balancing tolerance or allergy to food proteins. (2016) Trends in Immunology 37: 659–667
4.Sampson, Hugh (2016) Food allergy: past, present and future. Allergology international 65: 363–369 *An overivew by the guru of food allergy*
5.Mowat, Allan McI (2018) To respond or not to respond – a personal perspective of intestinal tolerance. Nat Rev. Immunol 18:405–15 *A personal view but very instructive overview from a renowned researcher in oral tolerance*
6.Cassani BA, Villablanca EJ, Callisto JD, Wang S, Mora JR. (2012) Vitamin A and immune regulation: role of retinoic acid in gut associated dendritic cell education, immune protection and tolerance. Mol Aspects Med 33:63–76
7.Breiteneder H, Mills ENC (2005) Molecular properties of food allergens. J Allergy Clin Immunol 115:14–23
8.McKenzie ANJ (2014) Type-2 innate lymphoid cells in asthma and allergy. Ann Am Thorac Soc 11: S263–270

9.Smith PK, Masilamani M, Li X-M, Sampson HA. (2017) The false alarm hypothesis: food allergy is associated with a high dietary advanced glycation end-products and proglycating dietary sugars that mimic alarmins. J Allergy Clin Immunol 139:429–37

10.Ruffner M, Spergel JM (2016) Non-IgE mediated food allergy syndromes. Ann Allergy Asthma Immunol 117: 452–454

Chapter 8

1.Tye-Din JA, Galpeau HJ, Agardh D. (2018) Celiac Disease: a review of current concepts in pathogenesis, prevention and novel therapies. Frontiers in Paediatrics 6:350: 1–19

2.Setty M, Discepolo V, Abadie V, Kamhawi S, Mayassi T et al. (2015) Distinct and synergistic contributions of epithelial stress and adaptive immunity to functions of intraepithelial killer cells and active celiac disease. Gastroenterology 149: 681–691

3.Maiuri L, Ciacci C, Ricciardelli I, Vacca L, Raia V, Auricchio S, Picard J, Osman M, Quaratino S, Londei M. (2003) Association between innate response to gliadin and activation of pathogenic T cells in coeliac disease. Lancet 362 30-37 *The first ideas about innate mechanisms in celiac disease*

4.Abadie V, Kim SM, Lejeune T, Palanski BA, Ernest JD et al. (2020) IL-15, gluten and HLA DQ8 drive tissue destruction in coeliac disease. Nature 578: 600-604 *A landmark paper*

5.Lammers KM, Lu R, Brownley J, Lu B, Gerard C et al. (2008) Gliadin induces an increase in intestinal permeability and zonulin release by binding to the chemokine receptor CXCR3. Gastroenterology 135: 194–204

6.Verdu EF, Galipeau HJ, Jabri B (2015) Novel players in coeliac disease pathogenesis: role of the gut microbiota. Nat Rev. gastroenterol Hepatol 12: 497–506

7.Dieterich W, Ehnis T, Bauer M, Donner P, Volta U, riecken EO, Schuppan D. (1997) Identification of tissue transglutaminase as the autoantigen of celiac disease. Nat Med 3: 797–801 *An important discovery*

8.Cellier C, Patey N, Mauvieux L, Jabri B, Delabasse E et al. (1998) Abnormal intestinal intraepithelial lymphocytes in refractory sprue gastroenterology 114: 471–481

9.Ettersperger J, Montcuquet N, Malamut G, DiSanto JP, Cerf-Bensussan N, Meresse B. (2016) Interleukin-15-dependent T-cell-like innate intraepithelial lymphocytes develop in the intestine and transform into lymphomas in celiac disease. Immunity 45: 610–625

10.Uhde M, Indart AC, Yu XB, Jang SS, de Girogio R, Green PHR, Volta U, Vernon SD, Alaedini A. (2019) Markers of non-coeliac wheat sensitivity in patients with myalgic encephalomyelitis/ chronic fatigue syndrome. Gut 68:377

Chapter 9

1. Cleynen I, Boucher G, Jostins L, Schumm LP, Zeissig S et al (2016) Inherited determinants of Crohn's disease and ulcerative colitis phenotypes: a genetic association study. Lancet 387: 156-167 *3 types of IBD – at least!*
2. Verstockt B, Smith KGC, Lee JC (2018) Genome-wide association studies in Crohn's Disease: past, present and future. Clin Trans Immunol 7: e101 *A pragmatic overview*
3. Mizoguchi E, Low D, Ezaki Y, Okada T. (2020) Recent updates on the basic mechanisms and pathogenesis of inflammatory bowel diseases in experimental animal models. Intest Res 18: 151-167 *Essential reading for anyone trying to understand the causes of IBD*
4. Ramos GP, Papadakis K (2019) Mechanisms of disease: inflammatory bowel diseases. Mayo Clin Proc 94: 155–165
5. Porter RJ, Kalla R, Ho G-T (2020) Ulcerative colitis: recent advances in the understanding of disease pathogenesis. F1000research:294
6. Palmela C, Chevarin C, Xu Z, Torres J, Sevrin G, Hirten R, Barnich N, Ng SC, Colombel J-F. (2018) Adherent-invasive Escherichia coli in inflammatory bowel disease. Gut 67: 574–587
7. Kaser A, Niederreiter L, Blumber RS (2011) Genetically determined epithelial host dysfunction and its consequences for microflora-host interactions. Cell Mol Life Sci 68: 364303649
8. Larabi A, Barnich N, Nguyen HTT (2020) New insights into the interplay between autophagy, gut microbiota and inflammatory responses in IBD. Autophagy 16: 38–51
9. Neurath MF (2017) Current and emerging therapeutic targets for IBD. Nat Rev. Gastroenterol Hepatol 14: 269–278
10. Powell JD, Pollizzi KN, Helkamp EB, Horton MR. (2012) Regulation of immune responses by mTOR. Annu Rev. Immunol 30: 39-68 *A comprehensive review of the immune effects of this master switch*

Printed in the United States
by Baker & Taylor Publisher Services